3D Videocommunication

3D Videocommunication

Algorithms, concepts and real-time systems in human centred communication

EDITED BY

Oliver Schreer
Fraunhofer Institute for Telecommunications
Heinrich-Hertz-Institut, Berlin, Germany

Peter Kauff
Fraunhofer Institute for Telecommunications
Heinrich-Hertz-Institut, Berlin, Germany

Thomas Sikora
Technical University Berlin, Germany

John Wiley & Sons, Ltd

Other Wiley Editorial Offices

John Wiley & Sons Inc., 111 River Street, Hoboken, NJ 07030, USA

Jossey-Bass, 989 Market Street, San Francisco, CA 94103-1741, USA

Wiley-VCH Verlag GmbH, Boschstr. 12, D-69469 Weinheim, Germany

John Wiley & Sons Australia Ltd, 33 Park Road, Milton, Queensland 4064, Australia

John Wiley & Sons (Asia) Pte Ltd, 2 Clementi Loop #02-01, Jin Xing Distripark, Singapore 129809

John Wiley & Sons Canada Ltd, 22 Worcester Road, Etobicoke, Ontario, Canada M9W 1L1

Wiley also publishes its books in a variety of electronic formats. Some content that appears in print may not
be available in electronic books.

Library of Congress Cataloging in Publication Data

(to follow)

British Library Cataloguing in Publication Data

A catalogue record for this book is available from the British Library

ISBN-13 978-0-470-02271-9 (HB)
ISBN-10 0-470-02271-X (HB)

Typeset in 10/12pt Times by Integra Software Services Pvt. Ltd, Pondicherry, India

FSC
Mixed Sources
Product group from well-managed
forests and other controlled sources

Cert no. SGS-COC-2953
www.fsc.org
© 1996 Forest Stewardship Council

Contents

List of Contributors

EDITORS

Peter Kauff
Fraunhofer Institute for Telecommunications
Heinrich-Hertz-Institut
Image Processing Department
Einsteinufer 37
D-10587 Berlin
Germany

Oliver Schreer
Fraunhofer Institute for Telecommunications
Heinrich-Hertz-Institut
Image Processing Department
Einsteinufer 37
D-10587 Berlin
Germany

Thomas Sikora
Technical University of Berlin
Communication Systems Group
Sekr. EN1
Einsteinufer 17
D-10587 Berlin
Germany

AUTHORS

Yousri Abdeljaoued
Ecole Polytechnique Fédérale de Lausanne - EPFL
CH-1015 Lausanne
Switzerland

Nicole Atzpadin
Fraunhofer Institute for Telecommunications
Heinrich-Hertz-Institut
Image Processing Department
Einsteinufer 37
D-10587 Berlin
Germany

Sandra Brix
Fraunhofer Institute for Digital Media Technology
Langewiesenerstr. 22
D-98693 Ilmenau
Germany

Christos Conomis
Fraunhofer Institute for Telecommunications
Heinrich-Hertz-Institut
Interactive Media Department
Einsteinufer 37
D-10587 Berlin
Germany

Touradj Ebrahimi
Ecole Polytechnique Fédérale de Lausanne - EPFL
CH-1015 Lausanne
Switzerland

Peter Eisert
Fraunhofer Institute for Telecommunications
Heinrich-Hertz-Institut
Image Processing Department
Einsteinufer 37
D-10587 Berlin
Germany

Jan-Friso Evers-Senne
Institute of Computer Science
Christian-Albrechts-University Kiel
Olshausenstr. 40
D-24098 Kiel
Germany

Christoph Fehn
Fraunhofer Institute for Telecommunications
Heinrich-Hertz-Institut
Image Processing Department
Einsteinufer 37
D-10587 Berlin
Germany

Andrea Fusiello
Dipartimento di Informatica
Università degli Studi di Verona
Strada Le Grazie, 15
I-37134 Verona
Italy

João G.M. Gonçalves
Nuclear Safeguards Unit
Institute for the Protection and Security of the Citizen (IPSC)
   European Commission – Joint Research Centre
I-21020 Ispra (VA)
Italy

Oliver Grau
Senior R&D Engineer
BBC R&D
Kingswood Warren
Tadworth
Surrey
KT20 6NP
UK

Wijnand A. IJsselsteijn
Human-Technology Interaction Group
Department of Technology Management
Eindhoven University of Technology
P.O. Box 513
5600 MB Eindhoven
The Netherlands

Spela Ivekovic
Electrical, Electronic and Computer Engineering
School of Engineering and Physical Sciences
Heriot-Watt University
Riccarton
Edinburgh
EH14 4AS
UK

Peter Kauff
Fraunhofer Institute for Telecommunications
Heinrich-Hertz-Institut
Image Processing Department
Einsteinufer 37
D-10587 Berlin
Germany

Reinhard Koch
Institute of Computer Science
Christian-Albrechts-University Kiel
Olshausenstr. 40
D-24098 Kiel
Germany

David Marimon i Sanjuan
Ecole Polytechnique Fédérale de Lausanne - EPFL
CH-1015 Lausanne
Switzerland

Lydia M.J. Meesters
Human-Technology Interaction Group
Department of Technology Management
Eindhoven University of Technology
P.O. Box 513
5600 MB Eindhoven
The Netherlands

Jane Mulligan
Department of Computer Science
Room 717, UCB 430
Engineering Center Office Tower
University of Colorado at Boulder
Boulder
CO 80309-0430
USA

Peter Noll
Technische Universität Berlin
Communications Systems Group
Sekr. EN1
Einsteinufer 17
D-10587 Berlin
Germany

Siegmund Pastoor
Fraunhofer Institute for Telecommunications
Heinrich-Hertz-Institut
Interactive Media Department
Einsteinufer 37
D-10587 Berlin
Germany

Oliver Schreer
Fraunhofer Institute for Telecommunications
Heinrich-Hertz-Institut
Image Processing Department
Einsteinufer 37
D-10587 Berlin
Germany

Markus Schwab
Technische Universität Berlin
Communications Systems Group
Sekr. EN1
Einsteinufer 17
D-10587 Berlin
Germany

Vítor Sequeira
Nuclear Safeguards Unit
Institute for the Protection and Security of the Citizen (IPSC)
    European Commission – Joint Research Centre
I-21020 Ispra (VA)
Italy

Pieter J.H. Seuntiëns
Human-Technology Interaction Group
Department of Technology Management
Eindhoven University of Technology
P.O. Box 513
5600 MB Eindhoven
The Netherlands

Thomas Sikora
Technical University of Berlin
Communication Systems Group
Sekr. EN1
Einsteinufer 17
D-10587 Berlin
Germany

Aljoscha Smolic
Fraunhofer Institute for Telecommunications
Heinrich-Hertz-Institut
Image Processing Department
Einsteinufer 37
D-10587 Berlin
Germany

Thomas Sporer
Fraunhofer Institute for Digital Media Technology
Langewiesener Strasse 22
D-98693 Ilmenau
Germany

Masayuki Tanimoto
Department of Electrical Engineering and Computer Science
Graduate School of Engineering
Nagoya University
Furo-cho
Chikusa-ku
Nagoya 464-8603
Japan

Emanuele Trucco
School of Engineering and Physical Sciences
Heriot-Watt University
Riccarton
Edinburgh
EH14 4AS
UK

Symbols

General

Scalar values x, y in italic lowercase. Coordinate values are scalars.
Vectors $\mathbf{X}$ as bold capitals (3D) or $\mathbf{x}$ as bold lowercase (2D).
Matrices M as italic capitals.

Specific symbols

t_c	Interaxial distance of a stereo camera
z_c	Convergence distance of a stereo camera
h	Sensor shift of a stereo camera
$\mathbf{M} = (x, y, z)^T$	Euclidean 3D point
$\mathbf{m} = (x, y)^T$	Euclidean 2D point
$\mathbf{M} = (x, y, z, w)^T$	Homogeneous 3D point
$\mathbf{m} = (u, v, w)^T$	Homogeneous 2D point
$\mathbb{R}^2$	Euclidian space in 2D
$\mathbb{R}^3$	Euclidian space in 3D
$\mathbb{P}^2$	Projective plane in 2D
$\mathbb{P}^3$	Projective space in 3D
l	Line is a vector in 2D
l_∞	Line at infinity
K	Camera calibration matrix
R	Rotation matrix
I	Identity matrix
$\mathbf{t}$	Translation vector
f	Focal length
$h_{1/r}$	Sensor shift, l = left, r = right
α_u, α_v	Focal length in multiples of pixels, horizontal and vertical
$\mathbf{C}$	Camera projection centre
$P = K[R^T \mid -R^T\mathbf{C}]$	Camera projection matrix
$I_{1,r}$	Image plane of l = left and r = right camera
$\mathbf{m}_1, \mathbf{m}_r$	Corresponding 2D points
$[\mathbf{t}]_\times$	Skew-symmetric matrix of vector $\mathbf{t}$
E	Essential matrix
F	Fundamental matrix
$\mathbf{l}_1, \mathbf{l}_r$	Two corresponding epipolar lines

$\mathbf{e}_l, \mathbf{e}_r$	Epipoles
H	Projective transformation/homography
$\mathbf{m}_1, \mathbf{m}_2, \ldots, \mathbf{m}_N$	Corresponding 2D points
$\mathcal{T}_i^{jk} = a_i^j b_4^k - a_4^j b_i^k$	Trifocal tensor
$[T_1, T_2, T_3]$	Trifocal tensor in matrix notation
$\mathbf{d}$	Disparity
$\prod$	3D plane
$\prod_\infty$	Plane at infinity
H_π	Homography related in a plane $\prod$
Q	Sound source
cov	Covariance
Δp	Error of localization
$\Delta \tau$	Time difference of arrival (TDOA)
c	Speed of sound in the air (Chapter 10) or speed of light
V	Volume
S	Surface
P_A	Sound pressure in the Fourier domain
p_0	Air density
V_n	Particle velocity
Δt	Pulse travel time
$\Delta \phi$	Phase difference
λ_a	Ambiguity interval

Abbreviations

1D-II	1D Integral imaging
2D	Two-dimensional
3D	Three-dimensional
3DAV	3D audio/visual
3DDAC	3D display with accomodation compensation
3D TV	Three-dimensional broadcast television
AAC	Advanced audio coding
AC	Alternating current
AED	Adaptive eigenvalue decomposition
AFX	Animation framework extension
ALS	Audio lossless coding
AM	Amplitude modulation
AP	Affine projection
AR	Augmented reality
ATTEST	Advanced three-dimensional television system (European IST project)
AVC	Advanced video coding
BAP	Body animation paramters
BDP	Body definition parameters
BIFS	Binary format for scenes
BMLD	Binaural masking level difference
BOOM	Binocular omni-oriented monitor
CAD	Computer-aided design
CAM	Computer-aided manufacturing
CAVE	Automatic Virtual Environment
CCD	Charge coupled device
CCIR	Comité Consultatif International du Radiodiffusion
CIF	Common intermediate format
CG	Computer graphics
CMOS	Complimentary metal–oxide semiconductor
CRT	Cathode ray tube
CSCW	Computer-supported collaborative work
CT	Computer tomography
CV	Computer vision
DC	Direct current
DCT	Discrete cosine transform
DIBR	Depth-image-based rendering
DMD	Digital mirror device

DMIF   Delivery multimedia integration framework
DOE   Diffractive optical elements
DOF   Degree of freedom
DPD   Displaced pixel difference
DSB   Delay-and-sum beam former
DSP   Digital signal processor
DVB   Digital video broadcasting
DVD   Digital video disc
DWT   Discrete Wavelet Transform
EPI   Epipolar plane image
ERLE   Echo return loss enhancement
ES   Elementary stream
ETSI   European Telecommunication Standards Institute
FAP   Facial animation parameters
FDP   Facial definitions parameters
FLOATS   Fresnel-lens-based optical apparatus for touchable distance stereoscopy
FM   Frequency modulation
FPA   Focal plane arrays
FPGA   Field programmable gate arrays
FS   Frame store
FTV   Free-viewpoint television
GCC   Generalized cross-correlation
GOP   Group of pictures
GPU   Graphical processing unit
GSC   Generalized sidelobe canceler
GUI   Graphical user interface
HCI   Human–computer interaction
HDTV   High-definition television
HMD   Head-mounted display
HOE   Holographic optical elements
HRM   Hybrid recursive matching
HRSE   High-resolution spectral estimation
HRTF   Head-related transfer function
HUD   Head-up display
IBR   Image-based rendering
ICT   Image cube trajectory
IEEE   Institute of Electrical and Electronics Engineers
IETF   Internet Engineering Task Force
IID   Interaural intensity difference
ILD   Interaural level difference
ILED   Infrared light-emitting diode
INS   Inertial navigation system
IP   Integral photography
IPD   Inter-pupillary distance
IPR   Intellectual property rights
ISDN   Integrated services digital network
IST   Information society technologies, area of European research programme

IT	Information technology
ITD	Interaural time difference
ITU	International telecommunications union
JPEG	Joint photographic experts group
LBE	Location-based environment
LCD	Liquid crystal display
LDI	Layered depth image
LED	Light-emitting diode
LFE	Low-frequency enhancement
LFVC	Light field video camera
Mbit	Megabit
MC	Motion compensation
MCOP	Multiple centre of projection
MCRA	Minimum controlled recursive averaging
MEMS	Microelectric mechanical systems
ML	Maximum likelihood
MMR	Mobile mixed reality
MMSE	Minimum mean-square error
MNCC	Modified normalized cross-correlation
MP	Main profile
MPEG	Moving Pictures Expert Group
MP-MPEG	Main profile of MPEG-standard
MR	Mixed reality
MRI	Magnetic resonance imaging
MUD	Multi-user dungeon
MUSIC	Multiple signal classification
MV	Motion vector
MVB	Multi-variance beamformer
MVP	Multiview profile
NCC	Normalized cross-correlation
NLMS	Normalized least-mean-square
OLED	Organic light-emitting diode
OM-LSA	Optimally modified log spectral amplitude
PC	Personal computer
PCI	Peripheral component interconnect, standard computer interface
PDA	Personal digital assistent
PF	Plenoptic function
PIP	Personal interaction panel
PS	Plenoptic sample
PSNR	Peak signal-to-noise ratio
QUXGA	Quad ultra-extended graphics array
RAID	Redundant array of independent disks
RGB	Red–green–blue colour space
R/D	Rate/distortion
RLS	Recursive least-squares
RMS	Root-mean-square
SAD	Sum of absolute difference

SCM	Spatial covariance matrix
SMNCC	Sum of modified normalized cross-correlation
SNR	Signal-to-noise ratio
SRP	Steered response power
SRP-PHAT	Steered response power with phase transform
SSD	Sum of squared differences
SSSD	Sum of sum of squared differences
SVD	Singular value decomposition
SVE	Shared virtual environment
SVTE	Shared virtual table environment
TDOA	Time difference of arrival
TUI	Tangible user interface
TV	Television
VAD	Voice activity detector
VAMR	Variable accommodation mixed reality
VB	Video buffer
VDTM	View-dependent texture mapping
VE	Virtual environment
VGA	Video graphics array
VLBC	Variable length binary code
VLC	Variable length coding
VLD	Variable length decoding
VLSI	Very-large-scale integration
VR	Virtual reality
VRD	Virtual retinal display
WFS	Wave field synthesis
WUXGA	Wide ultra-extended graphics array
ZPS	Zero parallax setting

Introduction

Oliver Schreer, Peter Kauff and Thomas Sikora

The migration of immersive media towards telecommunication applications continues to advance. Impressive progress in the field of media compression, media representation, and the larger and ever-increasing bandwidth available to the customer, will foster the introduction of these services in the future. It is widely accepted that this trend towards immersive media is going to have a strong impact on our daily life.

The ability to evoke a state of 'being there' and/or of 'being immersed' into media applications will no longer remain the domain of the flight simulators, CAVE systems, cyberspace applications, theme parks or IMAX theatres. It will arise in offices, venues and homes and it has the potential to enhance quality of life in general.

First steps in this direction have already been observed during the last few years:

- Video conferencing has become more and more attractive for various lines of business. Today video conferencing enhances distributed collaboration in an emerging global market. It is therefore regarded as a high-return investment for decision-making processes. Today, high-end videoconferencing systems already offer telepresence capabilities to achieve communication conditions as natural as possible. This business sector will benefit from the future advent of immersive systems providing improved realism of scene reproduction.
- The market of team collaboration systems grows drastically. The first synchronous collaboration tools are being sold today. They meet the demands of an increasing competition in costs, innovation, productivity and development cycles. Most of them still rely on the screen-sharing principle and suffer from a lack of natural communication between the collaborating partners. Emerging teleimmersion systems will go beyond these limitations of conventional collaborative team software. They will employ collaborative virtual environments (CVE) with intuitive interaction and communication capabilities.
- In the entertainment sector we are now beginning to see the viable economics of high-definition broadcasting of live events in sports or culture to cinemas, halls and large

3D Videocommunication — Algorithms, concepts and real-time systems in human centred communication
Edited by O. Schreer, P. Kauff and T. Sikora © 2005 John Wiley & Sons, Ltd

group venues. Applications such as e-theatres, d-cinemas, home theatres and immersive televisions are envisioned and/or being investigated by many R&D departments around the world. Television, computer games, sports arenas, live events or cinema as we know them today will inevitably develop into new immersive applications to satisfy consumer demands during the coming decades.

It is very difficult to predict developments in the field of immersive media beyond the topics discussed today. But the examples pointed out above already indicate a shift of paradigms in the way we will capture, transmit and consume media information in the future. Due to falling prices and advancing quality, large-screen displays, audio-visual 3D scene representation and intuitive human–machine interfaces will become more and more established in daily use, especially in offices and in home environments. Immersive systems will leave its experimental state and immersive portals will become ubiquitous in business and entertainment. The impact for consumers as well as for business processes and value chains will be drastic.

The development from two-dimensional (2D) towards three-dimensional (3D) audiovisual communications is generally seen as one of the key components for the envisioned applications. Scientific challenges in this field are manifold. They range from high-quality 3D analysis of audio and video and arbitrary view and sound synthesis to encoding of 3D audio and video. Understanding of real-time implementation issues, as well as system architectures and network aspects will be essential for the success of these applications. The introduction of many of these services will require new standards for the representation and coding of 3D audiovisual data. Since many of these services will change the way of how we consume and interact with media applications, it is important to take human factors research into account. The ultimate goal is to develop applications with sufficient service quality and user acceptance. The presence research community contributes to many aspects of this kind of user-centred communications.

This book presents a comprehensive overview of the principles and concepts involved in the fascinating field of 3D audiovisual communications. It offers a practical step-by-step walk through the various challenges, concepts, components and technologies involved in the development of applications and services. Researchers and students interested in the field of 3D audiovisual communications will find this book a valuable resource, covering a broad overview of the current state of the art. Practical engineers from industry will find this book useful in envisioning and building innovative applications.

The book is divided in four major parts. The first part introduces to the challenging field of 3D video communications by presenting the most important applications in this domain, namely 3D television, free view point video and immersive videoconferencing. The second part covers the theoretical aspects of 3D video and audio processing. Following the logical order of a common signal processing chain, the third part is related to 3D reproduction of audio-visual content. In the last part, several aspects of 3D data sensors are discussed.

The aim of Section I *Applications of 3D Videocommunication* is to give a comprehensive overview on the state of the art, the challenges and the potential of 3D videocommunication. This part opens with a chapter on *History of Telepresence* by W.A. IJsselsteijn. It presents the foundation and justification for this new field of research and development. A historical review describes how the term tele-presence emerged. The following chapter on *3D TV Broadcasting* by C. Fehn presents in detail one of the key applications in the field of 3D videocommunications. The history of television, the concept for a next generation television system and an elaborate description of the end-to-end stereoscopic video chain are discussed. These new emerging technologies also have a drastic impact on content creation and postproduction. To this end the chapter on *3D in Content Creation and Post-production* by O. Grau discusses new trends and perspectives in this domain. Since the ultimate goal of 3D audiovisual communication is to provide the user with a free view point, the chapter *Free Viewpoint Systems* by M. Tanimoto describes the challenges in this area and presents first results of experimental systems. Bidirectional 3D videocommunication is embedded in the concept of immersive videoconferencing. A chapter on *Immersive Videoconferencing* by P. Kauff and O. Schreer presents the history and the current state of the art of such systems for telepresence. Visionary approaches implemented in prototypes of immersive videoconferencing systems and immersive portals are outlined and discussed.

Section II of the book addresses the question of how 3D audiovisual data may be represented and processed. Our prime goal is to provide the reader with a complete overview on all aspects related to processing of audio and video. The chapter *Fundamentals of Multiple-view Geometry* by S. Ivekovic, A. Fusiello and E. Trucco outlines the theory relevant for understanding the imaging process of a 3D scene onto a single camera. The chapter focuses on the pinhole camera model, a stereo camera system by explaining the key issues of epipolar geometry and finally the three-view geometry based on the trifocal tensor. Several aspects of rectification and reconstruction are covered as well. This chapter provides the theoretical foundation for the following chapters of the section.

Stereo analysis provides implicit depth information from few camera images of the same scene. The chapter *Stereo Analysis* by N. Atzpadin and J. Mulligan illustrates in detail current approaches in stereo processing using two or three cameras. The fundamental challenges of disparity analysis are discussed and an overview on different algorithms is presented. More precise 3D models can be generated based on multiple views of a scene or an object. The chapter *Reconstruction of Volumetric 3D Models* by P. Eisert focuses on the relevant approaches in this domain — important for many new 3D multimedia services.

A next important step in the 3D processing chain consists of rendering novel views. In Chapter 9 *View Synthesis and Rendering Methods*, R. Koch and J.-F. Evers-Senne provide a classification of existing rendering methods. The subdivision in methods without geometry information, methods with implicit and explicit geometry gives a comprehensive insight into recently developed approaches in this new field of research.

The chapter *3D Audio Capture and Analysis* by M. Schwab and P. Noll covers aspects of the 3D acquisition process of human speech. This includes echo-control, noise reduction and 3D audio source localization to support convincing rendering of audiovisual scene content.

Standardization is a key issue for interoperability of systems from different vendors. Chapter 11 *Coding and Standardization* by A. Smolic and T. Sikora outlines the basic coding strategies frequently used for audio, image and video storage and transmission. International coding standards such as ITU, MPEG-2/4 and MP3 are discussed in this context. The most recent standardization activity relevant to 3D videocommunications is the MPEG-4 AdHoc-Group work on 3D Audio/Visual. The reader is provided with a detailed overview on these activities.

Section III covers different aspects of 3D reproduction of audiovisual content. It is introduced by a chapter on *Human Factors of 3D Displays* by W.A. IJsselsteijn, P.J.H. Seuntiens and L.M.J. Meesters. The authors discuss several aspects of stereoscopic viewing. The human factors aspect is essential for the development of convincing 3D video systems. The basics of human depth perception are presented since knowledge in this domain is fundamental for stereoscopic viewing. Principles of stereoscopic reproduction and the impact on stereoscopic image quality are discussed. The chapter *3D Displays* by S. Pastoor discusses core technologies for existing 3D displays. Aided viewing as well as autostereoscopic viewing approaches are addressed and discussed in the context of existing display prototype systems and products.

In conjunction with mixed reality applications, head-mounted displays (HMD) play an important role for visualization of virtual content in a real scene. Developments in this field are outlined in Chapter 14 on *Mixed Reality Displays* by S. Pastoor and C. Conomis. After a brief description of challenges of mixed reality displays and some aspects of human spatial vision in this field, a comprehensive overview of different technologies and systems is given.

Even in situations where vision is dominant, the auditory sense helps to analyse the environment and creates a feeling of immersion. The correct or at least plausible reproduction of spatial audio becomes an important topic. T. Sporer and S. Brix present the fundamentals of *Spatialized Audio and 3D Audio Rendering* in Chapter 15.

Section IV covers the field of active 3D data sensors. Active techniques enable the creation of very detailed 3D models with high accuracy. In Chapter 16 *Sensor-based Depth Capturing* by J.G.M. Goncalves and V. Sequeira, various active techniques for the capture of range images are outlined. The authors discuss limitations, accuracies and calibration aspects of these methods.

Chapter 17 *Tracking and User Interface for Mixed Reality* by Y. Abdeljaoued, D. Marimon; Sanjvan and T. Ebrahimi completes the book. The authors discuss the tracking of objects for the purpose of accurate registration between the real and virtual world. Furthermore, the importance of interaction technologies for the creation of convincing mixed reality systems is emphasized.

Section I
Applications of
3D Videocommunication

1

History of Telepresence

Wijnand A. IJsselsteijn

Eindhoven University of Technology, Eindhoven, The Netherlands

1.1 INTRODUCTION

The term *telepresence* was first used in the context of teleoperation by Marvin Minsky (suggested to him by his friend Pat Gunkel) in a bold 1979 funding proposal *Toward a Remotely-Manned Energy and Production Economy*, the essentials of which were laid down in his classic 1980 paper on the topic (Minsky 1980). It refers to the phenomenon that a human operator develops a sense of being physically present at a remote location through interaction with the system's human interface, that is, through the user's actions and the subsequent perceptual feedback he/she receives via the appropriate teleoperation technology.

The concept of presence had been discussed earlier in the context of theatrical performances, where actors are said to have a 'stage presence' (to indicate a certain strength and convincingness in the actor's stage appearance and performance). Bazin (1967) also discussed this type of presence in relation to photography and cinema. He writes:

> Presence, naturally, is defined in terms of time and space. 'To be in the presence of someone' is to recognise him as existing contemporaneously with us and to note that he comes within the actual range of our senses – in the case of cinema of our sight and in radio of our hearing. Before the arrival of photography and later of cinema, the plastic arts (especially portraiture) were the only intermediaries between actual physical presence and absence. Bazin (1967), p. 96, originally published in *Esprit* in 1951.

Bazin noted that in theatre, actors and spectators have a reciprocal relationship, both being able to respond to each other within shared time and space. With television, and any other broadcast medium, this reciprocity is incomplete in one direction, adding a new variant of 'pseudopresence' between presence and absence. Bazin:

> The spectator sees without being seen. There is no return Flow . . . . Nevertheless, this state of not being present is not truly an absence. The television actor has a sense of the million of ears and eyes virtually present and represented by the electronic camera. Bazin (1967), p. 97, footnote.

3D Videocommunication — Algorithms, concepts and real-time systems in human centred communication
Edited by O. Schreer, P. Kauff and T. Sikora © 2005 John Wiley & Sons, Ltd

The sense of *being together* and interacting with others within a real physical space can be traced back to the work of Goffman (1963), who used the concept of *co-presence* to indicate the individual's sense of perceiving others as well as the awareness of others being able to perceive the individual:

> The full conditions of co-presence, however, are found in less variable circumstances: persons must sense that they are close enough to be perceived in whatever they are doing, including their experiencing of others, and close enough to be perceived in this sensing of being perceived. Goffman (1963), p. 17.

This mutual and recursive awareness has a range of consequences on how individuals present themselves to others. Note, however, that Goffman applied the concept of co-presence only to social interactions in 'real' physical space. In our current society, the sense of co-presence *through a medium* is of significant importance as a growing number of our human social interactions are mediated, rather than co-located in physical space.

Since the early 1990s onwards, presence has been studied in relation to various media, most notably virtual environments (VEs). Sheridan (1992) refers to presence elicited by a VE as 'virtual presence', whereas he uses 'telepresence' for the case of teleoperation that Minsky (1980) was referring to. From the point of view of psychological analysis, a distinction based on enabling technologies is unnecessary and the broader term *presence* is used in this chapter to include both variations.

A number of authors have used the terms 'presence' and 'immersion' interchangeably, as they regard them as essentially the same thing. However, in this chapter, they are considered as different concepts, in line with, for instance, Slater and Wilbur (1997) and Draper *et al.* (1998). Immersion is a term which is reserved here for describing a set of physical properties of the media technology that may give rise to presence. A media system that offers display and tracking technologies that match and support the spatial and temporal fidelity of real-world perception and action is considered immersive. For an overview of criteria in the visual domain, see IJsselsteijn (2003). In a similar vein, Slater and Wilbur (1997) refer to immersion as the objectively measurable properties of a VE. According to them it is the 'extent to which computer displays are capable of delivering an inclusive, extensive, surrounding, and vivid illusion of reality to the senses of the VE participant' (p. 604). Presence can be conceptualised as the *experiential counterpart* of immersion — the human response. Presence and immersion are logically separable, yet several studies show a strong empirical relationship, as highly immersive systems are likely to engender a high degree of presence for the participant.

Lombard and Ditton (1997) reviewed a broad body of literature related to presence and identified six different conceptualizations of presence: realism, immersion, transportation, social richness, social actor within medium, and medium as social actor. Based on the commonalities between these different conceptualizations, they provide a unifying definition of presence as the *perceptual illusion of non-mediation*, that is, the extent to which a person fails to perceive or acknowledge the existence of a medium during a technologically mediated experience. The conceptualizations Lombard and Ditton identified can roughly be divided into two broad categories – *physical* and *social*. The physical category refers to the sense of being physically located in mediated space, whereas the social category refers to the feeling of being together, of social interaction with a virtual or remotely located communication partner. At the intersection of these two categories, we can identify *co-presence* or a sense of being together in a shared space at the same time, combining significant characteristics of both physical and social presence.

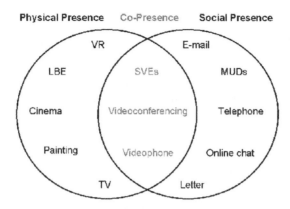

Figure 1.1 A graphical illustration of the relationship between physical presence, social presence and co-presence, with various media examples. Abbreviations: VR = virtual reality; LBE = location-based entertainment; SVEs = shared virtual environments; MUDs = multi-user dungeons. Technologies vary in both spatial and temporal fidelity

Figure 1.1 illustrates this relationship with a number of media examples that support the different types of presence to a varying extent. The examples vary significantly in both spatial and temporal fidelity. For example, while a painting may not necessarily represent physical space with a great degree of accuracy (although there are examples to the contrary, as we shall see), interactive computer graphics (i.e., virtual environments) have the potential to engender a convincing sense of physical space by immersing the participant and supporting head-related movement parallax. For communication systems, the extent to which synchronous communication is supported varies considerably. Time-lags are significant in the case of letters, and almost absent in the case of telephone or videoconferencing.

It is clear that physical and social presence are distinct categories that can and should be meaningfully distinguished. Whereas a unifying definition, such as the one provided by Lombard and Ditton (1997), accentuates the common elements of these different categories, it is of considerable practical importance to keep the differences between these categories in mind as well. The obvious difference is that of *communication* which is central to social presence, but unnecessary to establish a sense of physical presence. Indeed, a medium can provide a high degree of physical presence without having the capacity for transmitting reciprocal communicative signals at all. Conversely, one can experience a certain amount of social presence, or the 'nearness' of communication partners, using applications that supply only a minimal physical representation, as is the case, for example, with telephone or internet chat.

This is not to say, however, that the two types of presence are unrelated. There are likely to be a number of common determinants, such as the immediacy of the interaction, that are relevant to both social and physical presence. As illustrated in Figure 1.1, applications such as videoconferencing or shared virtual environments are in fact based on providing a mix of both the physical and social components. The extent to which shared space adds to the social component is an empirical question, but several studies have shown that as technology increasingly conveys non-verbal communicative cues, such as facial expression, gaze direction, gestures, or posture, social presence will increase.

In the remainder of this introductory chapter the historical development of a number of relevant presence technologies is described, with particular emphasis on their psychological impact. Though most media discussed here are relatively recent, the desire to render the real and the magical, to create illusory deceptions, and to transcend our physical and mortal existence may be traced back tens of thousands of years, to paleolithic people painting in a precisely distorted, or anamorphic, manner on the natural protuberances and depressions of cave walls in order to generate a three-dimensional appearance of hunting scenes. These paintings subtly remind us that, in spite of the impressive technological advances of today, our interests in constructing experiences through media are by no means recent. It is beyond the scope of this chapter to provide an exhaustive historical analysis; rather we want to inform our current endeavors and place them in a somewhat more humbling perspective.

1.2 THE ART OF IMMERSION: BARKER'S PANORAMAS

On June 17, 1787 Irish painter Robert Barker received a patent for a process under the name of 'la nature à coup d'oeil' by means of which he could depict a wide vista onto a completely circular surface in correct perspective. The *Repertory of Arts* which published the patent specifications in 1796 noted: 'This invention has since been called Panorama' (Oettermann 1997). Today, the term *Panorama* is used to denote a view or vista from an elevated lookout point, or, more metaphorically, to refer to an overview or survey of a particular body of knowledge, such art or literature. In the late 18th century, however, it was in fact a neologism created from two Greek roots, *pan*, meaning 'all', and *horama*, meaning 'view', to specifically describe the form of landscape painting which reproduced a 360-degree view. Its common usage today reflects some of the success of this art form at the time of its introduction.

The aim of the panorama was to convincingly reproduce the real world such that spectators would be tricked into believing that what they were seeing was genuine. Illusionistic or *trompe l'oeil* paintings had been a well-known phenomenon since Roman times, and such paintings would create the illusion of, for instance, walls containing a window to the outside world or a ceiling containing a view to the open sky. However, an observer's gaze can always move beyond the frame, where the physical surroundings often contradict the content of the painted world.

With panoramas, any glimpse of the real physical environment is obscured as the painting completely surrounds the viewer. Often, an observation platform with an umbrella-shaped roof (velum) was constructed such that the upper edge of the unframed canvas would be obscured from view (see Figure 1.2). The bottom edge of the painting would be obscured through either the observation platform itself or by means of some *faux terrain* stretching out between the platform and the canvas.

For example, the well-known *Panorama Mesdag*, painted in 1881 by Hendrik Willem Mesdag, offers a mesmerising view of the Dutch coast at Scheveningen. In the foreground, a real sandy beach with seaweed, fishing nets, anchors, and other assorted sea-related paraphernalia is visible and connects seamlessly to the beach in the painting. The top of the canvas is obscured through the roof of the beach tent one enters as one ascends the staircase, and emerges onto the viewing platform, surrounded by a balustrade. The viewer is completely surrounded by the illusionistic painting, which becomes particularly convincing as one looks out into the distance, where neither stereoscopic vision nor head-related movement parallax can provide conflicting information about the perceptual reality of what one is seeing.

Barker's first panoramic painting was a 21-metre-long 180-degree view of Edinburgh, the city where Barker worked as a drawing teacher. His 'breakthrough' piece, however, was the *Panorama of London*, first exhibited in 1792. After a successful tour of the English provinces,

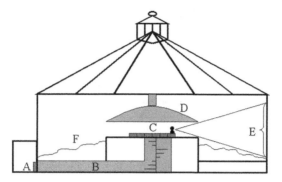

Figure 1.2 Cross-section of a panorama, consisting of: (A) entrance and box office, (B) darkened corridor and stairs, (C) observation platform, (D) umbrella-shaped roof, (E) observer's vertical angle of view, and (F) false terrain in the foreground

the panorama was shipped to the continent in 1799, to be first exhibited in Hamburg, Germany. A local paper, the *Privilegirte Wöchentliche Gemeinnützige Nachrichten von und für Hamburg* wrote in a review:

> It is most admirable. The visitor finds himself at the same spot on which the artist stood to make his sketch, namely on the roof of a mill, and from here has a most felicitous view of this great city and its environs in superb perspective. I would estimate that the viewer stands at a distance of some six paces from the exquisitely fashioned painting, so close that I wanted to reach out and touch it - but could not. I then wished there had been a little rope ladder tied to the railing on the roof of the mill, so I could have climbed down and joined the crowds crossing Blackfriar's Bridge on their way into the city. Quoted in Oettermann (1997), p. 185.

Seeing the same painting exhibited in Paris another reviewer commented for the German *Journal London und Paris*:

> No one leaves such a panorama dissatisfied, for who does not enjoy an imaginary journey of the mind, leaving one's present surroundings to rove in other regions! And the person who can travel in this manner to a panorama of his native country must enjoy the sweetest delight of all. Quoted in Oettermann (1997), p. 148.

Over the 19th century the panorama developed into a true mass medium, with millions of people visiting various panoramic paintings all across Europe, immersing themselves in the scenery of various great battles, admiring famous cities, or significant historic events. The panorama had many offshoots, most notably perhaps Daguerre's *Diorama* introduced in the 1820s, as well as the late 19th century *Photorama* by the Lumière brothers, and Grimoin-Sanson's *Cinéorama*, both of which applied film instead of painting. Fifty years later, when Hollywood needed to counter dropping box office receipts due to the introduction of television, attention turned again to a cinematographic panorama.

1.3 CINERAMA AND SENSORAMA

Cinerama, developed by inventor Fred Waller, used three 35 mm projections on a curved screen to create a 146-degree panorama. In addition to the impressive visuals, Cinerama also included a seven-channel directional sound system which added considerably to its psychological impact. Cinerama debuted at the Broadway Theatre, New York in 1952, with the independent production *This Is Cinerama*, containing the famous scene of the vertigo-inducing roller coaster ride, and was an instant success. The ads for *This is Cinerama* promised: 'You won't be gazing at a movie screen — you'll find yourself swept right into the picture, surrounded with sight and sound.' The film's program booklet proclaimed:

> You gasp and thrill with the excitement of a vividly realistic ride on the roller coaster . . . You feel the giddy sensations of a plane flight as you bank and turn over Niagara and skim through the rocky grandeur of the Grand Canyon. Everything that happens on the curved Cinerama screen is happening to you. And without moving from your seat, you share, personally, in the most remarkable new kind of emotional experience ever brought to the theater. Belton (1992), p. 189.

Interestingly, a precursor of the Cinerama system from the late 1930s — a projection system known as *Vitarama* — developed into what can be regarded as a forerunner of modern interactive simulation systems and arcade games. Vitarama consisted of a hemispherical projection of eleven interlocked 16 mm film tracks, filling the field of vision, and was adapted during the Second World War to a gunnery simulation system. The *Waller Flexible Gunnery Trainer*, named after its inventor, projected a film of attacking aircraft and included an electromechanical system for firing simulation and real-time positive feedback to the gunner if a target was hit. The gunnery trainer's displays were in fact already almost identical to the Cinerama system, so Waller did not have to do much work to convert it into the Cinerama system.

The perceptual effect of the widescreen presentation of motion pictures is that, while we focus more locally on character and content, the layout and motion presented to our peripheral visual systems surrounding that focus very much control our visceral responses. Moreover, peripheral vision is known to be more motion-sensitive than foveal vision, thereby heightening the impact of movement and optic flow patterns in the periphery, such as those engendered by a roller coaster sequence.

As Belton (1992) notes, the widescreen experience marked a new kind of relation between the spectator and the screen. Traditional narrow-screen motion pictures became associated, at least from an industry marketing point of view, with passive viewing. Widescreen cinema, on the other hand, became identified with the notion of *audience participation* — a heightened sense of engagement and physiological arousal as a consequence of the immersive wraparound widescreen image and multitrack stereo sound. The type of visceral thrills offered by Cinerama was not unlike the recreational participation that could be experienced at an amusement park, and Cinerama ads (Figure 1.3) often accentuated the audience's participatory activity by depicting them as part of the on-screen picture, such as sitting in the front seat of a roller coaster, 'skiing' side by side with on-screen water skiers, or hovering above the wings of airplanes (Belton 1992).

Unfortunately however, Cinerama's three projector system was costly for cinemas to install, seating capacity was lost to accommodate the level projection, required and a staff of seventeen people was needed to operate the system. In addition, Cinerama films were expensive to produce, and sometimes suffered from technical flaws. In particular, the seams where the three images were joined together were distractingly visible, an effect accentuated by variations in projector illumination (Belton 1992). Together these drawbacks prevented Cinerama from capitalizing on its initial success.

Following Cinerama, numerous other film formats have attempted to enhance the viewer's cinematic experience by using immersive projection and directional sound, with varying success. Today, the change to a wider aspect ratio, 1.65:1 or 1.85:1, has become a cinematic standard. In addition, some very large screen systems have been developed, of which IMAX, introduced at the World Fair in Osaka, Japan in 1970, is perhaps the best-known. When

Figure 1.3   Advertisement for Cinerama, 1952

projected, the horizontally run 70 mm IMAX film, the largest frame ever used in motion pictures, is displayed on screens as large as 30 × 22.5 m, with outstanding sharpness and brightness. By seating the public on steeply raked seats relatively close to the slightly curved screen, the image becomes highly immersive. As the ISC publicity says, 'IMAX films bring distant, exciting worlds within your grasp . . . It's the next best thing to being there' (Wollen 1993). IMAX has also introduced a stereoscopic version, *3-D Imax*, and a hemispherical one known as *Omnimax*. IMAX and other large-format theaters have been commercially quite successful, despite its auditoria being relatively few and far apart. (There are some 350 theaters worldwide that are able to project large-format movies.)

Meanwhile, cinematographer and inventor Morton Heilig was impressed and fascinated by the Cinerama system, and went on to work out a detailed design for an *Experience Theater* in 1959 that integrated many of the ideas previously explored, and expanded on them considerably. His goal was to produce total cinema, a complete illusion of reality engendering a strong sense of presence for the audience. To Heilig, Cinerama was only a promising start, not a conclusion. He wrote:

> If the new goal of film was to create a convincing illusion of reality, then why not toss tradition to the winds? Why not say goodbye to the rectangular picture frame, two-dimensional images, horizontal audiences, and the limited senses of sight and hearing, and reach out for everything and anything that would enhance the illusion of reality? Heilig (1998), pp. 343–344.

This is exactly what he aimed to do by designing the Experience Theater and subsequently the *Sensorama Simulator* (Heilig 1962) and *Telesphere Mask*, possibly the first head-mounted display. With building the Sensorama Simulator, Heilig tried to stimulate as much as possible all the different senses of the observers through coloured, widescreen, stereoscopic, moving images, combined with directional sound, aromas, wind and vibrations (Figure 1.4). The patent application for the Experience Theater explicitly mentions the presence-evoking capacity of this system:

> By feeding almost all of man's sensory apparatus with information from the scenes or programs rather than the theater, the experience theater makes the spectator in the audience feel that he has been physically transported into and made part of the scene itself. Heilig (1971).

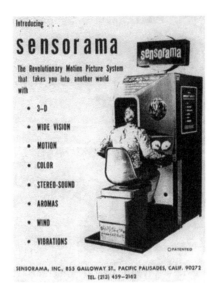

Figure 1.4 Advertisement for Sensorama, 1962

An interesting point illustrated by this quotation is the importance of receiving little or no sensory information that conflicts with the mediated content, such as incongruent information from one's physical surroundings (in this case the theatre). This signals the mediated nature of the experience and thus acts as a strong negative cue to presence — something that Robert Barker had already understood very well in the 18th century. Because of Sensorama's ability to completely immerse the participant in an alternate reality, the system is often cited as one of the precursors of modern virtual environment (VE) systems (Coyle 1993; Rheingold 1991). However, despite the considerable accomplishments of Heilig's prototypes, they were still based on a passive model of user perception, lacking the possibility of user action within the mediated environment, but rather offering completely predetermined content. In both Cinerama and Sensorama the participant was strictly a passenger. As we shall see, virtual environments derive their strength precisely from allowing participants to jump in the driver's seat — to *interact* with content in real-time.

1.4 VIRTUAL ENVIRONMENTS

Virtual environments allow users to interact with synthetic or computer-generated environments, by moving around within them and interacting with objects and actors represented there. Virtual environments are sometimes also referred to as *virtual reality* or VR. While both terms are considered essentially synonymous, the author agrees with Ellis (1991) who notes that the notion of an *environment* is in fact the appropriate metaphor for a head-coupled, coordinated sensory experience in three-dimensional space. In its best-known incarnation, VEs are presented to the user via a head-mounted display (HMD) where the (often stereoscopic) visual information is presented to the eyes via small CRTs or LCDs, and auditory information is presented by headphones. Because of weight and size restrictions, the resolution and angle of view of most affordable HMDs are quite poor. An HMD is usually fitted with a position tracking device which provides the necessary information for the computer to calculate and render the appropriate visual and auditory perspective, congruent with the user's head and body movements. The support of head-slaved motion parallax allows for the correct viewpoint-dependent transformations of the visual and aural scene, both of which are important for engendering a sense of presence in an environment (Biocca and Delaney 1995; Brooks 1999; Burdea and Coiffett 1994; Kalawsky 1993; Stanney 2002). A detailed description of current technologies in this field can be found in Chapter 14.

An alternative interface to the HMD is the BOOM (binocular omni-oriented monitor) where the display device is not worn on the head but mounted onto a flexible swivel arm construction so that it can be freely moved in space. Because a BOOM is externally supported and not worn on the head, heavier and hence higher resolution and larger angle-of-view displays can be used. Viewpoint position can be calculated by knowing the length of the swivel arms and measuring the angles of its joints. Moving a BOOM needs to be done manually, thereby occupying one of the hands. Tactile and force feedback is also sometimes provided through various devices ranging from inflatable pressure pads in data gloves or body suits to force-feedback arms or exoskeleton systems. Although there is an increasing interest in engineering truly multisensory virtual environments, such systems are still rather the exception.

A second common design of immersive virtual environments is through multiple projection screens and loudspeakers placed around the user. A popular implementation of such a projection system is known as the CAVE (Cruz-Neira *et al.* 1993), a recursive acronym for CAVE Automatic Virtual Environment, and a reference to *The Simile of the Cave* from Plato's *The Republic*, in which he discusses about inferring reality from projections (shadows) thrown on the wall of a cave. The 'standard' CAVE system, as it was originally developed at the Electronic Visualization Laboratory at the University of Illinois at Chicago, consists of three stereoscopic rear-projection screens for walls and a down-projection screen for the floor. A six-sided projection space has also been recently developed (at KTH in Stockholm, Sweden), allowing projections to fully surround the user, including the ceiling and the floor. Participants entering such room-like displays are surrounded by a nearly continuous virtual scene. They can wear shutterglasses in order to see the imagery in stereo, and wearing a position tracker is required to calculate and render the appropriate viewer-centred perspective. Although more than one person can enter a CAVE at any one time, only the participant controlling the position tracker will be able to perceive the rendered view in its correct perspective.

A spherical variation of CAVE-style wall-projection systems is known as the CyberSphere (Figure 1.5). The principle here is to project on the sides of a transparent sphere, with the participant being located on the inside. The movement of the participant is tracked by sensors at the base of the sphere and the projected images are updated accordingly. By integrating the display and locomotion surfaces, this type of display offers an interesting solution to the problem of limited locomotion in projection-based VEs, as any fixed display or projection surface will define the boundaries of physical locomotion.

Other less immersive implementations of virtual environments are gaining in popularity because they do not isolate the user (like an HMD) or require a special room (like a CAVE or CyberSphere) and are thus more easily integrated with daily activities. Such systems include stationary projection desks (e.g., the ImmersaDesk), walls, or head-tracked desktop systems (Fisher 1982). The latter is sometimes referred to as fish-tank virtual reality (Arthur *et al.* 1993; Ware *et al.* 1993).

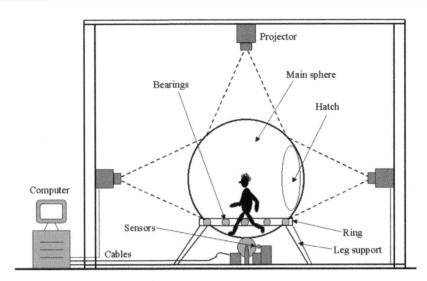

Figure 1.5 The CyberSphere allows for a participant to walk inside a sphere with projections from all sides

The first virtual environment display system where a totally computer-generated image was updated according to the user's head movements and displayed via a head-referenced visual display was introduced in 1968 by Ivan Sutherland, who is now generally acknowledged as one of the founding fathers of virtual reality. The helmet device Sutherland built was nicknamed 'Sword of Damocles' as it was too heavy to wear and had to be suspended from the ceiling, hanging over the user's head. In his classic 1965 paper *The Ultimate Display* he describes his ideas of immersion into computer-generated environments via new types of multimodal input and output devices. He concludes his paper with the following vision:

> The ultimate display would, of course, be a room within which the computer can control the existence of matter. A chair displayed in such a room would be good enough to sit in. Handcuffs displayed in such a room would be confining, and a bullet displayed in such a room would be fatal. With appropriate programming such a display could literally be the Wonderland into which Alice walked. Sutherland (1965), p. 508.

Although today we are still clearly a long way removed from such an ultimate display, current virtual environments, through their multisensory stimulation, immersive characteristics and real-time interactivity, already stimulate participants' perceptual–motor systems such that a strong sense of presence within the VE is often experienced. To date, the majority of academic papers on presence report on research that has been performed in the context of virtual environments, although a significant portion has also been carried out in the context of (stereoscopic) television and immersive telecommunications.

1.5   TELEOPERATION AND TELEROBOTICS

Interactive systems that allow users to control and manipulate real-world objects within a remote real environment are known as *teleoperator systems*. Remote-controlled manipulators (e.g., robot arms) and vehicles are being employed to enable human work in hazardous or challenging environments such as space exploration, undersea operations, minimally invasive surgery, or hazardous waste clean-up. The design goal of smooth and intuitive teleoperation triggered a considerable research effort in the area of human factors — see, for example, Johnson and Corliss (1971), Bejczy (1980), Sheridan (1992), and Stassen and Smets (1995). Teleoperation can be considered as one of the roots of today's presence research. In teleoperation, the human operator continuously guides and causes each change in the remote manipulator in real-time (master-slave configurations). Sensors at the remote site (e.g., stereoscopic camera, force sensors) provide continuous feedback about the slave's position in relation to the remote object, thereby closing the continuous perception–action loop that involves the operator, the master system through which the interaction takes place and the remote slave system.

Telerobotic systems are slightly different from teleoperation systems in that the communication between the human operator and slave robot occurs at a higher level of abstraction. Here, the operator primarily indicates goals, which the slave robot subsequently carries out by synthesizing the intermediate steps required to reach the specified goal. Where teleoperation systems aim to offer real-time and appropriately mapped sensory feedback, telerobotic systems have an intrinsic delay between issuing the command and carrying out the necessary action sequence.

As discussed earlier, the term telepresence was first used by Marvin Minsky in his 1979 proposal, which was essentially a manifesto to encourage the development of the science and technology necessary for a remote-controlled economy that would allow for the elimination of many hazardous, difficult or unpleasant human tasks, and would support beneficial developments such as the creation of new medical and surgical techniques, space exploration, and teleworking. He writes:

> The biggest challenge to developing telepresence is achieving that sense of 'being there'. Can telepresence be a true substitute for the real thing? Will we be able to couple our artificial devices naturally and comfortably to work together with the sensory mechanisms of human organisms? Minsky (1980), p. 48.

A critical issue in creating a sense of telepresence for the human operator at the remote site is the amount of time delay that occurs between the operator's actions and the subsequent feedback from the environment. Such delays may occur as a consequence of transmission delays between the teleoperation site and the remote worksite or local signal processing time in nodes of the transmitting network. Transmission at the speed of light becomes a significant limiting factor only at extraordinary distances, e.g., interplanetary teleoperation applications such as the remote control of unmanned vehicles on Mars. One solution that has been proposed to overcome this limitation is the use of predictive displays, where computer-generated images of the remote manipulator may be precisely superimposed over the returning video image (in fact, a form of video-based augmented reality). The computer-generated image will respond instantly to the operator commands, where the video image of the actual telerobotic system follows after a short delay.

The questions Minsky raised in 1980 are still valid today. Although the remote-controlled economy did not arrive in the way he envisioned, the development of telepresence technologies has significantly progressed in the various areas he identified (Hannaford 2000), and remains an active research area pursued by engineers and human factors specialists worldwide. In addition, the arrival and widespread use of the internet brings us remote access to thousands of homes, offices, street corners, and other locations where webcameras have been set up (Campanella 2000). In some cases, because of the two-way nature of the internet, users can log on to control a variety of telerobots and manipulate real-world objects. A well-known example is Ken Goldberg's *Telegarden*, where a real garden located in the Ars Electronica Museum in Austria was connected to the internet via a camera and where remote users could control a robotic arm to plant and water seeds, and subsequently watch their plants grow and flourish in real-time.

1.6 TELECOMMUNICATIONS

If, as it is said to be not unlikely in the near future, the principle of sight is applied to the telephone as well as that of sound, earth will be in truth a paradise, and distance will lose its enchantment by being abolished altogether. Arthur Mee, *The Pleasure Telephone*, 1898.

Media technologies have significantly extended our reach across space and time. They enable us to interact with individuals and groups beyond our immediate physical surroundings. An increasing proportion of our daily social interactions is mediated, i.e., occurs with representations of others, with virtual embodiments rather than physical bodies. The extent to which these media interactions can be optimised to be believable, realistic, productive, and satisfying has been the topic of scholarly investigations for several decades — a topic that is only increasing in relevance as new communication media emerge and become ubiquitous.

In their pioneering work, Short *et al.* (1976) conceptualized social presence as a way to analyse mediated communications. Their central hypothesis is that communication media vary in their degree of social presence and that these variations are important in determining the way individuals interact through the medium. Media capacity theories, such as social presence theory and media richness theory, are based on the premise that media have different capacities to carry interpersonal communicative cues. Theorists place the array of audiovisual communication media available to us today along a continuum ranging from face-to-face interaction at the richer, more social end and written communication at the less rich, less social end.

The majority of tele-relating studies to date have focused on audio- and videoconferencing systems in the context of professional, work-related meetings and computer-supported collaborative work (CSCW). Using such systems, participants typically appear in video-windows on a desktop system, or on adjoining monitors, and may work on shared applications that are shown simultaneously on each participant's screen. Examples include the work of Bly *et al.* (1993), and Fish *et al.* (1992). As more bandwidth becomes available (e.g., Internet2), the design ideal that is guiding much of the R&D effort in the telecommunication industry is to mimic face-to-face communication as closely as possible, and to address the challenges associated with supporting non-verbal communication cues such as eye contact, facial expressions and postural movements. These challenges are addressed in recent projects such as the National Tele-Immersion Initiative (Lanier 2001), VIRTUE (Kauff *et al.* 2000), im.point (Tanger *et al.* 2004) and TELEPORT (Gibbs *et al.* 1999), where the aim of such systems is to provide the remotely located participants with a sense of being together in a shared space, i.e., a sense of co-presence. In Chapter 5, a comprehensive overview on the current state-of-the-art is given.

The emergence and proliferation of email, mobile communication devices, internet chatrooms, shared virtual environments, advanced teleconferencing platforms and other telecommunication systems underlines the importance of investigating the basic human need of communication from a multidisciplinary perspective that integrates media design and engineering, multisensory perception, and social psychology. Add to this the increasingly social nature of interfaces and the increase in mediated communications with non-human entities (avatars, embodied agents), it becomes abundantly clear that we need to develop a deeper understanding, both in theory and in practice, of how people interact with each other and virtual others through communication media. The experience of social presence within different contexts and through different applications thus becomes a concept of central importance.

1.7 CONCLUSION

> When anything new comes along, everyone, like a child discovering the world, thinks that they've invented it, but you scratch a little and you find a caveman scratching on a wall is creating virtual reality in a sense. What is new here is that more sophisticated instruments give you the power to do it more easily. Virtual reality is dreams. Morton Heilig, quoted in Hamit (1993), p. 57.

André Bazin, pioneer of film studies, saw photography and cinema as progressive steps towards attaining the ideal of reproducing reality as nearly as possible. To Bazin, a photograph was first of all a reproduction of 'objective' sensory data and only later perhaps a work of art. In his 1946 essay, *The Myth of Total Cinema* (reprinted in his 1967 essay collection *What is Cinema?*), Bazin states that cinema began in inventors' dreams of reproducing reality with absolute accuracy and fidelity, and that the medium's technical development would continue until that ideal was achieved as nearly as possible.

The history of cinema and VR indeed appears to follow a relentless path towards greater perceptual realism, with current technologies enabling more realistic reproductions and simulations than ever before. But as Morton Heilig's quote at the beginning of this section echoes, the vision behind these developments is age-old. The search for the 'ultimate display', to use Sutherland's (1965) phrase, has been motivated by a drive to provide a perfect illusory deception, as well as the ancient desire for physical transcendence, that is, escaping from the confines of the physical world into an 'ideal' world dreamed up by the mind (Biocca *et al.* 1995).

Whereas arts and entertainment enterprises were behind much of the developments in cinema and television, the initial development of virtual environments, teleoperation, telecommunication and simulation systems has mainly been driven by industrial and military initiatives. Despite these different historical roots, a major connection between work in virtual environments, aircraft simulations, telerobotics, location-based entertainment, and other advanced interactive and non-interactive media is the need for a thorough understanding of the human experience in real environments (Ellis 1991). It is the perceptual experience of such environments that we are trying to create with media displays (de Kort *et al.* 2003).

Another common theme linking the above areas is that presence research offers the possibility to engineer a better user experience: to optimize pleasurability, enhance the impact of the media content, and adapt media technology to human capabilities, thereby optimizing the effective and efficient use of media applications. A more fundamental reason for studying presence is that it will further our theoretical understanding of the basic function of mediation: How do media convey a sense of places, beings and things that are not here? How do our minds and bodies adapt to living in a world of technologies that enhance our perceptual, cognitive and motor reach? Presence research provides a unique and necessary bridge between media research on the one hand and the massive interdisciplinary programme on properties of perception and consciousness on the other (Biocca 2001).

Finally, feeling present in our physical surroundings is regarded as a natural, default state of mind. Presence in the 'real world' is so familiar, effortless and transparent that it is hard to become aware of the contributing processes, to appreciate the problems the brain must solve in order to feel present. Immersive technologies now offer a unique window onto the presence experience, enabling researchers to unravel the question of how this complicated perception comes into being. In this way, technology helps us to know ourselves. This is perhaps the most important reason of all for studying presence.

REFERENCES

Arthur K, Booth K and Ware C 1993 Evaluating 3D task performance for fish tank virtual worlds. *ACM Transactions on Information Systems* **11**, 239–265.

Bazin A 1967 *What is Cinema? Vol. 1.* University of California Press, Berkeley, CA.

Bejczy A 1980 Sensors, controls, and man-machine interface for advanced teleoperation. *Science* **208**, 1327–1335.

Belton J 1992 *Widescreen Cinema.* Harvard University Press, Cambridge, MA.

Biocca F 2001 Presence on the M.I.N.D. Presented at the European Presence Research Conference, Eindhoven, 9–10 October 2001.

Biocca F and Delaney B 1995 Immersive virtual reality technology In *Communication in the Age of Virtual Reality* (ed. Biocca F and Levy M) Lawrence Erlbaum Hillsdale, NJ pp. 57–124.

Biocca F, Kim T and Levy M 1995 The vision of virtual reality In *Communication in the Age of Virtual Reality* (ed. Biocca F and Levy M) Lawrence Erlbaum Hillsdale, NJ pp. 3–14.

Bly S, Harrison S and Irwin S 1993 Mediaspaces: Bringing people together in video, audio, and computing environments. *Communications of the ACM* **36**, 29–47.

Brooks F. 1999 What's real about virtual reality. *IEEE Computer Graphics and Applications* November/December, 16–27.

Burdea G and Coiffett P 1994 *Virtual Reality Technology.* Wiley, New York.

Campanella T 2000 Eden by wire: Webcameras and the telepresent landscape In *The Robot in the Garden* (ed. Goldberg K) MIT Press Cambridge, MA pp. 22–46.

Coyle R 1993 The genesis of virtual reality In *Future Visions: New Technologies of the Screen* (ed. Hayward P and Wollen T) British Film Institute London pp. 148–165.

Cruz-Neira C, Sandin D and DeFanti T 1993 Surround-screen projection-based virtual reality: The design and implementation of the CAVE. *Computer Graphics: Proceedings of SIGGRAPH* pp. 135–142.

de Kort Y, IJsselsteijn W, Kooijman J and Schuurmans Y 2003 Virtual laboratories: Comparability of real and virtual environments for environmental psychology. *Presence: Teleoperators and Virtual Environments* **12**, 360–373.

Draper J, Kaber D and Usher J 1998 Telepresence. *Human Factors* **40**, 354–375.

Ellis S 1991 Nature and origins of virtual environments: A bibliographical essay. *Computing Systems in Engineering* **2**, 321–347.

Fish R, Kraut R, Root R and Rice R 1992 Evaluating video as a technology for informal communications. *Proceedings of the CHI '92* pp. 37–48.

Fisher S 1982 Viewpoint dependent imaging: An interactive stereoscopic display. *Proceedings of the SPIE* **367**, 41–45.

Gibbs S, Arapis C and Breiteneder C 1999 Teleport — towards immersive copresence. *Multimedia Systems* **7**, 214–221.

Goffman E 1963 *Behavior in Public Places: Notes on the Social Organisation of Gatherings.* The Free Press, New York.

Hamit F 1993 *Virtual Reality and the Exploration of Cyberspace.* SAMS Publishing, carmel, IN.

Hannaford B 2000 Feeling is believing: History of telerobotics technology In *The Robot in the Garden* (ed. Goldberg K) MIT Press Cambridge, MA pp. 246–274.

Heilig M 1962 Sensorama simulator U.S.Patent 3050870.

Heilig M 1971 Experience theater U.S.Patent 3628829.

Heilig M 1998 Beginnings: Sensorama and the telesphere mask In *Digital Illusion* (ed. Dodsworth Jr. C) ACM Press New York, NY pp. 343–351.

IJsselsteijn W 2003 Presence in the past: What can we learn from media history? In *Being There: Concepts, Effects and Measurements of User Presence in Synthetic Environments* (ed. Riva G, Davide F and IJsselsteijn W) IOS Press Amsterdam, The Netherlands pp. 17–40.

Johnson E and Corliss W 1971 *Human Factors Applications in Teleoperator Design and Operation.* Wiley-Interscience, New York.

Kalawsky R 1993 *The Science of Virtual Reality and Virtual Enviornments.* Addison-Wesley, Reading, MA.

Kauff P, Schäfer R and Schreer O 2000 Tele-immersion in shared presence conference systems Paper presented at the International Broadcasting Convention, Amsterdam, September 2000.

Lanier J 2001 Virtually there. *Scientifc American* pp. 52–61.

Lombard M and Ditton T 1997 At the heart of it all: The concept of presence. *Journal of Computer-Mediated Communication.*

Minksy M 1980 Telepresence. *Omni* pp. 45–51.

Oettermann S 1997 *The Panorama: History of a Mass Medium.* Zone Books, New York.

Rheingold H 1991 *Virtual Reality.* Martin Secker & Warburg, London.

Sheridan T 1992 Musings on telepresence and virtual presence. *Presence: Teleoperators and Virtual Environments* **1,** 120–126.

Short J, Williams E and Christie B 1976 *The Social Psychology of Telecommunications.* Wiley, London.

Slater M and Wilbur S 1997 A framework for immersive virtual environments (FIVE): Speculations on the role of presence in virtual environments. *Presence: Teleoperators and Virtual Environments* **6,** 603–616.

Stanney K 2002 *Handbook or Virtual Environments: Design, Implementation, and Applications.* Lawrence Erlbaum, Mahwah, NJ.

Staseen H and Smets G 1995 Telemanipulation and telepresence In *Analysis, Design and Evaluation of Man-Machine Systems 1995* (ed. Sheridan T) Elsevier Science, Oxford pp. 13–23.

Sutherland I 1965 The ultimate display. *Proceedings of the IFIP Congress* **2,** 506–508.

Tanger R, Kauff P and Schreer O 2004 Immersive meeting point (im.point) – an approach towards immersive media portals. *Proceedings of Pacific-Rim Conference on Multimedia.*

Ware C, Arthur K and Booth K 1993 Fish tank virtual reality. *Proceedings of INTERCHI '93 Conference on Human Factors in Computing Systems* pp. 37–42.

Wollen T 1993 The bigger the better: From CinemaScope to Imax In *Future Visions: New Technologies of the Screen* (ed. Hayward P and Wollen T) British Film Institute London pp. 10–30.

2

3D TV Broadcasting

Christoph Fehn

Fraunhofer Institute for Telecommunications/Heinrich-Hertz-Institut, Berlin, Germany

2.1 INTRODUCTION

Three-dimensional broadcast television (3D TV) is believed by many to be the next logical development towards a more natural and life-like visual home entertainment experience. Although the basic technical principles of stereoscopic TV were demonstrated in the 1920s by John Logie Baird, the step into the third dimension still remains to be taken. The many different reasons that have hampered a successful introduction of 3D TV so far could now be overcome by recent advances in the area of 3D display technology, image analysis and image-based-rendering (IBR) algorithms, as well as digital image compression and transmission. Building on these latest trends, a modern approach to three-dimensional television is described in the following, which aims to fulfil important requirements for a future broadcast 3D TV system: (1) backwards-compatibility to today's digital 2D color TV; (2) low additional storage and transmission overhead; (3) support for autostereoscopic, single- and multiple-user 3D displays; (4) flexibility in terms of viewer preferences on depth reproduction; (5) simplicity in producing sufficient, high-quality 3D content.

This chapter is organized into four distinct parts. First of all, a historical overview is provided, which covers some of the most significant milestones in the area of 3D TV research and development. This is followed by a description of the basic fundamentals of the proposed new approach to three-dimensional television as well as a comparison with the 'conventional' concept of an end-to-end stereoscopic video chain. Thereafter, an efficient depth-image-based rendering (DIBR) algorithm is derived, which is used to synthesize 'virtual' stereoscopic views from monoscopic colour video and associated per-pixel depth information. Some specific implementation details are also given in this context. Finally, the backwards-compatible compression and transmission of 3D imagery using state-of-the-art video coding standards is described.

3D Videocommunication — Algorithms, concepts and real-time systems in human centred communication
Edited by O. Schreer, P. Kauff and T. Sikora © 2005 John Wiley & Sons, Ltd

2.2 HISTORY OF 3D TV RESEARCH

The history of stereoscopy (from the Greek στερεός 'solid' and σκοπιά 'seeing') can be traced back to the year 1838, when Sir Charles Wheatstone, a british researcher and inventor, developed a mirror device that enabled the viewer to fuse two slightly different perspective images into a single, three-dimensional percept (Wheatstone 1838). His rather bulky and unhandy original *stereoscope* was further developed and minimized by Sir David Brewster in 1844 and at the end of the 19th century many families in both Europe and the US enjoyed watching stereoscopic still photographs of current events, scenic views or popular personalities (Lipton 1982).

With the advent of motion pictures, the popularity of the stereoscope began to decline, but proposals and concepts for stereoscopic cinema (early 1900s) and stereoscopic television (1920s) were present at the dawn of their monoscopic counterparts (IJsselsteijn *et al.* 2002). At the world fair in Paris in 1903, the Lumière brothers showed the first three-dimensional short movies. These screenings, however, could be watched by only one viewer at a time — with a modified stereoscope. The first full-length 3D feature film, *The Power Of Love*, was shown simultaneously to a large group of viewers at the Ambassador Hotel in Los Angeles at the end of the silent movie era in 1922. The projected left- and right-eye images were separated by the anaglyphic process. In 1928, John Logie Baird, a well-known british television pioneer, also applied the principle of the stereoscope to an experimental TV setup based on Paul Nipkow's early perforated disc technique (Tiltman 1928).

In spite of these very successful early demonstrations, it was not until the 1950s that Hollywood turned to 3D as the 'next big thing'. The motivations for this, however, were economic rather than artistic. In addition to the persecution of actors, writers and directors during Senator McCarthy's anti-communist crusade, the movie industry had to counteract the dropping box office receipts that occurred as a consequence of the increasing popularity of a competing technology — television (IJsselsteijn *et al.* 2002). On 26 November 1952, a low-budget independent 3D feature film called *Bwana Devil* opened in Los Angeles to capacity audiences. The overwhelming success of this movie – it earned nearly $100 000 in one week at this single theatre — started a 3D craze. At the height of the boom, between 1952 and 1954, Hollywood produced over sixty-five 3D movies, including Alfred Hitchcock's *Dial M For Murder* (1954) starring Grace Kelly and Ray Milland as well as Jack Arnold's classic horror B movie *Creature from the Black Lagoon* (1954). Publicity ads from this time (Figure 2.1) show how the movie industry tried to win back audiences by emhpasizing the increased sensation of the 3D experience with catch-phrases such as: '3-Dimension – Fantastic Sights Leap at You!', *It Came from Outer Space* (1953), 'Its Thrills Come Off the Screen Right at You!', *House of Wax* (1953) and 'The Shocking Chills of the Sensational Suspense Novel Leap From the Screen . . . in 3-Dimensions!', *The Maze — The Deadliest Trap in the World* (1953).

However, despite the immense first ardor, 3D cinema failed to fulfil its great expectations. Pourly produced footage caused by a lack of photographic experience with the new technology, low-quality 3D exploitation films, inadequate quality control in the laboratories and badly operated projection systems in the movie theatres quickly earned 3D films a bad reputation. It all added up in headaches, eyestrain and the demise of 3D movies in little more than a year (Lipton 1982). The few new 3D releases after 1954 couldn't revitalize the boom; most of them were never shown in 3D to the public. Today, large-scale three-dimensional

Figure 2.1 Publicity ads for 3D movies during the boom from 1952 to 1954. *It Came from Outer Space* (1953); *House of Wax* (1953); *The Maze – The Deadliest Trap in the World* (1953)

productions have found a small niche market in specially equipped cinemas in theme and amusement parks as well as in the currently commercially relatively successful IMAX 3D theaters.

Broadcast television also had a rather ambivalent relationship with 3D. While the first known experimental 3D TV broadcast in the US was on 29 April 1953, it took another 27 years until the first 'non-experimental' 3D broadcast was aired over SelectTV, a Los Angeles Pay-TV system, on 19 December 1980. The programme was a 3D feature film from 1953, *Miss Sadie Thompson*, and a 3D short starring the Three Stooges. Since that time, 3D TV has received some attention every now and then, with a few feature films and a number of TV shows such as *Family Matters*, *The Drew Carey Show* or *America's Funniest Videos* broadcasted using either the mentioned anaglyphic process or the time-parallax format. The hope, however, that this could be 'the TV of the future' was dashed by the poor quality attainable with these specific transmission and display technologies (Sand 1992).

With the foreseeable transition from analogue to digital TV services, new hope for 3D television arose in the early 1990s, and especially in Europe, a number of European-funded projects (e.g., COST 230, RACE DISTIMA, RACE PANORAMA, ACTS MIRAGE, ACTS TAPESTRIES) were set up with the aim to develop standards, technologies and production facilities for 3D TV (IJsselsteijn 2003). This work was added to by other groups, who focused on the investigation of the human factors requirements for high-quality stereoscopic television (Pastoor 1991).

Motivated by this revived interest in broadcast 3D services, the Moving Pictures Expert Group (MPEG) developed a compression technology for stereoscopic video sequences as part of the successful MPEG-2 standard (Imaizumi and Luthra 2002). The chosen approach, generally known as the multiview profile (MVP), can be regarded as an extension of the temporal scalability tool. It encodes the left-eye view as a base layer in conformance with the MPEG-2 Main Profile (MP), thus providing backwards-compatibility to 'conventional' 2D digital TV receivers. The right-eye view is encoded as an enhancement layer using the scalable coding tools with additional prediction from the base layer. While MPEG-2 has become the underlying technology for 2D broadcast TV worldwide, the multiview profile (MVP) has not found its application in any commercially available services so far.

Nevertheless, some promising attempts have been made to integrate stereoscopic TV and HDTV (high-definition television) into a new, high-quality 3D entertainment medium. In Japan, a 3D HDTV relay system developed by NHK was used in 1998 to transmit stereoscopically recorded live images of the Nagano Winter Games via satellite to a large-scale demonstration venue in Tokyo (Yuyama and Okui 2002) and a similar experiment was conducted in Korea/Japan during the 2002 FIFA World Cup, using a terrestrial and satellite network (Hur *et al.* 2002).

In recent years, a new paradigm for 3D TV has emerged that is based on the use of efficient image analysis and synthesis algorithms. The Australian company Dynamic Digital Depth (DDD), for example, has developed a proprietary (off-line) system that can convert monoscopic video sequences to stereoscopic 3D by appropriately displacing individual image segments (objects) in a captured scene (Dynamic Digital Depth 2004). Their technology has already been used successfully for the generation of high-quality 3D content for display in large IMAX 3D theatres.

For the future, further stimulation for the 3D market can be expected from two recent initiatives. (1) Initiated by the five major Japanese electronics companies Itochu, NTT Data, Sanyo Electric, Sharp and Sony, a 3D consortium was founded in March 2003, with the goal to enhance the potential market for 3D technologies (3D Consortium 2003). Subcommittees within this consortium, which at present unites over 200 members worldwide, discuss issues such as spreading image formats appropriate for various applications and I/O devices as well as developing guidelines and authoring tools for 3D content creation. (2) The MPEG committee established a new *ad hoc* group on 3D audio/visual coding (3DAV). This activity, started in December 2001, aims at the development of improved compression technologies for novel 3D applications such as omnidirectional video, free viewpoint television (FTV), or the new approach to 3D TV described in the remainder of this chapter (Smolic and McCutchen 2004).

2.3   A MODERN APPROACH TO 3D TV

A new approach to three-dimensional broadcast television has been proposed by the European Information Society Technologies (IST) project 'Advanced Three-Dimensional Television System Technologies' (ATTEST). In contrast to former proposals, which usually relied on the basic concept of an end-to-end stereoscopic video chain, i.e., on the capturing, transmission and display of two separate video streams, one for the left eye and one for the right eye, this novel idea is based on a more flexible joint distribution of monoscopic colour video and associated per-pixel depth information. From this 3D data representation, one or more 'virtual' views of a real-world scene can then be generated in real-time at the receiver side by means of so-called depth-image-based rendering (DIBR) techniques. The required backwards-compatibility to today's 2D digital TV infrastructure is achieved by compressing the video and depth data with already-standardized, state-of-the-art video coding algorithms such as MPEG-2 (ISO/IEC 1996), MPEG-4 Visual (ISO/IEC 1999) or the latest Advanced Video Coding (AVC) (ISO/IEC 2003).

For a better understanding of the fundamental ideas, the signal processing and data transmission chain of the ATTEST 3D TV concept is illustrated in Figure 2.2. It consists of five different functional building blocks: (1) 3D content creation; (2) 3D video coding; (3) transmission; (4) 'virtual' view synthesis; (5) 3D display.

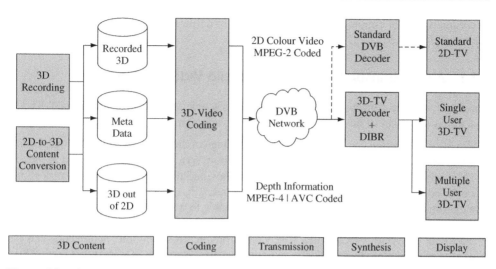

Figure 2.2   The ATTEST signal processing and data transmission chain consisting of five different functional building blocks: (1) 3D content generation; (2) 3D video coding; (3) transmission; (4) 'virtual' view synthesis; (5) 3D display

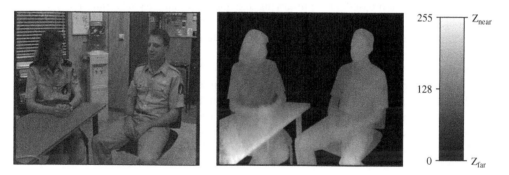

Figure 2.3   The ATTEST data representation format consisting of: (1) regular 2D colour video in digital TV format; (2) accompanying 8-bit depth-images with the same spatio-temporal resolution

For the generation of 3D content, a number of different (active and/or passive) approaches are described in Chapters 6–8 as well as Chapter 16. These methods can be used to create an advanced 3D representation consisting of regular 2D colour video in digital TV format (720 × 576 luminance pels, 25 Hz, interlaced) as well as an accompanying depth-image sequence with the same spatio-temporal resolution. Each of these *depth-images* stores depth information as 8-bit grey values with grey level 0 specifying the furthest value and grey level 255 defining the closest value (Figure 2.3). To translate this data format to real, metric depth values, required for the 'virtual' view generation, and to be flexible with respect to 3D scenes with different depth characteristics, the grey values are specified in reference to a near- and a far-depth clipping plane. The near clipping plane Z_{near} (grey level 255) defines the smallest metric depth value Z that can be represented in the particular depth-image.

Accordingly, the far clipping plane Z_{far} (grey level 0) defines the largest representable metric depth value.

2.3.1   A Comparison with a Stereoscopic Video Chain

Compared with the concept of an end-to-end stereoscopic video chain, the outlined 3D TV system has a number of advantages, with the most important virtues being the following:

- The 3D reproduction can be adjusted to a wide range of different stereoscopic displays and projection systems. As the required left- and right-eye views are only generated at the 3D TV receiver, their appearance in terms of parallax (and thus perceived depth impression) can be adapted to the particular viewing conditions. This allows one to provide the viewer with a customized 3D experience that is comfortable to watch on any kind of stereoscopic or autostereoscopic, single- or multiple-user 3D TV display.
- 2D-to-3D conversion techniques based on 'structure from motion' approaches can be used to convert already recorded monoscopic video material to 3D (Faugeras *et al.* 2001; Hartley and Zisserman 2000). This is a very important point, as it seems clear that the success of any future 3D TV broadcast system will depend to a great extent on the timely availability of sufficient interesting and exciting 3D video material.
- The chosen data representation format of monoscopic colour video plus associated per-pixel depth information is ideally suited to facilitate 3D post-processing. It enables automatic object segmentation based on depth-keying and allows for an easy integration of synthetic 3D objects into 'real-world' sequences (augmented reality) (Gvili *et al.* 2003). This is an important prerequisite for advanced television features such as virtual advertisement as well as for all kinds of real-time 3D special effects.
- Owing to the local smoothness characteristics of most 'real-world' object surfaces, the per-pixel depth information can be compressed much more efficiently than an additional colour video channel (which would be required to represent the second view in a conventionally recorded stereoscopic image). This feature makes it possible to introduce a 3D TV service, building on the proposed concept with only a very small transmission overhead (below 10–20% of the basic colour video bitrate) compared to today's conventional 2D digital TV system.
- The system allows the viewer to adjust the reproduction of depth to suit his or her personal preferences — much like every 2D TV set allows the viewer to adjust the colour reproduction by means of a (de-)saturation control. This is a very important system feature taking into account the evidence that there is a difference in depth appreciation over age groups. A study by (Norman *et al.* 2000) demonstrated that older adults were less sensitive than younger people to perceiving stereoscopic depth, in particular when the rendered parallax was higher.
- Photometric asymmetries, e.g., in terms of brightness, contrast or colour, between the left- and the right-eye view, which can destroy the stereoscopic sensation (Pastoor 1991), are eliminated from the first, as both views are effectively synthesized from the same original monoscopic colour image.
- Taking into account the noticeable recent trend to the development of multiview autostereoscopic 3D TV displays, it is of particular importance to note that these systems can also be driven very efficiently and — even more important — very flexible with 'virtual' images created in real-time from monoscopic colour video and associated per-pixel depth information (Bates *et al.* 2004).

Despite these advantages, it must not be ignored that the proposed concept also has a number of disadvantages that must be taken into careful consideration during the implementation of a commercially successful 3D TV system:

- The quality of the 'virtual' stereoscopic views depends on the accuracy of the per-pixel depth values of the original imagery. Therefore, it must be examined very carefully how different depth-image artifacts, which might be either due to the 3D content generation or result from the compression and transmission, translate into visually perceivable impairments in the synthesized images.
- An inherent problem of this new approach on 3D TV is that some areas that should be visible in the stereoscopic left- and right-eye views are occluded in the original (centre) image. This requires the development of suitable concealment techniques as well as an experimental assessment of the typical artifacts that these methods introduce during the 'virtual' view synthesis.
- Atmospheric effects such as fog or smoke, semi-transparent objects such as certain types of glass or plastic as well as such view-dependent effects as shadows, reflections and refractions can not be handled adequately by this concept. However, recent work on image-based rendering (IBR) as well as experiences from traditional computer graphics (CG) seem to indicate that this might not be a problem in most situations.

2.4 STEREOSCOPIC VIEW SYNTHESIS

Depth-image-based rendering (DIBR) is defined as the process of synthesizing 'virtual' views of a real-world scene from still or moving images and associated per-pixel depth information (Mark 1999; McMillan 1997). Conceptually, this can be understood as the following two-step process. At first, the original image points are reprojected into the three-dimensional world, utilizing the respective depth values. Thereafter, these intermediate space points are projected into the image plane of a 'virtual' camera, which is located at the required viewing position. The concatenation of reprojection (2D-to-3D) and subsequent projection (3D-to-2D) is usually called 3D image warping in the computer graphics (CG) literature and will be derived mathematically in the following paragraph.

2.4.1 3D Image Warping

Consider a system of two pinhole cameras and an arbitrary 3D point $\mathbf{M} = (x, y, z, 1)^{\mathrm{T}}$ with its projections $\mathbf{m}_l = (u_l, v_l, 1)^{\mathrm{T}}$ and $\mathbf{m}_r = (u_r, v_r, 1)^{\mathrm{T}}$ in the first and second view. Under the assumption that the world coordinate system equals the camera coordinate system of the first camera (this special choice of the world coordinate system does not limit the universality of the following expressions, it just simplifies the mathematical formalism, Faugeras 1995), the two perspective projection equations (see also Chapter 6) result in:

$$z_l \mathbf{m}_l = K_l \, [I|0] \, \mathbf{M} \tag{2.1}$$

$$z_r \mathbf{m}_r = K_r \, [R|\mathbf{t}] \, \mathbf{M} \tag{2.2}$$

where the 3×3 matrix R and the 3×1 vector $\mathbf{t}$ define the rotation and translation that transform the space point from the world coordinate system into the camera coordinate system of the second view, the two upper triangular 3×3 matrices K_l and K_r specify the intrinsic parameters of the first and second camera and z_l and z_r describe the scene depth in each camera coordinate system.

Rearranging Equation (2.1) gives an affine representation of the 3D point **M** that is linear dependent on its depth value z_l:

$$\mathbf{M} = z_l K_l^{-1} \mathbf{m}_l \tag{2.3}$$

Substituting Equation (2.3) into Equation (2.2) leads to the classical affine *disparity equation*, which defines the depth-dependent relation between corresponding points in two images of the same scene:

$$z_r \mathbf{m}_r = z_l K_r R K_l^{-1} \mathbf{m}_l + K_r \mathbf{t} \tag{2.4}$$

The disparity equation can also be considered as a *3D image warping* formalism, which can be used to create an arbitrary novel view from a known reference image. This requires nothing but the definition of the position **t** and the orientation R of a 'virtual' camera relative to the reference camera as well as the declaration of this synthetic camera's intrinsic parameters K. Then, if the depth values z of the corresponding 3D points are known for every pixel of the original image, the 'virtual' view can, on principle, be synthesized by applying Equation (2.4) to all original image points.

2.4.2 A 'Virtual' Stereo Camera

The simple 3D image warping concept described in the last section can of course also be used to create pairs of 'virtual' views that together comprise a stereoscopic image. For that purpose, two 'virtual' cameras are specified, which are separated by the *interaxial distance* t_c. To establish the so-called *zero-parallax setting* (ZPS), i.e., to choose the *convergence distance* z_c in the 3D scene, one of two different approaches, which are both in regular use in modern stereo cameras, can be chosen. With the 'toed-in' method, the ZPS is adjusted by a joint inward rotation of the left- and the right-eye camera. In the so-called 'shift-sensor' approach, shown in Figure 2.4, a plane of convergence is established by a small shift h of the parallel positioned cameras' CCD sensors (Woods *et al.* 1993).

While, technically, the 'toed-in' approach is easier to realize in real stereo camera hardware, the 'shift-sensor' method is usually prefered, because it does not introduce any unwanted vertical differences, which are known to be a potential source of eyestrain (Lipton 1982), between the left- and the right-eye view. Fortunately, this method is actually easier to implement with depth-image-based rendering (DIBR) as all the required signal processing is purely one-dimensional. All that is needed is the definition of two 'virtual' cameras, one for the left eye and one for the right eye. With respect to the reference (center) view, these 'virtual' cameras are symmetrically displaced by half the interaxial distance t_c and their CCD sensors are shifted relative to the position of the lenses.

Mathematically, this sensor shift can be formulated as a displacement of a camera's principal point c (Xu and Zhang 1996). The intrinsic parameters of the two 'virtual' cameras are thus chosen to exactly correspond to the intrinsic camera parameters of the original

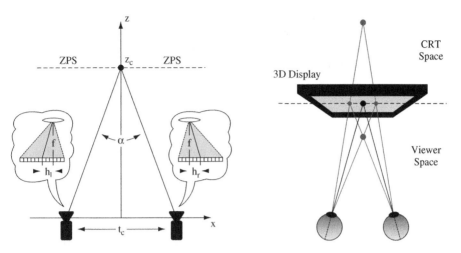

Figure 2.4   The 'shift-sensor' stereo camera and the respective three-dimensional reproduction on a stereoscopic or autostereoscopic 3D TV display (see Chapter 13)

view except for the horizontal shift h of the respective principle point. This can be written compactly as:

$$K_{l|r} = K + \begin{bmatrix} 0 & 0 & h_{l|r} \\ 0 & 0 & 0 \\ 0 & 0 & 0 \end{bmatrix} \tag{2.5}$$

where the $l|r$ symbol, which is used as a subscript here and in the following, should be substituted by either an l or an r, i.e., $K_{l|r}$ means either K_l or K_r, to denote that the respective parameter in the equation alludes to either the left or the right 'virtual' camera. (Note that for better readability no subscripts are used for the parameters that belong to the original view camera.)

Using the expression in Equation (2.5) and taking into consideration that the movement of the two 'virtual' cameras is restricted to be only translational with respect to the reference camera, i.e., $R = I$, the following simplifications can be made in the general 3D warping equation (2.4):

$$K_{l|r} R K^{-1} = K_{l|r} K^{-1} = \begin{bmatrix} 1 & 0 & h_{l|r} \\ 0 & 1 & 0 \\ 0 & 0 & 1 \end{bmatrix} = I + \begin{bmatrix} 0 & 0 & h_{l|r} \\ 0 & 0 & 0 \\ 0 & 0 & 0 \end{bmatrix} \tag{2.6}$$

Inserting the simplified expression in Equation (2.6) into Equation (2.4) then yields the following reduced form of the basic 3D warping formalism:

$$z_{l|r} \mathbf{m}_{l|r} = z \left(\mathbf{m} + \begin{bmatrix} h_{l|r} \\ 0 \\ 0 \end{bmatrix} \right) + K_{l|r} \mathbf{t}_{l|r} \tag{2.7}$$

This expression can be simplified even more by taking into account that the only non-zero translational component needed to create a 'shift-sensor' stereo camera is a horizontal translation t_x inside the focal plane of the original camera. Thus, with $t_z = 0$, it follows that the depth value of any 3D point is the same in the world coordinate system, which was chosen to equal the camera coordinate system of the original view, and in the coordinate system of the respective 'virtual' camera, i.e., $z_{l|r} = z$. Therefore Equation (2.7) further reduces to:

$$\mathbf{m}_{l|r} = \mathbf{m} + \frac{K_{l|r}\mathbf{t}_{l|r}}{z} + \begin{bmatrix} h_{l|r} \\ 0 \\ 0 \end{bmatrix} \quad \text{with } \mathbf{t}_{l|r} = \begin{bmatrix} t_{x,l|r} \\ 0 \\ 0 \end{bmatrix} \tag{2.8}$$

Thus, the affine position of each warped image point can simply be calculated as:

$$u_{l|r} = u + \Delta u_{l|r} \quad \text{and} \quad v_{l|r} = v$$
$$= u + \frac{\alpha_u t_{x,l|r}}{z} + h_{l|r} \tag{2.9}$$

where the horizontal translation $t_{x,l|r}$ results to the half of the selected interaxial distance t_c, with the direction of the movement given by:

$$t_{x,l|r} = \begin{cases} +\frac{t_c}{2} & \text{left-eye view} \\ -\frac{t_c}{2} & \text{right-eye view} \end{cases} \tag{2.10}$$

and α_u is the focal length of the original view camera (in multiples of the pixel width).

The size of the sensor shift $h_{l|r}$ depends only on the chosen convergence distance z_c and can be calculated from the requirement that for $z = z_c$, the horizontal component $u_{l|r}$ of any image point that is warped according to Equation (2.9) must be the same in the left and in the right view, i.e., $u_l = u_r$, which leads to the following expression:

$$h_{l|r} = -t_{x,l|r} \frac{\alpha_u}{z_c} \tag{2.11}$$

where $t_{x,l|r}$ is also defined by Equation (2.10).

The parallax between the 'virtual' left- and right-eye view then results in:

$$P_{r \to l} = u_r - u_l = \alpha_u t_c \left(\frac{1}{z_c} - \frac{1}{z} \right) \tag{2.12}$$

The expression in Equation (2.9) fully defines a simplified 3D warping equation that can be used to efficiently implement a 'virtual' stereo camera. Table 2.1 shows, how the resulting 3D reproduction is influenced by the choice of the three main system variables, i.e., by the choice of the interaxial distance t_c, the reference camera's focal length f as well as the convergence distance z_c. The respective changes in parallax, perceived depth and object size are qualitatively equal to what happens in a real stereo camera when these system parameters are manually adjusted *before* video capturing.

2.4.3 The Disocclusion Problem

An inherent problem of this stereoscopic view synthesis algorithm is due to the fact that scene parts, which are occluded in the original image, might become visible in any of the 'virtual' left- and right-eye views, an event referred to as *exposure* or *disocclusion* in the

Table 2.1 Qualitative changes in parallax, perceived depth and object size when varying the interaxial distance t_c, the focal length f or the convergence distance z_c of a real or 'virtual' stereo camera (adapted from Milgram and Krüger 1992)

Parameter	+/−	Parallax	Perceived depth	Object size
Interaxial distance t_c	+	Increase	Increase	Constant
	−	Decrease	Decrease	Constant
Focal length f	+	Increase	Increase	Increase
	−	Decrease	Decrease	Decrease
Convergence distance z_c	+	Decrease	Shift (forwards)	Constant
	−	Increase	Shift (backwards)	Constant

computer graphics (CG) literature (Mark 1999; McMillan 1997). The question arises as to, how these disocclusions should be treated during the view generation, as information about the previously occluded areas is neither available in the monoscopic colour video nor in the accompanying depth-images.

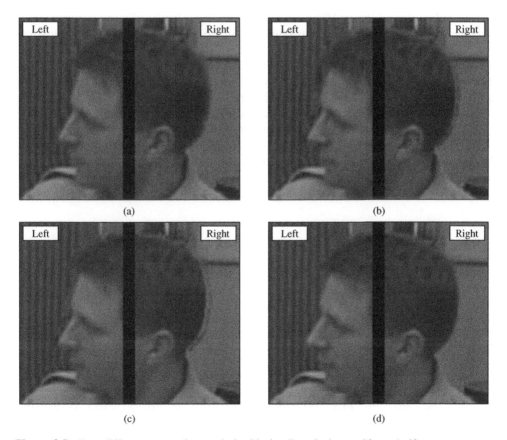

(a)

(b)

(c)

(d)

Figure 2.5 Four different approaches to deal with the disocclusion problem. Artifacts are more or less noticeable around the person's head and shoulders

One solution would be to rely on more complex, multi-dimensional data representations, however, this would again increase the overhead of the system. Therefore, techniques were developed and/or analyzed within the ATTEST project, which: (1) replace the missing image areas during the view synthesis with 'useful' colour information; or (2) preprocess the depth information in a way that no disocclusions appear in the 'virtual' views. Synthesis examples for four different approaches are provided in Figure 2.5 for a small display detail of the ATTEST test sequence 'Interview'.

In Figure 2.5, image (a) shows the 'rubber-sheets' that are typical for a linear colour interpolation between scene foreground (head) and background (wall). Image (b) visualizes the strip-like impairments that result from a simple extrapolation of the scene background. The artifacts in image (c) are due to a mirroring of background colours along the borders of the disocclusions. Finally, image (d) shows that visually less perceptible impairments can be achieved by preprocessing (smoothing) the depth information with a suitable Gaussian low-pass filter in a way that no disocclusions occur.

2.5   CODING OF 3D IMAGERY

The suitability of the different MPEG standard technologies for the efficient compression of depth-images was evaluated in a comparative coding experiment (Fehn 2004). The test group consisted of the following three codecs: (a) the MPEG-2 reference model codec (TM-5); (b) a rate-distortion (R/D) optimized MPEG-4 Visual codec developed at FhG/HHI; (d) the R/D optimized AVC reference model codec (v6.1a). The compression results for the two ATTEST test sequences 'Interview' and 'Orbi' are shown in Figure 2.6 for typical broadcast encoder settings, i.e., for a GOP (group of pictures) length equal to 12 with a GOP structure of IBBPBBP. . . , by means of rate–distortion curves over a wide range of different bitrates.

The provided results show that AVC as well as MPEG-4 Visual are very well suited for the compression of per-pixel depth information (with AVC being even more efficient). If a typical broadcast bitrate of 3 Mbit/s is assumed for the MPEG-2 encoded monoscopic colour information, it can be followed from the two graphs that the accompanying depth-images

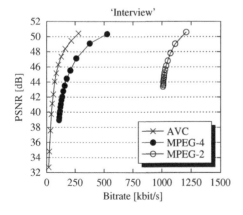 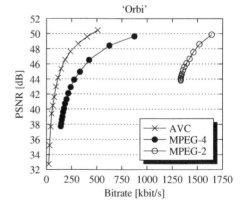

Figure 2.6   Rate–distortion curves for the compression of the two ATTEST test sequences 'Interview' and 'Orbi' using three different MPEG video coding standards

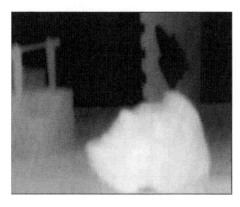 

Figure 2.7 Comparison of coding artifacts for 'Orbi' test sequence. Roughly the same bitrate of about 145 kbit/s was used for both codecs. (1) Rate-distortion (R/D) optimized MPEG-4 Visual codec - 37.74 dB (QP30); (2) AVC codec - 45.35 dB (QP30)

can be compressed to target rates significantly below 20% of this value. For example, AVC compression of the 'Interview' sequence at 105 kbit/s still leads to a very high PSNR of 46.29 dB. For the more complex 'Orbi' scene, this value can still be reached at a bitrate of approximately 184 kbit/s.

A better appreciation of the relative performance of the MPEG-4 Visual and the AVC codec can be attained from Figure 2.7, which compares two frames of the 'Orbi' scene each compressed at a bitrate of about 145 kbit/s. While in the case of MPEG-4 Visual encoding/decoding severe blocking artifacts are clearly noticeable in most areas of the picture, the visual quality of the AVC compressed image is only slightly degraded in comparison with the original. To illuminate the performance difference of the two coding standards from another perspective: The chosen AVC operating point (QP30) roughly corresponds to an MPEG-4 Visual quantization parameter of 8. With this QP, the MPEG-4 Visual encoder is able to achieve a comparable PSNR value of 44.94 dB (with roughly the same visual quality as with AVC encoding), the required bitrate, however, is more than twice as high (330 kbit/s compared with 145 kbit/s).

2.5.1 Human Factor Experiments

The 3D TV architecture described differs inherently from most consumer applications that utilize digital video compression algorithms in the sense that the outcome of the encoding/decoding process — in this case the MPEG-4 Visual or AVC compressed depth information — is never directly assessed by the viewer. Instead, it is processed together with the MPEG-2 encoded monoscopic colour video to generate 'virtual' stereoscopic images in real-time using the synthesis algorithm described in Section 2.4. For this reason, human factors experiments were carried out with the aim to develop an understanding of the relation between (a) the introduced depth coding artifacts on the one hand side; and (b) the perceptual attributes of the synthesised 3D imagery on the other hand side (Quante *et al.* 2003).

The evaluations were performed according to the double-stimulus subjective testing method specified in ITU-R BT.500-10 (ITU-R 2000). A group of 12 non-expert viewers was presented with stereoscopic stimulus material synthesized from the original and the

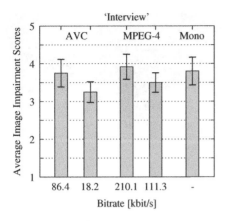 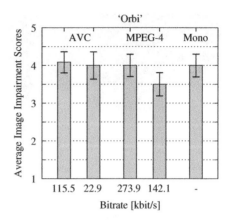

Figure 2.8 Average image impairment scores with 95% confidence intervals for different coded versions of the two ATTEST test sequences 'Interview' and 'Orbi'

encoded/decoded depth information. Based on their subjective impressions the participants had to rate the stimuli in terms of: (a) perceived image impairments; (b) depth quality. The viewers explored each stimulus as long as they wanted and marked their findings on a five-point double-stimulus impairment scale lettered with the semantic labels 'imperceptible' (5), 'perceptible, but not annoying' (4), 'slightly annoying' (3), 'annoying' (2) and 'very annoying' (1).

Some of the results are visualized in Figure 2.8. The plotted average image impairment scores show that for the chosen codecs and bitrates the artifacts perceived in the 'virtual' stereoscopic images are rated between 'perceptible, but not annoying' and 'slightly annoying'. However, as can be seen from the graphs, similar ratings also apply to the additionally evaluated monoscopic colour reference sequences, which indicates that the perceptual results not only reflect the introduced coding distortions, but also include image degradations that already appear in the original 2D imagery.

It should added that with regard to the perceived depth quality, the rating results (not reproduced here) verified that the proposed modern approach to 3D TV indeed leads to an added value when compared with conventional, monoscopic colour video material.

2.6 CONCLUSIONS

In summary, this chapter provided details of a modern approach to 3D TV that is based on depth-image-based rendering (DIBR). In this concept, 'virtual' stereoscopic views are generated in real-time at the receiver side from a 3D data representation format consisting of monoscopic colour video and associated per-pixel depth information. The main advantages of such a processing lie in the ability to customize the 3D reproduction with regard to different viewing conditions and/or user preferences as well as in the fact that the described scenario could be introduced with only a very minor transmission overhead (<20%) compared with today's conventional 2D digital TV.

On the other hand side, it must be noted that the quality of the synthesized 3D imagery of course depends on the accuracy of the provided depth information and even with perfect depth-images will be degraded to some extend by the mentioned disocclusion artifacts. These impairments could be avoided through the use of more complex 3D formats such as *multi-view images* (Cooke *et al.* 2003) or *layered depth-images* (LDIs) (Shade *et al.* 1998; Zitnick *et al.* 2004), however, this requires further research on the efficient compression and transmission of such data representations.

ACKNOWLEDGEMENTS

This work has been sponsored by the European Commission (EC) through their Information Society Technologies (IST) programme under proposal IST-2001-34396. The author would like to thank the project officers as well as all project partners (Philips Research, The Netherlands; TU Eindhoven, The Netherlands; FhG/HHI, Germany; KU Leuven, Belgium; CERTH/ITI, Greece; 3DV Systems, Israel; De Montfort University, United Kingdom; VRT, Belgium) for their support and for their input to this publication.

REFERENCES

3D Consortium 2003 [Online]. http://www.3dc.gr.jp/english/.

Bates R, Surman P, Sexton I, Craven M, Yow KC and Lee WK 2004 Building an Autostereoscopic Multiple-Viewer Television Display *Proceedings of 7th Asian Symposium on Information Display,* pp. 400–404, Nanjing, Jiangsu, China.

Cooke E, Feldmann I, Kauff P and Schreer O 2003 A Modular Approach to Virtual View Creation for a Scalable Immersive Teleconferencing Configuration *Proceedings of IEEE International Conference on Image Processing,* pp. 41–44, Barcelona, Spain.

Dynamic Digital Depth 2004 [Online]. http://www.ddd.com.

Faugeras O, Luong QT and Papadopoulo T 2001 *The Geometry of Multiple Images: The Laws That Govern the Formation of Multiple Images of a Scene and Some of Their Applications.* MIT Press, Cambridge, Massachussets, USA.

Faugeras OD 1995 Stratification of 3-D Vision: Projective, Affine, and Metric Representations. *Journal of the Optical Society of America* **12**(3), 465–484.

Fehn C 2004 Depth-Image-Based Rendering (DIBR), Compression and Transmission for a New Approach on 3D-TV *Proceedings of Stereoscopic Displays and Virtual Reality Systems XI,* pp. 93–104, San Jose, CA, USA.

Gvili R, Kaplan A, Ofek E and Yahav G 2003 Depth Keying *Proceedings of SPIE The Engineering Reality of Virtual Reality,* pp. 564–574, Santa Clara, CA, USA.

Hartley RI and Zisserman A 2000 *Multiple View Geometry in Computer Vision.* Cambridge University Press, Cambridge, UK.

Hur N, Ahn CH and Ahn C 2002 Experimental Service of 3DTV Broadcasting Relay in Korea *Proceedings of SPIE Three-Dimensional TV, Video, and Display,* pp. 1–13, Boston, MA, USA.

IJsselsteijn WA 2003 *Being There: Concepts, Effects and Measurement of User Presence in Synthetic Environments* Ios Press Amsterdam, The Netherlands chapter Presence in the Past: What can we Learn From Media History?, pp. 18–40.

IJsselsteijn WA, Seutiëns PJH and Meesters LMJ 2002 State-of-the-Art in Human Factors and Quality Issues of Stereoscopic Broadcast Television. Technical Report D1, IST-2001-34396 ATTEST.

Imaizumi H and Luthra A 2002 *Three-Dimensional Television, Video, and Display Technologies* Springer, Berlin, Germany, Chapter Stereoscopic Video Compression Standard "MPEG-2 Multiview Profile", pp. 169–181.

ITU-R 2000 Methodology for the Subjective Assessment of the Quality of Television Pictures. BT Series Recommendation BT.500–10.

Lipton L 1982 *Foundations of the Stereoscopic Cinema — A Study in Depth.* Van Nostrand Reinhold, New York, NY, USA.

Mark WR 1999 *Post-Rendering 3D Image Warping: Visibility, Reconstruction, and Performance for Depth-Image Warping* PhD thesis University of North Carolina at Chapel Hill Chapel Hill, NC, USA.

McMillan L 1997 *An Image-Based Approach to Three-Dimensional Computer Graphics* PhD thesis University of North Carolina at Chapel Hill, NC, USA.

Milgram P and Krüger M 1992 Adaptation Effects in Stero Due to On-line Changes in Camera Configuration *Proceedings of SPIE Stereoscopic Displays and Applications III,* pp. 122–134, San Jose, CA, USA.

Norman J, Dawson T and Butler A 2000 The Effects of Age Upon the Perception of Depth and 3-D Shape From Differential Motion and Binocular Disparity. *Perception* **29**(11), 1335–1359.

Pastoor S 1991 3D-Television: A Survey of Recent Research Results on Subjective Requirements. *Signal Processing: Image Communication* **4**(1), 21–32.

Quante B, Fehn C, Meesters LMJ, Seuntiëns PJH and IJsselsteijn WA 2003 Report of Perception of 3-D Coding Artifacts: A Non-Experts Evaluation of Depth Annotated Image Material. Technical Report D8a, IST-2001-34396 ATTEST.

Sand R 1992 3-DTV — A Review of Recent and Current Developments *IEE Colloquium on Stereoscopic Television,* pp. 1–4, London, UK.

Shade J, Gortler S, He LW and Szeliski R 1998 Layered Depth Images *Proceedings of ACM SIGGRAPH,* pp. 231–242, Orlando, FL, USA.

Smolic A and McCutchen D 2004 3DAV Exploration of Video-Based Rendering Technology in MPEG. *IEEE Transactions on Circuits and Systems for Video Technology* **14**(3), 349–356.

Tiltman RF 1928 How 'Stereoscopic' Television in Shown. Radio News.

Wheatstone C 1838 On Some Remarkable, and Hitherto Unobserved, Phenomena of Binocular Vision. *Philosophical Transactions of the Royal Society of London.*

Woods A, Docherty T and Koch R 1993 Image Distortions in Stereoscopic Video Systems *Proceedings of SPIE Stereoscopic Displays and Applications IV,* pp. 36–48, San Jose, CA, USA.

Xu G and Zhang Z 1996 *Epipolar Geometry in Stereo, Motion and Object Recognition.* Kluwer Academic Publishers, Dordrecht, The Netherlands.

Yuyama I and Okui M 2002 *Three-Dimensional Television, Video, and Display Technologies* Springer Press Berlin, Germany chapter Stereoscopic HDTV, pp. 3–34.

Zitnick CL, Kang SB, Uyttendaele M, Winder S and Szeliski R 2004 High-Quality Video View Interpolation Using a Layered Representation. *ACM Transactions on Graphics* **23**(3), 598–606.

3

3D in Content Creation and Post-production

Oliver Grau

BBC R&D, Tadworth, United Kingdom

3.1 INTRODUCTION

This chapter is dedicated to the use of 3D data for content creation in TV and film production. Computer graphics methods are used for a wide range of applications in content creation, such as the planning of productions, for pre-visualization on set and when 3D graphics appear in the final programme. The degree to which 3D graphics are involved ranges from their use in the planning phase of a production without any graphical elements in the final programme, to films that are entirely computer generated. For most types of programme it is in between, and typically involves a mixture of virtual and real scene content. Well-known examples are special effects in movies, which insert virtual objects or characters into real camera footage or virtual studios in TV productions that insert a presenter or actor into a completely virtual environment. Since the application of computer graphics is still very expensive usually only those components that can not be filmed will be generated virtually.

The final high-quality graphics for film and many TV productions are mostly generated off-line in a post-production phase. For on-set pre-visualization and many TV productions, that are either broadcast live or have a lower budget than feature film productions, the graphics are generated in real-time. The budget and the decision whether the image generation has to be in real-time constrain applicable techniques. For real-time TV productions usually only simple 2D scene composition can be used. In virtual studio systems for example, a camera image of the presenter or actor is taken in a (real) studio and then keyed out and overlayed onto a synthesized (virtual) image.

How convincing the composited result looks in terms of realism depends on the quality of the optical integration of the virtual and real scene components. In a full optical integration

3D Videocommunication — Algorithms, concepts and real-time systems in human centred communication
Edited by O. Schreer, P. Kauff and T. Sikora © 2005 John Wiley & Sons, Ltd

virtual and real objects are optically interacting with each other, that means they are occluding each other and cast shadows. With a 2D composition these optical interactions can be established in only a very limited way. For sophisticated optical interactions more 3D knowledge of the real (and virtual) scene is needed and the scene composition is done in the 3D domain.

The technical challenges of those productions that integrate virtual and real scene components are the subject of this chapter.

In order to achieve photo-realistic images of virtual scenes it is necessary to consider several optical phenomena. We here consider a *scene* as a set of observable objects and a set of relevant physical phenomena, that can explain the visual appearance of the scene.

Object types of particular interest include the following. A *body* as a coherent area in 3-space; often only opaque bodies are considered. For this case they can be represented equivalently as volumes or surface models. A *light source* sends light into the scene, so that bodies become visually apparent. A *real camera* is the sensor of the visual appearance of the real scene. Further a *virtual camera* is a model of a real camera that is used to generate images from a scene description, i.e., a virtual scene.

Both computer vision and computer graphics have developed models for these phenomena. For the creation of content that involves the integration of virtual content into real scenes (or vice versa) it is important that optical interactions between the integrated objects are harmonized.

A requirement for a believable integration is that both the virtual and the real scene are registered in the same coordinate space. In particular a *matching camera perspective* of both the real and virtual camera is very important for a convincing integration of real and virtual scene components. Depending on the distortions of the real camera, a more or less complex camera model has to be considered, i.e., at least camera pose and focal length and in some cases centre point shifts and radial distortions are used.

The most important *optical interactions* between scene objects are:

- *Occlusions*, which if not considered would destroy the realism of an integration.
- *Shadows* are important because they give a visual cue. Without shadows objects are appearing floating in space and look very unnatural.
- *Reflections* are important for shiny objects, but less often implemented because they can be avoided by careful selection of the scene elements.

Further to these phenomena there are many other physical effects that can be observed, such as refraction, transparency, particles (smoke, fog). These are modelled and implemented in computer graphics and also used in some productions. However, a harmonization between them for real and virtual scenes is rarely considered, mainly because of a lack of analysis tools and models for real scenes.

Not an optical interaction, but an often used feature is a *view interpolation* or *extrapolation*. In this case the virtual and real cameras are not matched, instead the virtual camera moves independently. This is often a desired feature for special effects or for visualization as in sports when a virtual camera flies over or around a scene. In order to be able to do this, more knowledge or data of the real scene must be available. Therefore, for a maximum of freedom, a 3D model of the real scene is required.

3.2 CURRENT TECHNIQUES FOR INTEGRATING REAL AND VIRTUAL SCENE CONTENT

The simplest integration of virtual and real content is a 2D/2D composition, as still in daily use in TV productions such as the weather forecast or standard news graphics. Normally the actor or presenter will be separated from the studio background by chroma-keying. This requires a specially equipped studio. Since there is no use of any 3D data the optical interactions are limited to carefully chosen scene layering: that means the presenter is always in front of the synthetic background. Camera moves are not possible.

The main innovation in the virtual studios of the 1990s was to attach a real-time camera tracking system to a studio camera. There are several tracking systems commercially available. One example of a camera tracking system developed by the BBC (Thomas *et al.* 1997) is based on coded targets mounted on the studio ceiling, as depicted in Figure 3.1. An auxiliary camera attached to the studio camera is mounted looking upwards. A real-time hardware system computes the accurate 3D position and orientation of the camera at 50 fps from those images. In addition zoom and focus can be retrieved through lens-attached sensors.

The measured parameters of the real camera are transfered to a virtual camera. That means the virtual camera parameters are cloned from the real camera, which is also called the master camera in this case. The virtual camera is implemented based on graphics hardware sufficiently fast to generate an image of the virtual scene with parameters changing in real-time. The composition is usually done with an external studio keyer that provides a chroma-key of the studio camera image and overlays it onto the virtual background. Since the virtual camera has no knowledge about the real scene it is not possible to implement

Figure 3.1 The BBC 'Free-D' camera tracking system is based on circular bar-coded targets on the ceiling and an auxiliary camera looking upwards. Real-time hardware computes the position and orientation of the camera at a 50 Hz update rate

optical interactions other than basic occlusions, although most keying systems provide the feature to pass the shadow of the studio floor onto the virtual image, giving the impression of having a shadow cast from the actor onto the virtual background. This only works for a flat virtual floor.

In TV productions virtual studios have found several applications, but conventional sets are still predominantly used. One of the reasons is that virtual studios require a chroma-keying facility that has to be installed permanently to be cost effective. Moreover, it requires expensive equipment and skilled operators and designers for the virtual sets. Therefore, virtual studios are mainly used for series with very short turn-around times, that means several different sets are used in the same studio, even on the same day, or if the content is already highly multi-media oriented and the programme benefits from the use of virtual techniques.

In the film industry the use of chroma-keying is a very common production technique, in particular for special effects. Techniques developed for virtual studios are used for pre-visualization on set. For example the BBC camera tracking system has been used in movies such as *A.I.* (Rosental *et al.* 2001) and others to give the director a pre-view of the camera framing with the action of actors and virtual backgrounds live on set. The final composition is done in post-production in a time-consuming off-line process that can take several days or even weeks to be finished. Therefore having the direct feedback available during the filming reduces the risk of a re-shoot of a scene. Unfortunately the actor is still working in a chroma-key environment and gets usually no direct feedback of the virtual scene.

A trend in recent TV productions is not to replace the entire environment with a virtual background. Instead only selected virtual scene components are integrated into a camera image. An example of such a production is depicted in Figure 3.2. The BBC programme is designed for children and is based on two teams. Each team designs its own small virtual robot and 'programs' it with a set of rules. The team's robots compete with each other in a virtual race run on the top of a large table. The robots are simulated in a games engine

Figure 3.2 *BAMZOOKi*

in real-time. A virtual image of the current scene is overlayed onto an image from a studio camera that is equipped with a tracking system. In order to cast a (virtual) shadow on the table, a model of the table in the form of a plane and the position of the studio key light is required. The shadow is then generated by the rendering software of the game engine.

An important part of the production is that children of both teams are present in the studio to 'support' their robot. A problem that occurs immediately is that the children need a feedback of the virtual robots, otherwise they can neither keep eye-contact nor react to the actions of the robots. Therefore an image of the virtual scene was projected onto the table in front of the children in the blanking phase of the studio camera, so that it is only visible to the children's eyes. The projection does not take into account the position of the viewers therefore the images are perspectively not absolutely correct, but they are sufficient for the visualization purpose, since the children only have to see where the virtual robots are and what they are doing.

Beyond the previous scenario many production situations ask for a bidirectional interface between the virtual and real world. For the previous given example that would mean that the children can not only see the virtual objects, but would also be able to touch them or give them a spin. This feature can be achieved by tracking the actors movements and by detecting collisions of the real and virtual world. Further, the position of actors can be used as a target for virtual characters to look at. Section 3.4 introduces a concept that is implementing a bidirectional interface.

Special effects found in feature film productions are normally not restricted to real-time and use more powerful rendering tools, such as ray-tracers for image synthesis. Therefore, special effects can afford to establish full optical interactions, that means occlusions, shadow casting and receiving, plus reflections if needed. Control over the lighting is very important, some effects even ask for control over the virtual camera.

In order to be able to implement these sophisticated optical interactions the virtual and the real scene have to be represented in 3D. Once both are described as a 3D model the scene composition can be done with production tools such as animation packages, providing a full optical integration.

How the creation of the 3D model of the real scene is done is very individual to each production and specific scene. Under many conditions the model is created interactively with an animation package, based on maps, measured dimensions of the set or based on images of the set or objects. If a higher level of detail or accuracy of the objects is required then an automatic modelling approach will be used.

The automated creation of 3D models of static objects makes use of active, sensor-based or passive, image-based methods. Active, sensor based 3D reconstruction systems, as described in Chapter 16, use an active illumination technique and are usually more robust compared with most passive techniques. There are several products available using active techniques from companies such as lasers or structured light. For the generation of both static objects and also complete sets laser-scanning systems are predominantly used by the film industry for special effects.

Passive, image-based 3D reconstruction techniques use only camera images without any active illumination. An example of this class of techniques is stereo vision, as described in Chapter 7. For the creation of 3D models of static objects or sets these methods are not much used in productions, mainly due to their restrictions in accuracy and robustness.

Concerning the estimation of the real lighting situation, it is common practice to take a 'light-probe', i.e., a picture with a white sphere in the scene. In an animation package a virtual white sphere is placed and virtual lights are placed interactively until the rendered image matches the image with the real sphere. In recent years this approach has been supplemented by image-based methods. In particular high-dynamic images of the scene, as described by Debevec (1998) can be used with many industrial rendering packages to re-create even complex lighting situations.

3.3   GENERATION OF 3D MODELS OF DYNAMIC SCENES

3D models of dynamic objects such as actors are usually generated with automated methods, since it requires a model update for every frame of a sequence. The same set of methods as discussed before for static objects can be considered, but only a few meet the requirements to capture the 3D shape and texture of an object in real-time.

Many active or sensor-based systems are not fast enough to capture at film or video frame rate. For this reason image-based methods are used or systems with active illumination based on cameras that can capture images of the light pattern and the texture. Such systems have been used for the real-time capture of faces. A structured light approach is commercially available from Eyetronics. In the movie *The Matrix Reloaded* a stereo reconstruction was used to generate 'clones' of persons.

For the 3D-capture of complete objects, in particular of actors in a studio environment, the computation of the visual hull from 2D silhouettes has been shown to be quite robust and suited for many practical object classes (Cheung *et al.* 2003; Grau and Thomas 2002; Matusik *et al.* 2000). The approach is based on the shape-from-silhouette method as described in Chapter 8, which is relatively simple to implement and can be computed in real-time in a specially equipped studio using chroma-keying or difference-keying techniques.

A disadvantage of the basic shape-from-silhouette algorithm is that no convex structures can be modelled. This problem was addressed by several extensions of the approach. The voxel colouring and shape carving technique (Kutulakos and Seitz 2000; Seitz and Dyer 1999) makes use of the colour information, i.e., the differences between the generated model and the camera images. A more robust and often computationally faster approach involves combining shape-from-silhouette with stereo matching (Starck and Hilton 2003). The application of stereo usually requires a different camera setup from a visual hull computation (pairs of relatively close grouped cameras). Another extension of the basic shape-from-silhouette algorithm is to consider the temporal changes, as in Vedula *et al.* (2002).

A different strategy, that fundamentally makes use of silhouette information, is the incorporation of high-level generic models of human bodies (Carranza *et al.* 2003; Hilton *et al.* 1999; Weik *et al.* 2000). These methods initialize a generic human shape model and then update only motion parameters over time. They give a good appearance, but are limited to human object classes.

Three-dimensional models of the actors are used for the real-time on-set visualization and off-line for post-production. In the first case a rough quality is sufficient since the purpose is to generate a pre-visualization, but it has to be in real-time. In the latter case the aim is to get a 3D quality that can be used for special effects in final programme quality. In both cases the shape-from-silhouette or visual hull concept can be used. For the real-time version the technique has to be computationally very fast. Several fast methods for the visual hull computation have been proposed (e.g., Matsuyama *et al.* 2004; Matusik *et al.* 2001; Niem 1994).

For use in the final programme the 3D models are usually computed off-line in post-production. Therefore more time consuming methods can be used in order to achieve more accurate and moreover temporally consistent 3D models. Most visual hull reconstruction algorithms compute a volumetric reconstruction in a first step and use an iso-surface generator, such as the marching-cubes algorithm (Lorensen and Cline 1987) to compute a 3D polygonal surface description in a second step. Possible artefacts in this reconstruction can be grouped into three categories:

- Remaining volume due to occlusions
- Bad volume approximation due to low number of cameras
- Sampling problems

Figure 3.3 illustrates the effects of these problems. The reconstruction in the left and middle was done using only six cameras and an octree-based data structure with hierarchical approach. The voxel resolution is $128 \times 128 \times 128$.

In the left and middle views of Figure 3.3 the problems due to occlusions can be seen. There are remaining bits on the feet and the separation of the two persons is quite poor. The approximation problems due to the low number of cameras are clearly visible in the top view (middle). In particular the person on the right shows a 'blocky' shape.

Both problems can be overcome by using more cameras: by upgrading the studio system from 6 to 12 cameras, the shape quality is improved, as can be seen in Figure 3.3 (right).

Unfortunately there are still very visible artefacts that are due to sampling problems. First, the 3D reconstruction looks bigger than the actual objects. This effect is caused by the 2D box test that gives a 3D voxel model that is on the average 0.5 voxels bigger than the real object. Therefore by increasing the voxel resolution the model can be better approximated. On the other hand an increase in voxel resolution also increases the number of triangles of the resulting surface description.

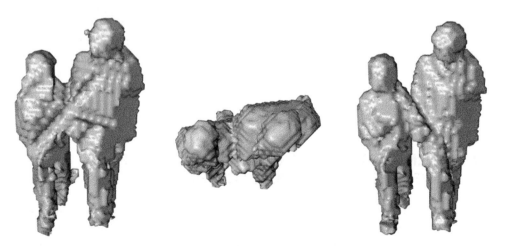

Figure 3.3   Result using six cameras and voxel-based 3D reconstruction ($128 \times 128 \times 128$) front view (left), top view (middle) and 12 cameras (right)

Figure 3.4 Result with voxel-based 3D reconstruction with super-sampling (left) and resulting frame of a video sequence with 3D model integrated into virtual background (right). 3D model of entrance provided by the Multimedia information processing group, Christian-Albrechts-University Kiel

Another visible problem in Figure 3.3 (right) is the fact that the surface looks 'voxelized', meaning that the voxel structure is visible due to another sampling problem: The used voxels are binary, they indicate only whether that voxel is foreground or background. The marching cubes algorithm that is used to generate a surface description from the voxel representation is introducing quantization noise for this kind of representation. One way to overcome this problem is using a line-based sampling method as described in Grau and Dearden (2003). In Grau (2004) we propose a method based on an octree-representation and super-sampling, that gives smooth 3D surfaces that can be used to generate video sequences. This approach reduces the sampling error that would otherwise be caused by a conventional volumetric reconstruction and the use of the marching cubes algorithm for the generation of a surface model. The new approach extends the accuracy of the volumetric shape reconstruction by super-sampling without increasing the number of triangles in the 3D model: The leaf nodes of an octree are further subdivided and the value of the original node is replaced by a counter of the number of sub-nodes that are found as belonging to the object. This value is than used in a standard marching cubes algorithm to compute a smoother 3D surface. Figure 3.4 (left) shows a resulting reconstruction using one super-sampling level. Further we apply Gaussian smoothing to the 3D models in order to suppress temporal artefacts that would be visible in a synthesized video sequence. Figure 3.4 (right) shows an integration of the model into a 3D model of an entrance with full optical interactions.

3.4 IMPLEMENTATION OF A BIDIRECTIONAL INTERFACE BETWEEN REAL AND VIRTUAL SCENES

The interface between the real and the virtual scene components plays an important role during 3D content production. In particular the on-set visualization of the virtual scene becomes an important issue for the actor, and the pre-visualization of the composited scene for the director and operators on set. If the actor or presenter has to interact with a virtual object, e.g., a virtual character, then without feedback there is often a noticeable difference between the direction in which the actor is actually looking and the position of the virtual character (the so-called eye-line problem). This can be quite disturbing in the final programme since human observers are very sensitive to wrong eye-lines.

On the other hand feedback from the real scene to the virtual scene is also very desirable in many production situations. It requires sensing of the real scene and can be used to generate precise event triggers or to create virtual objects that react to the real world.

An approach to provide the actor or presenter with a visual cue of objects in a virtual set is described in Tzidon *et al.* (1996). The system projects an outline of virtual objects onto the floor and walls. However, this method is restricted to show only the point of intersection of the virtual objects with the particular floor or wall. That means a virtual actor in the scene would be visualized as footprints only, and the eye-line problem persists.

Systems that do provide the required functionality are projection-based VR (virtual reality) systems, such as the CAVE (Cruz Neira *et al.* 1992) or the office of the future (Raskar *et al.* 1998). The main application of these systems is to provide an immersive, collaborative environment. Therefore, a CAVE system tracks the position of the viewer's head and computes an image for that particular viewpoint, which is then projected onto a large screen forming one wall of the environment. Several screens are usually used to provide a wide field-of-view, and some installations provide an all-round view using six projection screens that completely surround the viewer. Therefore it is possible to present objects that appear virtual in space, giving the viewer an immersive experience. The head position of the viewer is usually tracked using a special device, e.g., a helmet with an electromagnetic transmitter, and if stereo projection is desired, a pair of shuttered glasses must be worn. Such devices cannot be used in a production environment, because they would be visible in the final programme.

Although such collaborative projection-based VR systems would probably provide a good immersive feedback for an actor, they are not designed to create 3D models of the person. The main problem with the integration of a 3D-capturing component into a CAVE-like system is that there should not be any interference between both sub-systems. A concept of integrating 3D capture into a CAVE-like system was proposed in the blue-c system (Gross *et al.* 2003). This system uses back-projection onto a special screen that can be switched in three cycles between transparent and opaque modes. In the transparent mode images are taken in parallel from several cameras. In the opaque mode a projector projects an image for the left and right eye of the virtual scene in two separate cycles in order to give a stereo cue. The user has to wear shuttered glasses, that separate the images for both eyes. For the 3D reconstruction an approach similar to Matuski *et al.* (2000) based on difference keying is proposed.

The system discussed here was designed for the production of 3D content as part of the IST-ORIGAMI project (Grau *et al.* 2003). Therefore, the quality of the final 3D content has the highest priority. Furthermore the system is intended to be used in a standard studio environment. Taking these requirements into account a view-dependent front-projection system was developed that requires less space than a rear-projection system and can be fitted into most existing studios. Images are projected onto a special retro-reflective cloth that allows a robust chroma-key in conjunction with cameras equipped with a ring of monochrome (blue) LEDs.

The system is composed of a number of modular components, as depicted in Figure 3.5. The communication and data exchange between these components is predominantly based upon a local area network and standard IT components. This concept provides a cost-effective way of implementing a system which can be easily configured to suit particular requirements.

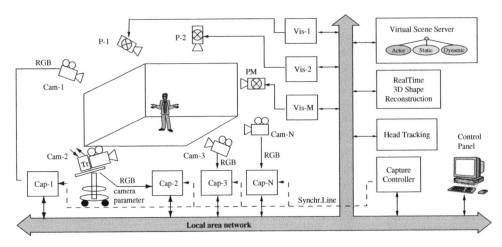

Figure 3.5 System overview. Each camera (Cam-1 ... Cam-N) is connected to a capturing server (Cap-1 ... Cap-N). Equally the data projectors (P-1 ... P-N) are driven by a projection server (Vis-1 ... Vis-N)

For example, the number of cameras can be varied depending on the available space in the studio and the specific production needs.

Each camera (Cam-1 ... Cam-N) is connected to a capturing server (Cap-1 ... Cap-N). The capturing servers are standard PCs, equipped with a frame grabber card. Each capturing server also has a RAID disc array. This allows the incoming video from the cameras to be saved to disc as uncompressed video (704 × 576 pixel, 24-bit colour resolution at 25 fps, progressive scan). This data is later processed in an off-line phase to produce high-quality 3D models for the final programme.

Usually the cameras are fixed and their parameters are determined with a calibration procedure. In addition, cameras equipped with a real-time tracking system can be used, as shown for Cam-2 in Figure 3.5. The real-time tracker delivers exact position and orientation and the internal camera parameters. We are using our previously developed 'Free-D' camera tracking system, as described in Section 3.2 for our experiments.

In addition to their function as an image sequence recorder, the capturing servers provide several online services over the network. On request they can send the latest grabbed image and provide a chroma-keying service, allowing the other software components to request alpha masks. One of the components making use of this is the real-time 3D shape reconstruction, that synchronously requests alpha masks from all the capture servers, and uses these to compute a 3D model of the actor using fast visual hull computations, as described in the previous section.

The 3D data is used by the head tracker and passed to the virtual scene server. The head position of the actor is used by the projection system to render a view-dependent image of the scene.

The virtual scene server provides a description of the virtual scene. This includes static scene elements, usually the virtual set, dynamic parts, for example any virtual characters involved in the scene, and the actor, provided by the 3D shape reconstruction module. The virtual scene server synchronizes all scene updates and distributes the scene updates to it.

The visualisation servers (Vis-1 ... Vis-M) are standard PCs equipped with OpenGL accelerated graphics cards that render the scene and drive the data projectors (P-1, ... P-M). They access the latest scene data from the scene server and the actor's head position from the head tracking module. Each server uses this data, together with the calibrated position of the corresponding projector, to calculate and render the scene for the viewpoint of the actor. The generated image is then projected onto the retro-reflective cloth.

The control panel is the main interface to the system. It consists of several components:

- The *camera remote control* allows the remote setting of camera parameters, such as focus, zoom, aperture and so on. This is particularly important because one or more cameras may be hanging from the ceiling and hence will not be accessible.
- The *capture control* is used to start and stop the capturing service on the servers.
- The *animation control* is able to start pre-defined 3D animations or control them live.
- The *3D preview* provides a preview of the production for the director and the camera operators. This allows the director to see a view of the scene from any position he or she chooses, allowing planning for camera shots. A textured 3D mesh of the actor is inserted in the virtual scene to give a preview of the final composited programme.

The control panel can be physically distributed over several workstations. This allows tasks to be delegated to different people or to be controlled from different locations. Moreover, the preview module can be instantiated several times, e.g., a view for the director, for a camera operator, for an animator and so on. The particular viewpoint of each preview can be set individually and is dynamic.

The capture controller synchronizes control events and the actual capturing. Therefore all capture servers are connected with a hardware line (indicated as 'Synchr.Line' in Figure 3.5) that allows frame-accurate synchronization of the capture.

A more detailed description of the studio set-up and the system components is given in Grau *et al.* (2004).

3.4.1 Head Tracking

The view-dependent actor feedback system requires information of the actor's head position. CAVE systems use mainly electromagnetic or acoustic tracking devices that are attached to the auxiliary glasses (e.g., shutter or polarization glasses), that are necessary for the stereo perception. For a production system this intrusion is not acceptable. Therefore a passive method is used. For this purpose a fast 3D template filter can be applied to the 3D volumetric representation generated by the visual hull reconstruction. The filter consists of two boxes, V_1 and V_2, as depicted in Figure 3.6. The filter computes a match $M(\mathbf{P})$ for all positions $\mathbf{P} = (m, n, k)^{\mathrm{T}}$ in a (discrete) volume:

$$M(\mathbf{P}) = 2a_2(\mathbf{P}) - a_1(\mathbf{P}) \tag{3.1}$$

$$a1(\mathbf{P}) = \mathrm{Match}(V_1(\mathbf{P})) \tag{3.2}$$

$$a2(\mathbf{P}) = \mathrm{Match}(V_2(\mathbf{P})) \tag{3.3}$$

$$\mathbf{P}_{Head} = \mathbf{P} : M(\mathbf{P}) \to \max \tag{3.4}$$

The match functions in (3.2) and (3.3) count the numbers of voxels inside the test box $V_1(\mathbf{P})$ and $V_2(\mathbf{P})$. In practical use the method has shown itself to be quite robust.

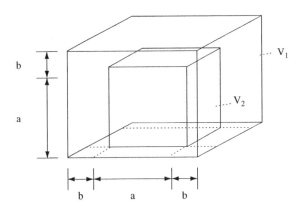

Figure 3.6 Template for head tracking

Besides the use in a view-dependent projection system the actor's head position can also be used as a feedback source for virtual scenes. For example an autonomous (virtual) character can react to the actor's motion. Because the head position is precisely known the virtual actor could track the actor and keep its eye-line correct.

3.4.2 View-dependent Rendering

The rendering engines in the visualization server for the projectors request the latest head position from the head tracker and use this to calculate and render the projected image from the point of view of the actor. This is done by setting a viewing frustum with the origin at the actor's head position. In addition a homographic distortion (8 parametric perspective image transform) is added to the projection matrix of the virtual camera, taking into account the relative positions of the projection screen (wall) and the projector position.

The view-dependent rendering allows the actor, if required, to keep looking at the face of the virtual character as he or she walks around him. That means the virtual scene components appear in space and the actor is 'immersed'.

The renderer also receives any updates in the scene from the virtual scene server. If the virtual character were as to move then the actor would see these movements. The combination of the scene updates and the viewpoint-dependent rendering thus allows complex interaction between the virtual and the real scene elements.

Figure 3.7 shows the setup during a demo production. The scene is taking place in the (virtual) entrance hall of a museum and includes a pterosaur flying through the hall from one end to the other. The flight was visualized using four projectors and allowed the boy to precisely track the pterosaur over 180°. The boy's eye-line was perfect in all takes that were recorded.

3.4.3 Mask Generation

A problem in a front-projection system is that there could be projected light falling onto an actor. One way to make this unwanted light invisible to the cameras is by operating

Figure 3.7 Flying pterosaur on the projection screen (in the top centre screen). The black area in front of the actor is caused by the shadow mask

the cameras in a shuttered mode, and projecting only during the time that the shutters are closed, as proposed in Tzidon *et al.* (1996). One disadvantage of this approach is that the light intensity of the projectors must be very high in order for the actors to be able to see the projected image under normal studio lighting conditions. Furthermore, the need to shutter the cameras may force the use of higher levels of studio illumination. Also, there remains the potential problem that the actors may be dazzled when looking towards a projector. An alternative approach, proposed here, is to generate a mask from the captured 3D shape and overlay it onto the projected image. This removes the need for high-intensity projectors with a rapid temporal response, so that conventional data projectors may be used; it also solves the problem of dazzle. The mask can be seen in Figure 3.7 as a black area on the projection screen just in front of the actor.

3.4.4 Texturing

For the director or camera operator, a synthesized view is provided that gives a pre-visualization of the final composited scene, i.e., the virtual and real scene elements. This renderer receives updates from the virtual scene server and grabs the latest 3D shape model of the actors. It can then generate a 3D representation of the scene and allows the director to view the scene from any position. This position can be dynamically updated to allow simulation of shots where the camera is moving.

In order to give a more realistic image the 3D shape model of the actor is textured with a view from one of the cameras. Therefore the renderer determines the studio camera that has the smallest angle to the virtual camera. The 3D shape model is stamped with the time-code of the alpha masks used to generate it, so the renderer requests the image from that time-code from the relevant capturing server, and uses it to texture the 3D shape model.

3.4.5 Collision Detection

Collision detection is a computer graphics technique to detect whether two 3D objects are in contact or are intersecting each other. Here it is used to establish another source of feedback for the virtual scene. In particular the collision detection allows one to determine whether an actor is in contact with a virtual object.

In order to keep the latency low the collision detection should take place after the visual hull computation. Here the test was implemented in the volumetric data representation: an updated bounding box of the virtual object of interest is compared with the result of the visual hull reconstruction. An event is generated if the intersection increases beyond a certain threshold of voxels, which avoids firing events caused by noise in the volumetric reconstruction.

The output signals are sent to the event server, which re-broadcasts them to its other clients. The use of these signals to trigger external events is designed into the functionality of the clients (i.e., the studio renderers).

Collision detection has been used in tests together with the projection system. Because it is intuitive to use and the event triggers are very precise, acceptance was very high.

3.5 CONCLUSIONS

3D data are increasingly used in content production, in particular in the movie industry for special effects, but also for TV productions. A driving force in this industry is the search for new effects or new kinds of programmes. Examples are movies such as *The Matrix* with excessive special effects or completely new television programmes like the BBC *BAMZOOKi* programme.

On the other hand new production methods can significantly decrease production costs. An example is the use for on-set visualization that has already changed the production flow for special effects in the film industry. The production system discussed here, with a bidirectional interface between real and virtual worlds, introduces a set of new tools for these kind of productions. It is worth mentioning that the degree of immersion in this system is not as high as in stereoscopic display systems, but it is a valuable tool to solve interaction problems between virtual and real actors, in particular the important eye-line problem.

For the future the increase in the use of 3D data and 3D scene composition will continue. To tap the full potential of cost savings, 3D planning and pre-visualization tools will be further developed.

In addition the final programme content will benefit from new production methods and the use of 3D data, mainly for special effects, but also for new emerging techniques that use 3D delivery, such as 3D TV and gaming platforms.

REFERENCES

Carranza J, Theobalt C, Magnor M and Seidel HP 2003 Free-viewpoint video of human actors. *ACM Transactions on Computer Graphics*.
Cheung K, Baker S and Kanade T 2003 Shape-from-silhouette of articulated objects and its use for human body kinematics estimation and motion capture. *Proceedings of the IEEE Conference on Computer Vision and Pattern Recognition*, pp. 77–84.

Cruz-Neira C, Sandin D, DeFanti T, Kenyon R and Hart J 1992 The cave: Audio visual experience automatic virtual environment. *Communications of the ACM* **35**(6), 65–72.

Debevec PE 1998 Rendering synthetic objects into real scenes: Bridging traditional and image-based graphics with global illumination and high dynamic range photography. *Proceedings of SIGGRAPH 98, Computer Graphics Proceedings, Annual Conference Series*, pp. 189–198, Orlando, USA.

Grau O 2004 3d sequence generation from multiple cameras. *Proceedings of the IEEE, International Workshop on Multimedia Signal Processing 2004*, Siena, Italy.

Grau O and Dearden A 2003 A fast and accurate method for 3d surface reconstruction from image silhouettes. *Proceedings of the 4th European Workshop on Image Analysis for Multimedia Interactive Services (WIAMIS)*, London, UK, pp. 395–404.

Grau O *et al.* 2003 New production tools for the planning and the on-set visualisation of virtual and real scenes. *Conference Proceedings of the International Broadcasting Convention*, Amsterdam, NL.

Grau O and Thomas GA 2002 Use of image-based 3d modelling techniques in broadcast applications. *2002 Tyrrhenian International Workshop on Digital Communications*, Capri, Italy, pp. 177–183.

Grau O, Pullen T and Thomas GA 2004 A combined studio production system for 3-d capturing of live action and immersive actor feedback. *IEEE Transactions on Circuits and Systems for Video Technology* **14**(3), 370–380.

Gross M, Wuermlin S, Naef M, Lamboray E, Spagno C, Kunz A, Koller-Meier E, Svoboda T, Gool LV, Lang S, Strehlke K, Moere AV and Staadt O 2003 blue-c: A spatially immersive display and 3d video portal for telepresence. *Proceedings of ACM SIGGRAPH 2003*, San Diego, USA, pp. 819–827.

Hilton A, Beresford D, Gentils T, Smith R and Sun W 1999 Virtual people: Capturing human models to populate virtual worlds. *IEEE Conference on Computer Animation*, pp. 174–185.

Kutulakos K and Seitz S 2000 A theory of shape by shape carving. *International Journal of Computer Vision* **38**(3), 197–216.

Lorensen WE and Cline HE 1987 Marching cubes: A high resolution 3d surface construction algorithm. *Proceedings of the 14th Annual Conference on Computer Graphics and Interactive Techniques*, ACM Press, pp. 163–169.

Matsuyama T, Wu X, Takai T and Wada T 2004 Real-time dynamic 3-d object shape reconstruction and high-fidelity texture mapping for 3-d video. *IEEE Transactions on Circuits and Systems for Video Technology* **14**(3), 357–369.

Matusik W, Buehler C and McMillan L 2001 Polyhedral visual hulls for real-time rendering. *Proceedings of the 12th Eurographics Workshop on Rendering*, pp. pages 116–126.

Matusik W, Buehler C, Raskar R, Gortler SJ and McMillan L 2000 Image-based visual hulls. In *Siggraph 2000, Computer Graphics Proceedings* (ed. Akeley K), ACM Press / ACM SIGGRAPH / Addison Wesley Longman, pp. 369–374.

Niem W 1994 Robust and fast modelling of 3d natural objects from multiple views. *SPIE Proceedings, Image and Video Processing II*, San Jose, vol. 2182, pp. 388–397.

Raskar R, Welch G, Cutts, M, Lake A, Stesin L and Fuchs H 1998 The office of the future: A unified approach to image-based modeling and spatially immersive displays. *Computer Graphics* **32**(Annual Conference Series), 179–188.

Rosenthal S, Griffin D and Sanders M 2001 Real-time compter graphics for on-set visualization: 'a.i.' and 'the mummy returns'. *Siggraph 2001, Sketches and Applications*.

Seitz S and Dyer C 1999 Photorealistic scene reconstruction by voxel coloring. *International Journal of Computer Vision* **35**(2), 151–173.

Starck J and Hilton A 2003 *Proceedings of ICCV*, pp. 915–922.

Thomas GA, Jin J, Niblett T and Urquhart 1997 A versatile camera position measurement system for virtual reality tv production. *Conference Proceedings of the International Broadcasting Convention*, pp. 284–289.

Tzidon *et al.* 1996 Prompting guide for chroma keying. *United States Patent*, 08/595,311, February 1, 1996.

Vedula S, Baker S and Kanada T 2002 Spatio-temporal view interpolation. *Proceedings of the 13th ACM Eurographics Workshop on Rendering*, pp. 65–76.

Weik S, Wingbermhle J and Niem W 2000 Automatic creation of flexible antropomorphic models for 3d videoconferencing. *Journal of Visualization and Computer Animation* **11**, 145–154.

4

Free Viewpoint Systems

Masayuki Tanimoto

Nagoya University, Japan

4.1 GENERAL OVERVIEW OF FREE VIEWPOINT SYSTEMS

Free viewpoint is the most general and challenging application scenario of 3D video communication. Television has a long history since it realized a human dream of seeing a distant world in real-time. It maintains its position as the most important and useful visual information system. However, TV shows us the same scene, even if we move our viewpoint in front of the display. This is quite different from what we experience in the real world. From TV, users can get only a limited view of a real 3D world they want to see. The view is determined not by the user, but by a camera placed in the 3D world. Although many important technologies have been developed, this function of TV has never changed. If TV could develop a new function such that we can view a 3D scene freely from any viewpoint as if we were there, it would be an epoch-making change in the history of television.

Many systems have been proposed to realize this function. It is common to all systems that a scene is captured with a set of cameras. Therefore, these systems are classified based on the methods of free view image generation as shown in Table 4.1. The methods of free view image generation are closely related to camera density. If the camera density is very high, view generation is very simple. It is performed by selecting a camera image or by collecting pixels from camera images. This is referred to as 'integral photography' in Table 4.1. If the camera density is not very high, view generation needs some processing. If it is moderately high, intermediate views can be generated precisely by interpolation of camera images using camera parameters. This is referred to as 'ray-space' in the table. Although Light field and Lumigraph belong to this category, precise interpolation was not reported in these methods. If intermediate views are generated, not by precise interpolation, but by warping or projection of camera images, generated views are not precise. This is referred to as 'image domain' in the table. If the camera density is low, intermediate views can be generated by detecting objects in the scene. This is referred to as 'model-based' in the table. Generated views are not precise in this method. Detection of objects in natural

3D Videocommunication — Algorithms, concepts and real-time systems in human centred communication
Edited by O. Schreer, P. Kauff and T. Sikora © 2005 John Wiley & Sons, Ltd

Table 4.1 Methods for free view image generation and their features

Method	Data acquisition	Data conversion	View generation	Quality of generated view image	Transparency
Image domain	Direct acquisition	No	Warping/ projection	Not precise	High
Integral pho- tography	Direct acquisition (precisely aligned)	Optical	Optical	Precise	High
Ray-space	Calibration and registration	Coordinate transform	Memory access/ interpolation	Precise	High
Surface light field	Calibration and registration	Decomposition approximation	Texture mapping/ pixel-by-pixel multiplication	Not precise in the case of approximation	Low
Model-based	Calibration and registration	3D model texture	Texture mapping	Not precise	Low

scenes has been studied intensively in the field of computer vision. However, it is a very difficult problem and has not yet been solved. The boundary between camera densities of ray-space and Model-Based methods is not clear and strongly depends on the performance of the interpolation. We can extend camera intervals for ray-space by using high-performance interpolation (Kobayashi *et al.* 2000, Droese *et al.* 2004). There is a hybrid of the ray-space and model-based methods which is referred to as 'surface light field' in the table. It uses both the surface model of the object and the information from light rays. From the ray-space side, it is regarded as a limiting case where the reference plane of the ray-space converges to the surface of the object. From the model-based side, many textures are attached on the surface of the object to improve the quality of the generated views.

'Transparency' of these methods is also listed in Table 4.1. Free viewpoint systems will be used in wide application areas such as entertainment, nature observation, sightseeing, art, museum, archive, content production, education, medicine, security, surveillance, ITS (Intelligent Transportation Systems) etc. If the method has strong scene dependence, its application is very limited. Transparency of the method is low in this case. If the method has no scene dependency, it can be used in many applications and the transparency of the method is high. A typical example of the latter case is TV, which is transparent because it works well for any scene and content. Therefore, TV can be a platform of 2D visual information. Transparency is also a very important factor in 3D video communication system, too.

These representation methods are described in more detail in Chapter 9. According to the definitions in Chapter 9, integral photography and ray-space belong to 'rendering without geometry', image domain belongs to 'rendering with geometry compensation', and surface light field and model-based methods belong to 'rendering from approximate geometry'.

Figure 4.1 summarizes typical free viewpoint image generation in different camera density. Figure 4.1(a) shows the case of very dense or continuous camera configuration such as very

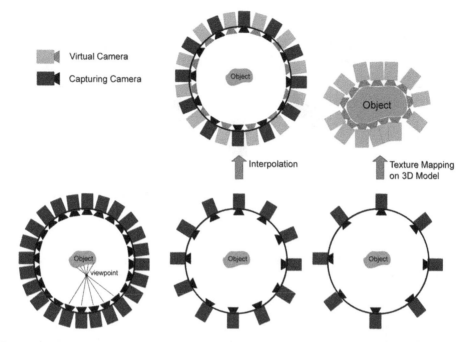

Figure 4.1 Free viewpoint image generation for different camera densities: (a) very dense or continuous (very dense ray-space); (b) dense (ray-space with interpolation); (c) sparse (model-based) (Reproduced with permission from Professor T. Kanade, CMU, USA)

dense ray-space. In this case, any viewpoint image can easily be obtained by collecting the rays that pass the viewpoint from different cameras. In the case of dense camera configuration, as shown in Figure 4.1(b), undetected rays are generated by interpolation and then the method shown in Figure 4.1(a) is applied to generate free viewpoint images. This is ray-space with interpolation. Figure 4.1(c) shows the case where camera configuration is so sparse that the interpolation of undetected rays is difficult. This is the model-based case. In this case, a 3D model of the object is made and texture is mapped on the surface of the object. This is expressed by many virtual cameras located on the surface of the object in Figure 4.1(c).

Typical free viewpoint systems due to image domain, ray-space, surface light field, model-based and integral photography methods are described in the following.

4.2 IMAGE DOMAIN SYSTEM

4.2.1 EyeVision

A system proposed by Carnegie Mellon University (CMU), USA, was given the name 'EyeVision' (http://www.ri.cmu.edu/events/sb35/tksuperbowl.html). Multiple video images of a dynamic event such as a football game were captured with multiple cameras placed at different angles. It was used to create special effects for the Super Bowl game. This approach is similar to those used for special effects in movies such as *The Matrix*. The difference is that the cameras of EyeVision can move tracking the object mechanically.

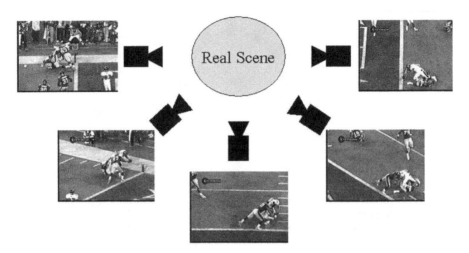

Figure 4.2 Virtual camera fly by EyeVision (Reproduced by permission from the Committee of 3D Conference, Japan)

The action at the game was captured by 30 cameras, each poised some 80 feet (~25 m) above the field at Raymond James Stadium in Tampa, Florida. Each camera, with computer-controlled zoom and focus capabilities, was mounted on a custom-built, robotic pan–tilt head, which could swing the camera in any direction at the command of a computer. These camera heads were controlled in concert so that cameras pointed, zoomed and focused at the same time on the same spot on the field, where a touchdown or fumble occurred.

Virtual camera flies were created along the predefined path of the real cameras around the stadium by switching the stored camera images in succession, as shown in Figure 4.2. The path is limited to the predefined original camera positions since no interpolation of intermediate views is performed.

4.2.2 3D-TV

A system was proposed by Mitsubishi Electric Research Laboratories (MERL) (Matusik and Pfister 2004), and called '3D-TV'. A 3D-TV prototype system was implemented with real-time acquisition, transmission and 3D display of dynamic scenes.

Figure 4.3 shows a schematic representation of the 3D-TV system. The acquisition stage consists of an array of hardware-synchronized cameras. Small clusters of cameras are connected to producer PCs. The producers capture live, uncompressed video streams and encode them using standard MPEG coding. The compressed video streams are then broadcast on separate channels over a transmission network, which could be digital cable, satellite TV, or the Internet. On the receiver side, individual video streams are decompressed by decoders. The decoders are connected by the network (e.g., gigabit Ethernet) to a cluster of consumer PCs. The consumers render the appropriate views and send them to a standard 2D, stereo-pair 3D, or multi-view 3D display.

View images are generated by assuming that all objects have the same depth. Therefore, the main processing in consumer side is a linear combination of the pixels of different views

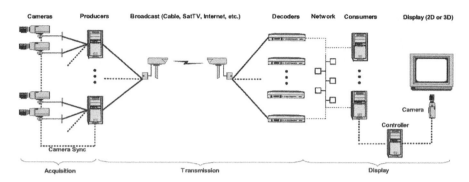

Figure 4.3 Configuration of 3D-TV system (Reproduced by permission of the ACM from Matusik and Pfister 2004)

for a given depth. Note that the system is not using multiple depth maps to synthesize the new view, so the generated view cannot be precise.

4.2.3 Free Viewpoint Play

Practical free viewpoint systems based on the image domain method have already been available in storage applications (http://www.sonymusic.co.jp/VisualStore/WonderZone/, http://www.ponycanyon.co.jp/eizoplay/). Scenes of entertainment are captured with many cameras, and these signals are stored in DVD. The number of cameras is 12 to 17, and recording time is about 40 minutes. Transition effect is added to the generated images at the change of viewpoint to increase the feeling of smoothness by using a function of Play Station 2. Several kinds of DVD software are available commercially and the user can change the viewpoint interactively.

4.3 RAY-SPACE SYSTEM

4.3.1 FTV (Free Viewpoint TV)

The system was proposed by Nagoya University and called 'Free viewpoint TV (FTV)' (Tanimoto and Fujii 2002; Fujii and Tanimoto 2002; Na Bangchang *et al.* 2003; Panahpour Tehrani *et al.* 2004). In FTV, the user can freely control the viewpoint of any dynamic 3D scene in real-time.

The configuration of FTV is shown in Figure 4.4. The major process of FTV consists of acquisition, interpolation, compression and display. At the transmitter side, the scene is acquired with many cameras. There are several types of camera arrangement, depending on how freely we want to see the scene. For example, if we want to see the scene from one side only, the cameras are placed on a line. If we want to see the scene from the backside, they are placed on a circle. If we want to see the scene from the top, they are placed on a dome. The dense camera arrangement makes the ray-space dense. In the case of line camera arrangement, the images of these cameras are placed vertically in parallel in

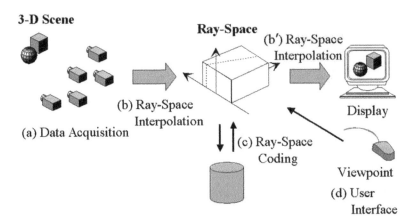

Figure 4.4 Configuration of FTV (Reproduced by permission of the ACM from Tanimoto and Fujii 2002; Fujii and Tanimoto 2002)

the ray-space, forming the FTV signal. The ray-space is originally four-dimensional, five-dimensional including the time parameter. If we place cameras in a limited region such as on a line or a circle, the rays obtained are limited and the ray-space constructed from these rays is a sub-space of the ray-space.

The sparse camera arrangement makes the ray-space sparse. A dense ray-space can be made from the sparse ray-space by interpolation using the special structure of the ray-space. This ray-space interpolation is done at the transmitter side or the receiver side depending on the application.

The FTV signal consists of multiple camera signals and camera parameters. Compression of the FTV signal is very important because the number of cameras is large. The simplest way is to apply conventional MPEG coding to each camera signal independently. However, the FTV signal can be compressed more effectively by considering the whole FTV signal and the structure of the ray-space because there is strong correlation between the camera images (Fujii *et al.* 1996, Tanimoto *et al.* 2001, Na Bangchang *et al.* 2004).

The display of FTV is very easy. When the user defines his or her viewpoint to the system, the view image is simply obtained as a section image of the ray-space. The view image can be displayed on a 2D or 3D display (Higashi *et al.* 2004). Two or more cross-sections are made for 3D or multi-view display. Figure 4.5 shows an experimental FTV system . Examples of the generated free viewpoint images are shown in Figure 4.6. Complicated natural scenes, including sophisticated objects such as small moving fish, bubbles and the reflection of light from the aquarium glass are reproduced very well in real-time.

FTV can acquire, store and transmit all the visual information of 3D space. Thus, the function of FTV is not only to generate free viewpoint images, but also to transport the 3D visual space. FTV is a good candidate for a platform of 3D visual information.

4.3.2 Bird's-eye View System

The system was proposed by Nagoya University and called the 'Bird's-Eye View System' (Sekitoh *et al.* 2001; Sekitoh *et al.* 2003). It can generate a bird's-eye view over an intersection

Figure 4.5 Experimental FTV system (Reproduced by permission of Professor M. Tanimoto, Nagoya University, Japan)

Figure 4.6 Examples of generated free viewpoint images in FTV (Panahpour Tehrani *et al.* 2004): (a) intermediate position; (b) forward position

in real-time. A 1/24-scale miniature car is driven on a diorama by wireless control. This scene is captured with 16 CCD cameras mounted on a dome structure. The 16 CCD cameras are set at each vertex of triangular patches, which compose a hemisphere dome with 50 cm radius over the intersection of the diorama. These cameras are arranged at about 20 cm intervals in the shape of an equilateral triangle.

The hardware of the Bird's-Eye View System is shown in Figure 4.7. This system consists of one set of 16 client PCs, and one host PC. The client PCs and the host PC are connected by gigabit Ethernet. Each PC is a general-purpose PC. The image capturing board is mounted in a PCI bus on each client PC. A 3D mouse is connected to RS232-C port and a virtual viewpoint can be inputted.

Since the camera arrangement of the Bird's-Eye View System is sparse compared with FTV, view images are generated using model-assisted interpolation. Silhouette images are obtained by taking the difference with the background for each input image. 2D projective transformation is carried out to the silhouette images from each view to virtual view.

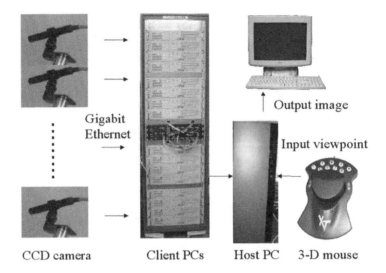

Figure 4.7 Hardware of the Bird's-eye View System (Sekitoh *et al.* 2003)

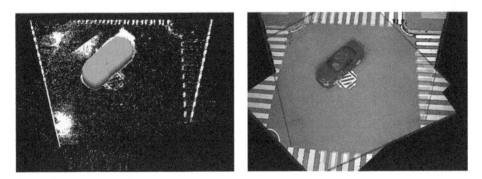

Figure 4.8 Examples of height map and generated bird's-eye view (Sekitoh *et al.* 2003): (a) height map; (b) generated bird's-eye view

The height map is created from the transformed images. Examples of a height map and generated bird's-eye view image are shown in Figure 4.8(a, b).

4.3.3 Light Field Video Camera System

This system was proposed by Stanford University and called the 'Light Field Video Camera System (LFVC system)' (Levoy and Hanrahan 1996; Goldluecke *et al.* 2002; Wilburn *et al.* 2002; Vaish *et al.* 2004). The LFVC system uses a camera array for capture, as shown in Figure 4.9. The array system is formed as 3×2 camera array, which leads for a sample hardware requirement. The camera heads are aligned in parallel. Camera nodes are connected to each other and to a host PC via an IEEE1394 High Speed Serial Bus. To keep the cameras synchronized, a common clock signal is distributed to all of the cameras. A reset signal and two general purpose triggers are also routed to all camera nodes.

Figure 4.9 Individual LFVC and LFVC system (Wilburn *et al.* 2002)

The system can display a dynamic scene from new viewpoints at 20 frames per second. However, this system is not real-time. First, this system captures a scene with these cameras and records the camera images. Then, disparity maps are made off-line from the recorded camera images. View images are generated by warping and blending the recorded camera images with the disparity maps. The quality of the generated view image largely depends on the accuracy of the disparity maps.

In the rendering step, a number of images with recalculated dense disparity maps are warped and blended to generate new view images. To speed up the processing, the polygon processing capabilities of OpenGL as well as hardware texturing and blending provided by modern graphics hardware are used.

This system uses special cameras called the 'Light Field Video Camera' (LFVC) to construct a camera array. The basic block diagram of the LFVC is shown in Figure 4.10. The LFVC consists of multiple CMOS imagers, each providing 640×480 pixel RGB resolution at 30 frames per second. The use of more than one camera inevitably leads to mismatches in hue, brightness and radial distortion among different camera images. These differences

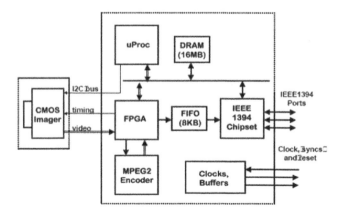

Figure 4.10 Block diagram of LFVC (Reproduced by permission of the SPIE from The Light Field Video Camera, *Proc. Media Processors*, Wilburn, B., Smulski, M., Kelin Lee H., Horowitz, M. (2002). Figure 1 and Figure 2)

need to be minimized by careful calibration prior to further processing. Custom-built driver boards enable on-line calibration as well as MPEG-2 compression of each video stream.

4.4   SURFACE LIGHT FIELD SYSTEM

This system was proposed by Intel and called the 'Surface Light Field System' (Chen *et al.* 2002). The surface light field is regarded as a limiting case that the reference plane of the ray-space converges to the surface of the object as described in Section 4.1. Therefore, it needs the shape information of the object. A light field parameterized on the surface offers a natural and intuitive description of the view-dependent appearance of scenes with complex reflectance properties.

In the Surface Light Field System, acquisition is off-line and view image generation is real-time. A compact representation suitable for an accelerated graphics pipeline is introduced. The light field data is approximated by partitioning it over elementary surface primitives and factorizing each part into a small set of lower-dimensional functions. After acquisition, it can be further compressed using standard image compression techniques leading to extremely compact data sets that are up to four orders of magnitude smaller than the input data. Finally, an image-based rendering method and light field mapping are used that can visualize surface light fields directly from this compact representation at interactive frame rates on a personal computer.

Figure 4.10 gives an illustration of the overall data acquisition process. From 200 to 400 images are captured with a hand-held digital camera as shown in Figure 4.11(a). Figure 4.11(b) shows one sample image. Observe that the object is placed on a platform designed for the purpose of automatic registration. The colour circles are first automatically detected on the images using a simple colour segmentation scheme. This provides an initial guess for the position of the grid corners that are then accurately localized using a corner finder. The precise corner locations are then used to compute the 3D position of the camera relative to the object. The object geometry is computed using a structured lighting system consisting of a projector and a camera. The two devices are visible in Figure 4.11(a). Figure 4.11(c) shows an example of camera image acquired during scanning. The projector is used to project a translating stripped pattern onto the object. A temporal analysis is used for accurate range sensing. In order to facilitate scanning, the object is painted with white removable paint. Between 10 and 20 scans are taken to completely cover the surface geometry of the object. Between two consecutive scans, the object is rotated in front of the

Figure 4.11   Data acquistion process (Reproduced by permission of the ACM from Chen *et al.* 2002): (a) capturing many images; (b) sample image; (c) painted object; (d) 3D triangular mesh

camera and projector by about 20°. The resulting cloud of points (approximately 500 000 points) is then fed into mesh editing software to build the final triangular surface mesh shown in Figure 4.11(d). Since the same calibration platform is used for both image and geometry acquisition, the resulting triangular mesh is naturally registered to the camera images.

4.5 MODEL-BASED SYSTEM

4.5.1 3D Room

The 3D Room system was proposed by writers at Carnegie Mellon University (CMU), USA, and it aims to capture and model a real time-varying event inside the viewing zone (Nayaranan *et al.* 1998; Vedula *et al.* 1998; Kanade *et al.* 1997; Saito *et al.* 1999). The 3D Room is 20 feet (L) × 20 feet (W) × 9 feet (H). As shown in Figure 4.12 and Figure 4.13, 49 cameras are distributed inside the room. A PC cluster system (17 PCs) is used to digitize all the video signals from the cameras simultaneously as uncompressed and lossless colour images.

Figure 4.13 shows the overview of the procedure for generating virtual view images from multiple image sequences collected in the 3D room. Depth image sequences are obtained from input image sequences by applying multiple baselines stereo frame by frame. The depth images of all cameras are merged into a sequence of 3D shape models, using a volumetric merging algorithm. This volumetric merging generates a 3D shape model of the object that is represented by a triangle mesh. The volume of interest is specified during the volumetric merging so that only the objects of interest can be extracted. For controlling the appearance-based virtual viewpoint in the 3D Room, two cameras are selected. The intermediate images between the selected two camera images are generated by interpolation of the selected images from the correspondence between the images. The corresponding points are computed by

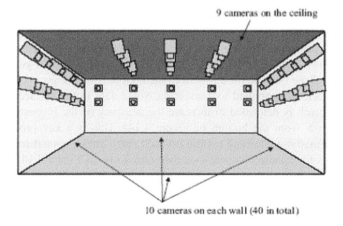

Figure 4.12   Camera placement in the 3D Room (Reproduced by permission of The IEEE, T. Kanade, from Saito *et al.* 1999)

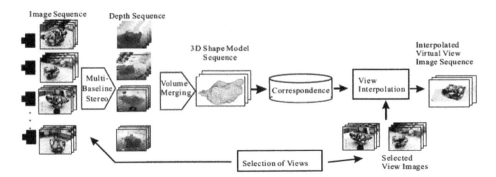

Figure 4.13 Overview of the procedure for generating virtual view images from multiple cameras in the 3D Room (Reproduced by permission of The IEEE, Professor T. Kanade from Saito *et al.* 1999)

using the 3D shape model. The weighting value between the images controls the appearance of the virtual viewpoint.

4.5.2 3D Video

This system was proposed by Kyoto University and called '3D Video' (Matsuyama and Takai 2002; Wu *et al.* 2003; Matsuyama *et al.* 2003; 2004). 3D video also aims to capture and model a real time-varying event inside the viewing zone as 3D Room. This system uses a parallel pipeline processing method for reconstructing a dynamic 3D object shape from multi-view video images, by which a temporal series of 3D voxel representation of the object can be obtained in real-time. However, the conversion of the voxel representation to the surface patch representation and the deformation of the surface patch to increase the accuracy of the reconstructed 3D shape are off-line.

Figure 4.14 illustrates the basic process of generating a 3D video frame. A set of multi-view object images are taken simultaneously by a group of distributed video cameras (see the top row of Figure 4.14). Background subtraction is applied to each captured image to generate a set of multi-view object silhouettes (see the second row from the top in Figure 4.14). Each silhouette is back-projected into the common 3D space to generate a visual cone encasing the 3D object. Then, such 3D cones are intersected with each other to generate the voxel data of the object shape (see the third row from the bottom row of Figure 4.14). The discrete marching cubes method is applied to convert the voxel data to the surface patch data, and then the surface patch is deformed to increase the accuracy of the reconstructed 3D shape (see the second row from the bottom of Figure 4.14). This is a key process to obtain a good 3D shape. Finally, colour and texture on each patch are computed from the observed multi-view images for texture mapping (see the bottom row of Figure 4.14).

By repeating the above process for each video frame, a live 3D motion picture can be created. It should be noted that since the surface patch deformation requires large computation time (approximately a few minutes per frame), the entire process of 3D Video cannot run in real-time.

The method is improved by introducing image-based rendering techniques to generate arbitrary view images based on 3D shape data reconstructed from multi-view images, where

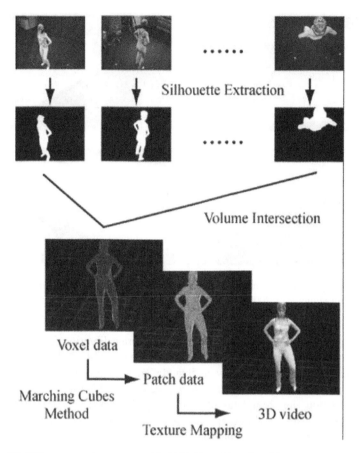

Silhouette Extraction

Volume Intersection

Voxel data

Marching Cubes
Method

Patch data

3D video

Texture Mapping

Figure 4.14 3D Video generation process (© 2004 IEEE. Reprinted by permission. Matsuyama T, Wu X, Takai T, and Wada T (2004) Real-time dynamic 3-D object shape reconstruction and high-fidelity texture mapping for 3-D Video, *IEEE Trans. on Circuits and Systems for Video Technology*, **14**(3), Figure 1)

a virtual view direction is specified to control the blending process of multi-view images. While such image-based rendering methods can avoid jitter in generated images, image sharpness is degraded due to the blending operation.

4.5.3 Multi-texturing

The system was proposed by Heinrich-Hertz-Institut (HHI) and called 'Multi-texturing' (Debevec *et al.* 1998; Müller *et al.* 2003a, b; Smolic *et al.* 2003). The feature of this system is to use image-based rendering techniques to improve generated view images as 3D Video.

Figure 4.15 shows examples of views generated by the multi-texturing for view-dependent rendering of graphics. A dynamic scene is captured by a number of static cameras. The approach consists of a model-based reconstruction method requiring *a priori* knowledge

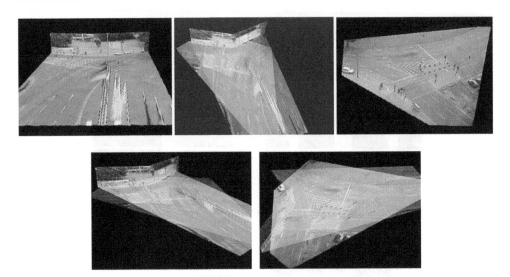

Figure 4.15 Examples of view images generated by multi-texturing (Reproduced by permission of Dr. A. Smolic from Müller *et al.* 2003a,b): (a) original; (b) blended; (c) original; (d) blended; (e) blended

about the scene geometry (e.g., ground plane) and camera calibration. The scene is separated into static parts (streets, sidewalks, buildings, etc.) that are modelled semi-automatically (i.e., those areas are chosen manually and modelled automatically) and dynamic objects (e.g., vehicles, pedestrians, bicycles, etc.), which are modelled automatically, using primitive geometry information (i.e., templates of vehicles, pedestrians, bicycles, etc.). All components are combined, using a 3D compositor that allows free navigation within the scene. After camera calibration and segmentation, a background image is extracted for each view. These are used to create a multi-texture model of the background using the multi-texture tool of DirectX. All background images are projected onto common geometric primitives, e.g., a ground plane. Individual normal vectors are assigned to each of the textures.

This enables view-dependent rendering by individual weighting of the contributing textures, as illustrated in Figure 4.15. Figure 4.15(a, c) show the scene from original camera viewpoints. Figure 4.15(b) shows a central view where both original views are equally weighted. Figure 4.15(d, e) show viewpoints closer to one of the original cameras, where the corresponding weight of the closer camera is set to a higher value, respectively. Objects such as flags that are not modelled are deformed in the blended images.

4.6 INTEGRAL PHOTOGRAPHY SYSTEM

4.6.1 NHK System

This system was proposed by NHK and called 'Integral Photography System' (Okui and Okano 2003). The viewer can see a 3D moving scene without wearing any special glasses. The images constructed from light rays appear real to the eyes of the observer.

Integral photography (IP) is a well-known technique of autostereoscopic photography (Lippmann 1908). Figure 4.16 shows the principle of the original IP method. In order to

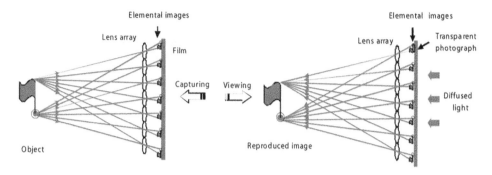

Figure 4.16 A baisc IP method (Reproduced by permission of Dr. Fumio Akano and Makoto Okui, NHK from Okui and Okano 2003): (a) pickup; (b) display

produce an IP image composed of many elemental images, a lens array with many convex microlenses is positioned immediately in front of the photographic film. Numerous small elemental images are focused onto the film, with their number corresponding to that of the microlenses. The film is then developed to obtain a transparent photograph. For reproduction, this transparent photograph is placed where the film was and is irradiated from behind by diffused light. The light beams passing through the photograph retrace the original routes and converge at the point where the photographed object was, forming an autostereoscopic image.

The basic IP method has the following problems. The first is that the reconstructed images are viewed from the opposite direction to that of the image pickup. The second problem is optical cross-talk between elemental images, which causes double or multiple image interference. The third problem is the resolution of the reproduced image.

NHK developed a real-time IP system as shown in Figure 4.17, solving these problems. The film and the transparent photograph in Figure 4.16 are replaced by a high-definition television (HDTV) camera and a LCD, respectively, to realize a real-time system. Furthermore, a gradient-index (GRIN) lens (Arai *et al.* 1998) is introduced as the pickup lens array to avoid the first and second problems. A depth control lens (Okano *et al.* 1999), which is a large-aperture convex lens, is employed to solve the third problem.

It should be noted that the IP System and FTV are based on the same principle because the elementary images of the IP System are the same as the ray-space of FTV. The difference

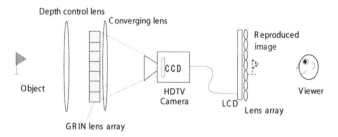

Figure 4.17 Configuration of a real-time IP system (Reproduced by permission of Dr. Fumio Akano and Makoto Okui, NHK from Okui and Okano 2003)

Table 4.2   Specifications of 1D-II 3D display system (Taira *et al.* 2004)

Panel	20.8-inch LCD	15.4-inch LCD
Panel Resolution	3200 × 2400 (QUXGA)	1920 × 1200 (WUXGA)
Number of Parallaxes	32	18
3D resolution	300 × 800	300 × 400
Viewing angle	±10°	±14°
Luminance	160 cd/m²	150 cd/m²
Contents	Still Images	Movies/interactive

between the two systems is the sampling density of light rays in the location and direction of the ray-space. In the IP System, the sampling density in the location is dense and that in the direction is sparse. On the other hand, in FTV, the sampling density in the location is sparse and that in the direction is dense. This is why the IP System does not need view interpolation and the resolution of the view image is low. The resolution of the view image has been improved by using a super HDTV camera instead of the HDTV camera.

4.6.2   1D-II 3D Display System

This system was developed by Toshiba and called the '1D Integral Imaging (1D-II) 3D Display System' (Fukushima *et al.* 2004; Saishu *et al.* 2004; Taira *et al.* 2004). Two types of 3D PC-based display systems were developed, which are 20.8 and 15.4 inches diagonal size, using one-dimensional integral imaging. The specifications of the systems are listed in Table 4.2.

Horizontal resolution of 300 pixels with 32 parallaxes in the 20.8-inch type and 18 parallaxes in the 15.4-inch type was achieved by mosaic pixel arrangement of the display panel. These systems have wide viewing area and a viewer can observe 3D images with continuous motion parallax, suppressing the flipping of 3D images observed in conventional multi-view display systems. By using lenticular sheets, these systems realized high brightness in spite of having many parallaxes.

Plug-ins of PC for creating 3D image for CG software was also developed. Movie of CG contents and real-time interaction are demonstrated in the 15.4-inch type. The processing flow of the real time interaction is shown in Figure 4.18.

4.7   SUMMARY

Free viewpoint systems have been the stuff of dreams. Now, many of such systems have become available as shown in this chapter. Their cost is still high because they need a large amount of memory and a large computational cost. However, the rapid progress of technologies will make such systems to be realized more easily soon.

The free viewpoint systems are very important not only in application, but also academically because such systems eventually need frontier technologies that treat rays one by one

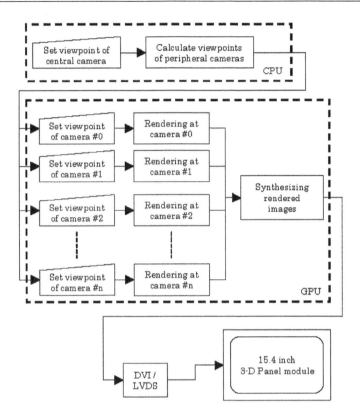

Figure 4.18 Processing flow of the real-time interaction of 1D-II 3D display system (Reproduced by permission of the Committee of 3D Conference, Japan, from Taira *et al.* 2004)

in acquisition, processing and display. In this chapter, we have seen some examples of such technologies as ray acquisition, ray processing and ray display. Other types of ray acquisition and ray display systems have also been reported (Endo *et al.* 2000; Fujii and Tanimoto 2004). This work will open a new frontier in image engineering, namely 'Ray-Based, Image Engineering'.

REFERENCES

Arai J, Okano F, Hoshino H and Yuyama I 1998 Gradient-index lens array method based on real-time integral photography for three-dimensional images. *Applied Optics*, **37**(11), 2034–2045.

Chen WC, Bouguet JY, Chu MH and Grzeszczuk R 2002 Light field mapping: efficient representation and hardware rendering of surface light fields. *ACM Transactions on Graphics* **21** (3), 447–456.

Debevec P, Yu Y and Borshukov G 1998 Efficient view-dependent image-based rendering with projective texture mapping. *Proceedings ACM SIGGRAPH'98*, USA, 105–116.

Droese M, Fujii T and Tanimoto M 2004 Ray-space interpolation constraining smooth disparities based on loopy belief propagation. *Proceedings of the 11th International Workshop on Systems, Signals and Image Processing ambient multimedia (IWSSIP)*, Poland, 247–250.

Endo T, Kajiki, Y, Honda T and Sato M 2000 Cylindrical 3-D video display observable from all directions. *Proceedings of Pacific Graphics*, 300–306.

Fujii T and Tanimoto M 2002 Free-viewpoint television based on the ray-space representation. *SPIE ITCom*, 175–189.

Fujii T and Tanimoto M 2004 Real-time ray-space acquisition system. *Proceedings of SPIE*, **5291**, USA, 179–187.

Fujii T, Kimoto T and Tanimoto M 1996 Ray space coding for 3D visual communication. *Proceedings of Picture Coding Symposium*, 447–451.

Fukushima R, Taira K, Saishu T and Hirayama Y 2004 Novel viewing zone control method for computer generated integral 3-D imaging. *Proceedings of SPIE*, **5291**(A62).

Goldlucke B, Magnor M and Wilburn B 2002 Hardware accelerated dynamic light field rendering. *VMV* 2002.

Higashi T, Fujii T and Tanimoto M 2004 Three-dimensional display system for user interface of FTV. *Proceedings of SPIE*, **5291**A(25), USA.

http://www.ponycanyon.co.jp/eizoplay/.

http://www.ri.cmu.edu/events/sb35/tksuperbowl.html.

http://www.sonymusic.co.jp/VisualStore/WonderZone/.

Kanade T, Rander PW and Nayaranan PJ 1997 Virtualized reality: constructing virtual worlds from real scenes. *IEEE Multimedia* **4**(1), 34–47.

Kobayashi T, Fujii T, Kimoto T and Tanimoto T 2000 Interpolation of ray-space data by adaptive filtering. *Proceedings of SPIE*, 3958, USA, 252–259

Levoy M and Hanrahan P 1996 Light field rendering. *Proceedings of SIGGRAPH*, 31–42.

Lippmann G 1908 *Comptes-Rendus Academie des Sciences*, **146**, 446–451.

Matsuyama T and Takai T 2002 Generation, visualization, and editing of 3D video. *Proceedings of the Symposium on 3D Data Processing Visualization and Transmission*, Italy, 234–245.

Matsuyama T, Wu X, Takai T and Nobuhara S 2003 Real-time generation and high fidelity visualization of 3D video. *Proceedings of MIRAGE2003*, 1–10.

Matsuyama T, Wu X, Takai T and Wada T 2004 Real-time dynamic 3-D object shape reconstruction and high-fidelity texture mapping for 3-D video *IEEE Transactions on Circuits and Systems for Video Technology*, **14**(3), 357–369.

Matusik W and Pfister H 2004 3D-TV a scalable system for real-time acquisition, transmission and autostereoscopic display of dynamic scenes. *Proceedings of ACM SIGGRAPH*, 814–824.

Müller K, Smolic A, Droese M, Voigt P and Wiegand T 2003a Multi-texture modelling of 3D traffic scenes. *Proceedings of ICME 2003*, USA, 6–9.

Müller K, Smolic A, Droese M, Voigt P and Wiegand T 2003b 3D modelling of traffic scenes using multi-texture surfaces. *Proceedings of PCS 2003*, France, 309–314.

Na Bangchang P, Fujii T and Tanimoto M 2003 Experimental system of free viewpoint television. *Proceedings of SPIE* 5006, USA, 554–563.

Na Bangchang P, Fujii T and Tanimoto M 2004 Ray-space data compression using spatial and temporal disparity compensation. *Proceedings of IWAIT*, Singapore, 171–175.

Nayaranan PJ Rander PW and Kanade T 1998 Constructing virtual worlds using dense stereo. *Proceedings of IEEE 6th International Conference on Computer Vision*, India, 3–10.

Okano F, Arai J, Hoshino H and Yuyama, 1999 Three-dimensional video system based on integral photography. *Optic Engineering*, **38**(6), 1072–1077.

Okui M and Okano F 2003 Real-time 3-D imaging method based on integral photography. *Report (for AHG on 3DAV), ISO/IEC JTC 1/SC 29/WG 11* m10088, Australia.

Panahpour Tehrani M, Na Bangchang P, Fujii T and Tanimoto M 2004 The optimization of distributed processing for arbitrary view generation in camera sensor networks. *IEICE Transactions on Fundamentals of Electronics, Communication and Computer Sciences*, E87-A, **8**, 1863–1870.

Saishu T, Taira K, Fukushima R and Hirayama Y 2004 Distortion control in a one-dimensional integral imaging autostereoscopic display system with parallel optical beam groups. *Proceedings of SID*, **53**(3), USA, 1438–1441.

Saito H, Baba S, Kimura M, Vedula S and Kanada T 1999 Appearance-based virtual view generation of temporally-varying events from multi-camera images in the 3D room. *Proceedings of 3-D Imaging and Modelling*, 516–525.

Sekitoh M, Fujii T, Kimoto T and Tanimoto M 2001 Bird's eye view system for ITS. *Proceedings of the IEEE Intelligent Vehicle Symposium*, 119–123.

Sekitoh M, Kutsuna T, Fujii T, Kimoto T and Tanimoto M 2003 Arbitrary view generation by model-based interpolation in the ray space. *Proceedings of SPIE* **4660**(58), USA, 465–474.

Smolic A, Mueller K, Droese M, Voigt P and Wiegand T 2003 Multiple view video streaming and 3D-scene reconstruction for traffic surveillance. *Proceedings of WIAMIS 2003, 4th European Workshop on Image Analysis for Multimedia Interactive Services*, UK, 427–432.

Taira K, Yanagawa S, Kobayashi H, Yamauchi Y and Hirayama Y 2004 Development of 3-D display system using one-dimensional integral imaging method. *Proceedings of the 3D Conference*, Japan, 21–24.

Tanimoto M and Fujii T 2002 FTV free viewpoint television. *ISO/IEC JTC1/SC29/WG11*, M8595.

Tanimoto M, Nakanishi A, Fujii T and Kimoto T 2001 The hierarchical ray-space for scalable 3-D image coding. *Proceedings of the Picture Coding Symposium*, 81–84.

Vaish V, Wilburn B, Joshi N and Levoy M 2004 Using plane + parallax for calibrating dense camera arrays. *Proceedings of the CVPR 2004*, 2–9.

Vedula S, Rander PW, Saito H and Kanade T 1998 Modelling, combining, and rendering dynamic real-world events from image sequences. *Proceedings of the 4th Conference on Virtual Systems and Multimedia*, Japan, **1**, 326–332.

Wilburn B, Smulski M, Kelin Lee H and Horowitz M 2002 The light field video camera. *Media Processors, Proceedings of SPIE*, **4674**, 29–36.

Wu X and Matsuyama T 2003 Real-time active 3D shape reconstruction for 3D video. *Proceedings of the 3rd International Symposium on Image and Signal Processing and Analysis*, Italy, 186–191.

5

Immersive Videoconferencing

Peter Kauff and Oliver Schreer

*Fraunhofer Institute for Telecommunications/Heinrich-Hertz-Institut,
Berlin, Germany*

5.1 INTRODUCTION

The recent developments in the area of videoconferencing are the best proof for an ongoing migration of immersive media to telecommunication. Today, videoconferencing is going to become more and more attractive for various lines of business. It encourages distributed collaboration in an emerging global market place and is therefore regarded as a high-return investment for decision-making. Corporate reasons for using videoconferencing systems include business globalization, increased competition, pressure for higher reactivity and shorter decision phases, increase in number of partners, and reduced time and travel costs.

Considering the positive market forecasts, one may assume that videoconferencing has been a killer application from the beginning. However, reviewing the advent of videoconferencing system shows that the introduction phase was difficult and stony. In fact, many people were concerned about usability and effectiveness when the first systems were brought to the market more then fifteen years ago. Early experiments even failed to show an advantage over simple audio applications such as phone conferences. According to a former Wainhouse report from the beginning of this decade, voice conferencing and e-mail were still preferred to videoconferencing for a long time. Key barriers were the high unit prices and costs of ownership as well as the concerns about integration, lack of training, and user friendliness. One reason for limited user acceptance was the restricted support of natural human-centred communication. Early implementations were mainly based on simple system architectures, reproducing single-user images either at small displays or in separate PC windows. Important communication modalities such as facial expressions, eye contact, gaze direction, bodily

3D Videocommunication — Algorithms, concepts and real-time systems in human centred communication
Edited by O. Schreer, P. Kauff and T. Sikora © 2005 John Wiley & Sons, Ltd

movement, postures and gestures, audiovisual coherence or spatial audio perception were often addressed in a dissatisfying manner.

Most of these former deficits were caused by limited bandwidth capabilities, inefficient video compression techniques and insufficient knowledge about presence engineering. Meanwhile this situation has changed dramatically. High-speed networks are more and more established in daily work and state-of-the-art multimedia standards provide efficient compression techniques for coding audiovisual data with high quality. Large plasma or LCD displays as well as large screen projections are available for a reasonable cost and, finally, recent efforts in presence research have defined design rules and recommendations for natural inter-human telecommunication services.

As a consequence, many high-end videoconferencing systems nowadays offer telepresence capabilities. They use large display technology to present conferees at life-size, and to support inter-human communication attributes to the greatest possible extent. Especially equipped conference rooms integrate large projection walls into real meeting venues to get the impression of a seamless transition between two physically separated locations.

Anticipating that these developments are only the beginning of new era in telecommunication, a lot of worldwide research efforts have investigated the potential of immersive videoconferencing that goes beyond today's telepresence systems. Presence research has therefore developed many concepts to exceed the limitations of traditional communication and have investigated how to utilize these concepts for an improvement of physical presence (spatial closeness despite of distance) in videoconferencing. Most of these activities were strongly linked to the concept of teleimmersion. The ultimate goal is to design immersive portals, which offer rich communication modalities and where conferees can meet and collaborate, as they would do in a real physical face-to-face event. In the ideal case they should not be able to perceive any difference to a real meeting venue.

It is the objective of this book chapter to present a snapshot of current developments in this business segment. For this purpose, the next section will outline the meaning of telepresence for available high-end videoconferencing products. Section 5.3 will then describe the concept of shared virtual table environment (SVTE), which is the basis of multi-party communication in immersive videoconferencing. An overview about related international research and experimental platforms for immersive videoconferencing will subsequently be given in Section 5.4. Finally, Section 5.5 gives an outlook to new perspectives and trends.

5.2 THE MEANING OF TELEPRESENCE IN VIDEOCONFERENCING

Telepresence is a relatively new term in communication science. It was first used by Minsky (1980). Since then the concept of presence has been keenly discussed among human factor researchers and the discussion is likely to continue. The structure and the effects of presence are to a large extent still unclear (IJsselsteijn *et al.* 2001). Especially during the 1990s several researchers emphasized the particular importance of telepresence for videoconferencing (Acker and Levitt 1987; Buxton 1992; Mühlbach *et al.* 1995). Various aspects of telepresence have been investigated on the basis of sophisticated test-beds and the researchers have given recommendations for the design of adequate videoconferencing systems. Meanwhile this knowledge has found its way into commercial products and videoconferencing sellers advertise high-end solutions with telepresence capabilities.

One of the most important requirements on a telepresence system is realism and, with it, a high-quality video reproduction of the remote conferees. In this context it is essential that conference partners appear in life-size and in correct perspective at a place where they would be expected in a real meeting. In principle, this can only be achieved by capturing a conferee with a camera that has been placed in the line-of-sight, preferably as close as possible to the assumed head position of the other conferee. Figure 5.1 illustrates this for the simple case of a face-to-face conversation and its reproduction through a two-party single-user video conference. Here, the camera that is capturing the partner on the left is placed at the assumed head position of the partner on the right and vice versa.

In practice, however, it is not easy to implement this viewing condition. The crucial point is that the camera has to be mounted physically or virtually behind the display. The optical solution to this problem is the usage of semitransparent displays. The classical representatives in this context, the so-called video tunnels, are optical systems with half-silvered mirrors which reflect a certain percentage of light, but which allow the rest of the light to pass through (Acker and Levitt 1988; Buxton 1992). Similar to teleprompter technology known from conference and news equipment, the half-silvered mirror is placed in an adequate manner (e.g., with a tilt angle of 45°) between display and viewer (Figure 5.2). It thus becomes possible to split the optical paths of display and camera. Hence, through a combination of further mirrors, the camera appears virtually behind the display and is effectively placed

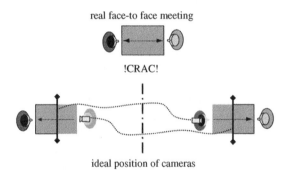

Figure 5.1   Line-of-sight method: reproduction of real face-to-face meeting by placing cameras in the line-of-sight at the assumed positions of the conferees' heads

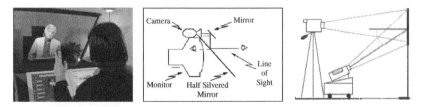

Figure 5.2   Telepresence systems with semi-transparent displays. *Left*: example of a commercial product. Reproduced by permission of Teleportec Inc. (Teleportec 2004). *Middle*: original EuroPARC illustration of reciprocal video tunnel using half-silvered mirror to deflect the optical camera path. Reproduced by permission of William Buxton (Buxton and Moran 1990). *Right*: projector and camera mounting for a half-transparent display using diffraction pattern (either Frensel- or HOE-based)

in the line-of-sight. Although this method was patented long ago by Rosenthal (1947) and has frequently been used in experimental systems since the term of telepresence appeared in videoconferencing (Acker and Levitt 1987; Buxton 1992, Okada *et al.* 1994), it took years for the related technology to be applied to commercial videoconferencing products (see examples in Figures 5.2, 5.3 and Figure 5.4). Some of these products also use alternative technologies for semitransparent displays that have turned up in the recent years. This especially applies to Fresnel lenticular displays which filter out light from a specific incident angle in one direction whereas light from the reverse direction passes through regularly. Hence, in contrast to the classical video tunnel, the cameras are mounted behind the display in this case and rear projection beams under a specific vertical angel adapted to the diffraction pattern of the fresnel structure (Figure 5.2). Further promising alternatives are displays with holographic optical elements (HOE) or shuttered LCD displays, but both are found only in experimental systems so far (Aoki *et al.* 1999, Gross *et al.* 2003).

Strictly speaking, the line-of-sight method from Figure 5.1 is limited to the special case of two single conferees talking to each other. Nevertheless, the method is also frequently used for team videoconferencing with small groups as well as one-to-many applications (teleteaching, board room meetings, etc.). Usually this can be accepted as long as the viewing distance between the two parties is sufficiently large (see examples in Figure 5.3). In this case the deviation of the individual viewing direction from the optimal line-of-sight remains in certain limits. The human perception is not very sensitive in this respect. The only relevant parameter is the sensitivity of eye contact that has intensively been studied in the past. Recent results in combination with an excellent survey on former investigations can be found in (Chen 2001; 2002).

A critical threshold is, however, exceeded if either the groups become too large or viewing distances become too low. In these cases the deviation of the individual viewing direction from the optimal line-of-sight becomes annoying. Several videoconferencing products have targeted this niche market. They use multi-projection wall and multi-camera systems to overcome these restrictions. One very interesting approach in this context is the Telepresence Wall shown in Figure 5.4. It connects two teleportals that can be placed in large venues such as corridors or lobbies with steady flow of personnel traffic. It is permanently available and allows for meetings of up to 20 persons. This is achieved by a large multi-projection system serving a curved panorama screen. Each projection unit is equipped with one video and one audio module. Similar to the reciprocal video tunnel from Figure 5.2, the optical camera path is deflected via semi-transparent and regular mirrors. The whole mirror system is designed such that all cameras are virtually placed at the same position, the common intersection point of all projector axes, but with different orientations coinciding with the

Figure 5.3 Examples of telepresence systems for teleteaching, board meetings and team conversation. Reproduced by permission of Teleportec Inc. (Teleportec 2004)

 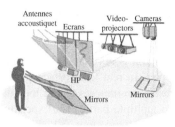

Figure 5.4 Telepresence Wall: portal for large group conversation. *Left*: running session. *Left*: illustration of mounting. Reproduced by permission of France Telecom RD (France Telecom 2003)

one of the corresponding projector. Due to this special mounting the camera images can be stitched seamlessly towards the large panoramic video beamed by the multi-projection system. Furthermore, specific acoustic antennas and spatial audio rendering enhance the reproduction of natural communication and reinforce the sensation of proximity to the conference partners at the other site. All these features together form a telepresence system where either two large groups can converse jointly or separated subgroups can talk to each other simultaneously without disturbing the conversation of the other group. The Telepresence Wall is based on the eConf technology, providing a totally customisable software solution which can be used for a wide range of video conference systems using immersive portals as well as mobile terminals (France Telecom 2004).

For the sake of completeness it should be mentioned that the usage of semi-transparent displays is only one possibility to implement the line-of-sight method from Figure 5.1. It is the most pragmatic one and the only one that has found its way onto the market so far. However, from research and experimental systems we know a couple of alternatives. In contrast to semi-transparent displays that are based on an optical solution, these alternatives utilize image processing to generate novel views referring to virtual cameras that are placed in the line-of-sight. Most approaches use stereo image processing to achieve this goal (Ohm *et al.* 1997; Mulligan *et al.* 2004; Yang and Zhang 2002; Atzpadin *et al.* 2004). Others are based on monocular approaches using gaze correction in combination with face models (Yip and Jin 2003a, b; Zitnick *et al.* 1999; Gemmel *et al.* 2000).

5.3 MULTI-PARTY COMMUNICATION USING THE SHARED TABLE CONCEPT

Available telepresence systems such as the ones mentioned in Section 5.1 clearly focus on two-party communication. Although most of them also offer multi-party options, they usually lose their telepresence capabilities in this mode. In fact, conversational structures with three and more parties are much more complicated than two-party conversations and natural reproduction of multi-party communication requires more sophisticated terminals and system architectures than the ones that are available from commercial products.

One crucial issue is that non-verbal communication cues that contain directive information (who is looking at whom, who is pointing on what, etc.) play a completely different role in multi-party communication. An exact reproduction of gaze direction, pointing gestures or directional sound has highest priority in this case. A simple example, which explains these circumstances in more detail, is the appearance of changing roles in conferencing. In a two-party communication the non-speaking party always feels responsible to take the lead as soon as the other party asks to do so, or simply stops talking. In multi-party communication, however, the situation is not that clear because there exist at least two parties that can take over the active part. In real face-to-face communication people often use their eye gaze or make a plain gesture to indicate whom they are addressing or to suggest whom they ask to go ahead. When multiple parties start speaking at the same time, gaze direction and body language often decide who stops talking and who continues. Moreover, side-conversations between sub-groups may happen in multi-party conferences. In natural face-to-face talking, the human perception is able to filter out side-talk and to concentrate on the voice of his direct interlocutor by selective listening (*cocktail-party effect*). However, that requires audiovisual congruence, meaning that the perceived audio source spatially coincides with the position at which the corresponding person appears visually.

Due to lacking gaze awareness and wrongly presented postures in multi-party modes, commercial videoconferencing systems are not able to support these features. Hence, conferees still need to address each other by using names and often a session chair is required to explicitly control the order and the discipline of speakers (Vertegaal *et al.* 2000). In this sense, multi-party communication in videoconferencing has in fact no real advantage over phone conferencing. The main restriction is that conventional conferencing systems are effectively limited to the one line-of-sight of a two-party conference shown in Figure 5.1, even if they are running in the multi-party mode using multiple windows presenting more than one remote party at one screen. This is not sufficient because a N-party conference certainly has more than just one line-of-sight. The example in Figure 5.5 shows this for the case of a three-party conference.

Each conferee has two lines of sights in this case and has to be captured from two different directions therefore. For instance, conferee C has to be captured by a first camera that is placed in the line-of-sight between A and C to obtain the correct perspective of C for A, and, the other way round, a second camera is positioned at the line-of-sight between B and C to pick up an adequate viewpoint of C for B. In total, this results in six cameras for three parties and $N(N-1)$ cameras in the general case of N parties. This method is often called the round-table concept.

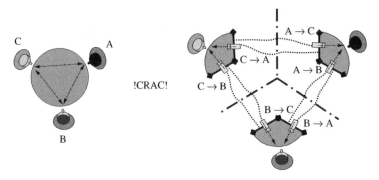

Figure 5.5 Round table concept: N parties are virtually grouped around a round table ($N = 3$). In accordance to Figure 5.1, cameras are placed in the lines of sight to achieve a natural reproduction of the N-party communication yielding $N(N-1)$ cameras in total

Following this concept, several experimental multi-party systems have been implemented over the years. Definitely the first one and — looking back from the 21st century – certainly the most humorous one is the system implementation reported by Negroponte (1995). It was developed in the 1970s under the political pressure of the cold war. The mission was to link five nuclear bunkers with a conference system that offered the most natural communication conditions for military decision purposes. The solution was an exact mechanical reproduction of a round table with five participants at each site. Every table consisted of one vacant seat for the local conferee and four seats occupied by life-sized physical replicas of the remote participants. All replicas provided transparent masks, shaped like a human face. During a conference session high-quality videos were projected into the facial masks. In addition, a remote control allowed for basic head movements such as turning, nodding or shaking.

Another early approach following a similar philosophy is the Hydra system proposed by Snellen and Buxton (1992). As shown in Figure 5.6, those places that would otherwise be occupied by conversational partners are held by a corresponding number of Hydra units, meaning that each remote partner is physically substituted through one technical module consisting of its own camera, monitor and loudspeaker. Thus, keeping unique round-table geometry at all participating sites, the Hydra concept is able to preserve gaze and sound direction to a certain extent. Although the original Hydra implementation was restricted to four parties, the concept behind it is open and can in principle be extended towards the general case of N parties. Thanks to these features, the Hydra system holds its position of having been a cutting-edge application in this area. Supposing that the system were updated with today's technology (e.g., using teleprompter terminals from Figure 5.2 instead of the dated Hydra units from Figure 5.6), it would meet most requirements of a state-of-the-art telepresence system for multi-party communication. The crucial disadvantage of the Hydra concept, however, is that the overall bit rate increases disproportionately with the number of multi-point connections. Each site of a N-party communication needs $(N-1)$ video links to the remote partners, resulting in $N(N-1)$ video links in total. Assuming that the bit rate for one efficiently compressed high-quality video stream ranges between 0.5 and 1.0 Mbit/s, a three-party conference yields a total bit rate of 3–6 Mbit/s. That is already beyond the limit being tolerated, even for professional use. A five-party conference would end up with unacceptable rate 10–20 Mbit/s and a conference with 10 parties would even require utopian 45–90 Mbit/s.

Figure 5.6   The Hydra concept: a local user is sitting in front of three Hydra units, substituting the remote conferees physically. Each unit consists of its own camera, monitor and loudspeaker. Reproduced by permission of William Buxton (Snellen and Buxton 1992)

Basically, the video views that are sent from each site to its $(N-1)$ remote partners contain extremely redundant information. Unfortunately, the Hydra concept, or any other round-table application representing the remote participants physically through separated terminals, cannot exploit this redundancy for compression purposes, because the video streams have to be transmitted separately using unicast network structures. Therefore, latest approaches are based on a new concept that integrates generic 3D video representations of the conferees into shared virtual table environment (SVTE) as shown in Figure 5.7. Hence, the round-table concept is no longer built physically, but it is simulated by means of virtual reality. The 3D video representations that are required for this purpose are typically extracted from multi-view capturing, using at least two cameras (Mulligan *et al.* 2004, Atzpadin *et al.* 2004).

In contrast to former round-table concepts such as Hydra, this approach allows for the usage of efficient multicast network structures, meaning that the same generic 3D video representation is sent to all $(N-1)$ remote stations. Thus, the required overall bit rate is proportional to the number of connected sites. Taking into account that depth information can usually be compressed much more efficiently than video data, the additional expenses of an adequate 3D representation format are marginal compared with a regular video stream. Thus, one video-plus-depth stream can be compressed to less than 1 Mbit/s and the total bit rate with N parties is lower than N Mbit/s. Obviously, this is much more efficient than the unicast approach as we know it, for example, from the Hydra concept.

Using a common scene description and integrating the $(N-1)$ received 3D video streams of the related remote partners into this scene, the virtual conference scene is composed and updated at each terminal. The individual view for the local conferee is then rendered through a virtual camera, which coincides with the user's particular head position. Both 3D scene composition and usage of virtual cameras give lot of space for flexible system design, especially for the adaptation to different terminal structures and futures extensions towards new display technology. For reason the SVTE concept often appears in connection with immersive system architectures. In this context the next section gives a survey of the most important implementations in this area.

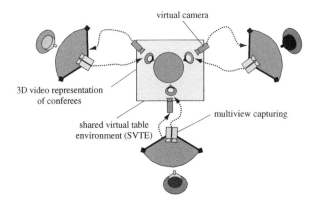

Figure 5.7 Shared virtual table environment (SVTE): generic 3D representations of the conferees are integrated into a shared virtual environment. Individual views of the conference scene are rendered by the means of a virtual camera. The video representations are extracted from multi-view capturing, using at least two cameras

5.4 EXPERIMENTAL SYSTEMS FOR IMMERSIVE VIDEOCONFERENCING

An early experimental multi-party conference combining the round-table concept with at least semi-immersive display technology was the MAJIC system (multi-attendance joint interface for collaboration), proposed at Keio University, Yokohama, in 1994 (Okada *et al.* 1994). At each site a large semi-transparent curved screen was mounted behind a normal computer terminal. Life-sized images of the remote partners appeared on this screen. Eye contact and gaze awareness was supported by careful placement of participants, cameras and projectors. A corresponding placement of microphones and loudspeakers was used to achieve spatial audio rendering as well. In this sense, it followed the Hydra concept from Figure 5.6, but used more sophisticated technology and more precise positioning of capturing and rendering devices.

Some years later the so-called MONJUnoCHIE system was proposed and investigated at the University of Tokyo (Aoki *et al.* 1999). In contrast to MAJIC, MONJUnoCHIE did not use separated screens for videoconferencing on one hand and shared applications on the other hand. The three-party terminal configuration from Figure 5.8 was developed for demonstration purposes. A special semi-transparent display based on holographic optical elements (HOE) was used to respect eye contact and gaze awareness.

The applied HOE display was a glass plate coated with a holographic film containing a diffraction pattern. Due to diffraction, images were projected from the rear under a vertical angle of about 30°, and a camera could therefore be positioned in the line-of-sight directly behind the display. Moreover, the HOE displays were equipped with touch sensors that allowed for shared applications and joint interaction with further objects at the screen.

Both, MAJIC and MONJUnoCHIE still used separated windows or screens and 2D video for representing the remote partner. Thus, they did not scale well in terms of the number of participants. Similar to the original Hydra concept, the total amount of capturing and rendering devices as well as the required overall bit rate increased with the square of the connected sites (see Section 5.2). The TELEPORT system from Figure 5.9 probably represents the first proposal that took a different approach to overcome these limitations (Gibbs *et al.* 1999). Rather than aligning separated displays or windows to the different remote partners, the TELEPORT moved somewhat further towards the SVTE concept from Figure 5.7 and implemented the round table in a shared virtual environment. All remote signals were

Figure 5.8 MONJUnoCHIE: the system name originates from a Japanese proverb meaning that two is better than one. *Left*: terminal for a three-party connection. *Middle*: image of a running session. *Right*: HOE display with touch screen sensors. Reproduced by permission of Yasuda-Aoki-Laboratory, Research Center for Advanced Science and Technology (RCAST), University of Tokyo (Aoki *et al.* 1999)

Figure 5.9 TELEPORT: immersive video conference system using shared virtual environment. *Left*: seamless transition between real workspace and projected image. *Middle*: rendered image as it appears at the full-wall projection screen. *Right*: example of a running session. Reproduced by permission of Fraunhofer Institute for Media Communication IMK (Gibbs *et al.* 1999)

composed jointly in a virtual conference room following a unique scene description script. For seamless integration of the videos into the virtual 3D environment the conferee's silhouette was segmented by means of chroma- or delta-keying. The TELEPORT display room is equipped with a full-wall projection of 3×2.25 m and allowed for stereoscopic viewing with passive glasses. In addition, a viewer tracking system determined the position of the local user to render the images from the right viewpoint and to enhance the illusion of a seamless transition between the real and the virtual world. To enforce this effect, elements in the real room mirrored or aligned with elements in the virtual extension, including shape, colour, geometry and lighting.

Although the TELEPORT used virtual 3D environments, the segmented videos that have been embedded into these environments were still limited to 2D representation. In contrast, the US National Tele-Immersion Initiative (NTII) simultaneously proposed another SVTE approach that applied 3D video representation to an experimental videoconferencing system based on so-called telecubicles (Chen *et al.* 2000; Towles *et al.* 2002). As shown in Figures 5.10 and 5.11, the remote participants appeared, life size on separate stereo displays arranged in accordance to the SVTE concept. The current head position of the local user was tracked through sensory devices, and the stereo rendering was carried out

Figure 5.10 Telecubicles: the NTII demonstrator at University of North Carolina in Chapel Hill (Chen *et al.* 2000). Local user is talking to the left remote partner. Reproduced by permission of Ketchum Photography, Chapel Hill, North Carolina

Figure 5.11 Telecubicles: the NTII demonstrator. *Left*: local user has turned to the right partner. *Right*: follow-up terminal with multi-view camera setup for 3D acquisition. Reproduced by permission of Department of Computer Science, University of North Carolina, Chapel Hill

with respect to the tracker data. Note that head tracking in combination with SVTE set up guarantees consistent eye contact, gaze awareness and gesture reproduction: everybody in the conference perceives consistently who is talking to whom or who is pointing at what (i.e., everybody perceives the same spatial arrangement) and in the correct perspective (i.e., the view is consistent with each individual viewpoint). In addition, the videos on the screen could be viewed in stereo through shuttered glasses. 3D video representation was based on voxel data derived from multiview images captured by the multi-baseline set up. As a result, dynamic voxel models could be sent as unique 3D video representations to all remote sites.

Although both TELEPORT and NTII have did some of the definitive spadework in immersive videoconferencing research, they were of experimental use only. They used complex display mountings, special tracking devices were needed to adapt 3D rendering to the user's head position and the user had to wear glasses to perceive 3D cues. Furthermore the NTII system still aligns separated windows to the remote partners and thus does not scale well, whereas TELEPORT did not use 3D video representations at all.

Based on these experiences, the European IST project VIRTUE (virtual team user environment) has proposed a semi-immersive and transportable desktop solution for immersive videoconferencing (Schreer *et al.* 2002). It used a 61-inch plasma display to present a virtual conference scene at life size at one common screen (Figure 5.12). The remote partners

Figure 5.12 Semi-immersive videoconferencing on its way to the market. *Left*: the experimental system of the European IST project VIRTUE using a 61-inch plasma display. *Right*: the immersive meeting point (im.point), a follow-up prototype of VIRTUE using a MPEG-4-based software implementation and standard PC technology

appeared as natural videos within the virtual environment. Similar to the TELEPORT, the local user got the impression of a seamless transition between real workspace and shared virtual environment. In contrast to TELEPORT, but similar to NTII, the VIRTUE system has used multi-view capturing to achieve 3D video representation of the conferees. Although the internal scene representation was based on an entire 3D data structure (including video), the display was restricted to 2D images to avoid the burden of using stereo glasses in a video conference application. Nevertheless, the 3D-to-2D rendering via a virtual camera was adapted to the tracked head position of the local user. The intention of this processing was to support depth perception by head motion parallax viewing instead of stereo. The system architecture, however, is open in this sense and autostereoscopic displays can be used as soon as they are available with sufficient size and quality.

Based on VIRTUE, a follow-up system called the immersive meeting point (im.point, Figure 5.12) has recently been developed with the ambition to provide a scalable system bridging the gap between a prototype with reduced complexity, suitable as basis for a marketable product, on the one hand and an experimental platform for the investigation of future extensions, especially in the field of 3D audio and video processing, on the other (Tanger *et al.* 2004). This implementation uses the state-of-the-art MPEG-4 standard for audiovisual coding, scene description and multimedia integration framework. The system has been implemented entirely in software on the basis of standard PC technology and can be scaled from simple 2D video to full 3D video processing, depending on application and performance requirements. Due to these features, the im.point belongs to a new system generation of semi-immersive videoconferencing focusing on compact solutions in order to bring these systems closer to the market. A further example for this evolution is the RealMeet room system shown in Figure 5.13. The technology behind is similar to the Telepresence Wall from Figure 5.4, but RealMeet targets three-party communication between small groups. A running system has recently been established between three different sites of France Telecom RD. Comparable solutions are offered by the US company TeleSuite (TeleSuite 2004). Other solutions, such as the Coliseum (Baker *et al.* 2003) or Gaze-2 (McLeod *et al.* 1999) are going in the opposite direction and have reduced the complexity of immersive videoconferencing to 20-inch TFT displays to integrate the conference application into usual desktop working environments.

Figure 5.13 The RealMeet room system. *Left*: the system mounting with two screens connected to two different remote sites. Each screen uses a semi-transparent mirror, like the Telepresence Wall from figure 5.4 *Right*: a running session with one remote site connected. Reproduced by permission of France Telecom RD

5.5 PERSPECTIVE AND TRENDS

The migration from early system implementations such as Hydra via first SVTE implementations (TELEPORT, NTII, VIRTUE) towards marketable prototype solutions (im.point, RealMeet) shows the enormous potential behind these developments. Especially the embedding of videoconferencing functionalities into shared virtual environments paves the way to a new era in telecommunications. It combines two fields of applications that have evolved separately in the past: the VR-based functionality of shared applications and the realism of videoconferencing. It is quite obvious that these hybrid solutions can also be used to interact with objects of the virtual scene. Thus, such systems are not limited to achieving realism and supporting physical presence. There are much more interesting things to do with the SVTE concept. Using existing experimental systems, this aspect has been instensively investigated in former studies. For instance, the NTII conference system from Figures 5.10 and 5.11 has been used for telecollaboration experiments where local and remote users could jointly manipulate 3D graphics objects that have been downloaded in advance and integrated into the shared virtual environment (Towles *et al.* 2002). Another example refers to early studies where two distant groups could collaborate in a TELEPORT-like constellation through a shared whiteboard that was partly located in the real workspace and partly integrated into the virtual environment (McLeod *et al.* 1999).

Many researchers believe that this evolution will end up with immersive portals that can be used in the same manner for communication and collaboration. As with the holo-deck in the *Star Trek* movies, a local user should not recognize any difference from reality while using an immersive portal for telecommunication purposes. An interesting and innovative research activity bringing this future vision closer to reality is the blue-C project at the Technical University of Zurich, Switzerland (Gross *et al.* 2003). It has a strong focus on three-dimensional video analysis and synthesis in a CAVE-like environment (Figure 5.14). Its scientific objectives are close to the original NTII work, but blue-C applies the latest technology and uses an experimental setup that is more sophisticated than the NTII system. It enables a number of participants to perceive photo-realistic 3D images of their remote partners in real-time, while interacting and collaborating inside an immersive virtual world. The key technology is a shuttered LCD screen that can be switched from an opaque state for display purposes (using rear projection) to a transparent state for acquisition purposes (through cameras behind the screen). A special synchronization unit accurately switches between screen, projectors, active illumination, stereo glasses and cameras.

Figure 5.14 Blue-C: an immersive portal for tele-collaboration. *Left*: the CAVE-like construction of blue-C. *Middle*: outside view of a blue-C session while active screen is transparent. *Right*: inside view while active screen is opaque (shop application). Reproduced by permission of the blue-C project, ETH Zürich, Switzerland

The blue-C system is still a prototype version, but it is the last development in an evolution chain, which indicates that immersive portals with complex real-time 3D analysis and display technologies will become reality in near future. In this sense, this section has emphasized the relevance of telepresence for true and realistic immersive videoconferencing. The decent review of state-of-the-art technology as well as prototype systems in research laboratories has drawn a clear picture of limits and challenges in this important field of application in 3D video communication.

ACKNOWLEDGEMENTS

The authors would like to thank all international collaboration partners for their fruitful discussions, the exchange of knowledge and the support for gathering all information for this survey paper. Particularly, we thank Teleportec Inc. for giving us the permission to include their application images, William Buxton for providing his former drawings on the reciprocal video tunnel and the photograph of the Hydra system, Olivier Gachignard from France Telecom RD for his material on the Telepresence Wall and the RealMeet room system, Hiroshi Yasuda and Terumasa Aoki from the Research Center for Advanced Science and Technology at University of Tokyo for their information and pictures on the MONJUnoCHIE system, the Fraunhofer Institute for Media Communication IMK in Sankt Augustin, Germany, for giving us the permission to include images of the TELEPORT system, Larry Ketchum from Ketchum Photography and the Computer Science Department of University of North Carolina in Chapel Hill for providing us with photographs of the experimental NTII system and — last, but not least – Markus Gross and his blue-C project team from ETH Zürich, Switzerland, for giving us permission to include pictures of their blue-C system. In addition, we thank all NTII members for their inspirations on our own research on immersive videoconferencing – in particular, Henry Fuchs, Greg Welch and Herman Towles from University of North Carolina in Chapel Hill, Kostas Daniilidis from University of Pennsylvania, Amela Sadagic and Jaron Lanier from Advanced Networks and Services and Jane Mulligan from University of Colorado, and all VIRTUE partners for our successful collaboration, in particular, Emanuele Trucco from Heriot-Watt University in Edinburgh, UK, John Stone from Sony UK, Michael Jewell from British Telecom, Emile Hendriks from Delft University of Technology, The Netherlands and Jan-Maarten Schraagen from TNO Human Factors, The Netherlands.

REFERENCES

Acker S and Levitt S 1987 Designing videoconference facilities for improved eye contact. *Journal of Broadcasting and Electronic Media,* **31**(2), 181–191.

Aoki T, Widoyo K, Sakamoto N, Suzuki K, Saburi T and Yasuda H 1999 MONJUnoCHIE system : videoconference system with eye contact for decision making. *International Workshop on Advanced Image Technology (IWAIT).*

Atzpadin N, Kauff P and Schreer O 2004 Stereo analysis by hybrid recursive matching for real-time immersive video conferencing. *IEEE Transactions on Circuits Systems and Video Technology, Special Issue on Immersive Telecommunications,* **14**(3) 321–334.

Baker H, Bhatti N, Tanguay D, Sobel I, Gelb D, Goss ME, MacCormick J, Culbertson WB and Malzbender T 2003 Computation and Performance Issues in Coliseum, An Immersive Videoconferencing System. *ACM Multimedia,* Berkeley, CA, USA.

Buxton W and Moran T 1990 EuroPARC's integrated interactive intermedia facility: early experience. In Gibbs S and Verrijn-Stuart AA (eds.), Multi-user interfaces and applications, *Proceedings of the IFIP WG 8.4 Conference on Multi-user Interfaces and Applications*, Heraklion, Crete. Amsterdam: Elsevier Science Publishers B.V. (North-Holland), 11–34.

Buxton WAS 1992 Telepresence: integrating shared task and person spaces. *Proceedings of Graphics Interface,* 123–129.

Chen M 2001 Design of a virtual auditorium. *Proceedings of ACM Multimedia 2001,* 19–28.

Chen M 2002 Leveraging the asymmetric sensitivity of eye contact for videoconferencing. *Proceedings of ACM CHI 2002,* 49–56.

Chen WC, Towles H, Nyland L, Welch G and Fuchs H 2000 Towards a compelling sensation of telepresence: demonstrating a portal to a distant (static) office. *Proceedings of IEEE Visualization.* Salt Lake City, UT, USA, IEEE Computer Science Press, 327–333.

France Telecom 2003 The Telepresence Wall which eradicates frontiers and distance. *http://www.rd.francetelecom.com/en/galerie/mur_telepresence/pdf/doc.pdf.*

France Telecom 2004 eConf: the videoconferencing software over IP. *http://www.rd.francetelecom. com/en/brevets/e-conf/econf.php*

Gemmell J, Zitnick L, Kang T and Toyama K 2000 Software-enabled gaze-aware videoconferencing. *IEEE Multimedia,* **7,**(4), 26–35.

Gibbs SJ, Arapis C and of Breiteneder CJ 1999 TELEPORT — towards immersive co-presence. *Multimedia Systems,* **7**(3), 214–221.

Gross M, Würmlin S, Naef M, Lamboray E, Spagno C, Kunz A, Koller-Meier E, Svoboda T, Van Gool L, Lang S, Strehlke K, Vande MA and Staadt O 2003 Blue-C: a spatially immersive display and 3D video portal for telepresence. *Proceedings of ACM SIGGRAPH 2003,* 819–827.

IJsselsteijn WA, Freeman J, De Ridder H 2001 Presence: where are we?. *CyberPsychology and Behaviour,* **4**(2), 179–182.

McLeod D, Neumann U, Nikias CL and Sawchuk AA 1999 Integrated media systems: the move towards media immersion. *IEEE Signal Processing Magazine* **16**(1), 33–76.

Minsky M 1980 Telepresence. *Omni,* 45–51.

Mühlbach L, Böcker M and Prussog A 1995 Telepresence in videocommunications: a study on stereoscopy and individual eye-contact. *Human Factors* **37,** 290–305.

Mulligan J, Isler V and Daniilidis K 2004 Trinocular stereo: a real-time algorithm and its evaluation. *International Journal of Computer Vision* **47**(1/2/3), 51–61.

Negroponte N 1995 *Being Digital.* New York: Vintage Books.

Ohm JR, Grüneberg K, Hendriks E, Izquierdo E, Kalivas D, Karl M, Papadimatos D and Redert A 1998 A realtime hardware system for stereoscopic videoconferencing with viewpoint adaptation. *Image Communication* **14,** 147–171.

Okada K, Maeda F, Ickikawaa Y and Matsushita Y 1994 multi-party videoconferencing at virtual social distance: MAJIC design. *Proceedings of CSCW'94,* 385–393.

Rosenthal AH 1947 Two-way television communication unit. *United States Patent 2,420,198.*

Schreer O, Hendriks E, Schraagen J, Stone J, Trucco E and Jewell M 2002 Virtual team user environments (VIRTUE) — a key application in telecommunication. *Proceedings of eBusiness and eWork,* Prague, Cech Republic, CD-ROM.

Snellen A and Buxton B 1992 Using spatial cues to improve videoconferencing. *Proceedings of CHI'92,* 651–652.

Tanger R, Kauff P and Schreer O 2004 Immersive meeting point (im.point) — an approach towards immersive media portals. *Proceedings of Pacific-Rim Conference on Multimedia,* Tokyo, Japan, Part 1, 89–96.

Teleportec 2004: The ultimate in global face-to-face communication. *www.teleportec.com.*

TeleSuite 2004. Virtually anywhere. *www.telesuite.com.*

Towles H, Chen WC, Yang R, Kum SU and Fuchs H 2002 3D Tele-collaboration over Internet 2. *Proceedings of the International Workshop on Immersive Tele-Presence (ITP),* Juan Les Pins, France, 28–31.

Vertegaal R, Van der Veer G and Vons H 2000 Effects of gaze on multi-party mediated communication. *Proceedings of Graphics Interface,* Montreal, Canada, 95–102.

Yang R and Zhang Z 2002 Eye gaze correction with stereovision for video-teleconferencing. *Microsoft Research, Technical Report MSR-TR-2001-119.*

Yip B and Jin JS 2003a An effective eye gaze correction operation for video conference using anti-rotation formulas. *Proceedings of the 4th IEEE Pacific-Rim Conference On Multimedia (PCM)* **2**, 699–703.

Yip B and Jin JS 2003b Face re-orientation using ellipsoid model in video conference. *Proceedings of the 7th IASTED International Conference on Internet and Multimedia Systems and Applications*, 245–250.

Zitnick L, Gemmell J and Toyama K 1999 Manipulation of video eye gaze and head orientation for video teleconferencing. *Microsoft Research, Technical Report MSR-TR-99-46*.

Section II
3D Data Representation and Processing

6

Fundamentals of Multiple-view Geometry

Spela Ivekovic[1], Andrea Fusiello[2] and Emanuele Trucco[1]

[1]*Heriot-Watt University, Edinburgh, United Kingdom*
[2]*University of Verona, Verona, Italy*

6.1 INTRODUCTION

This chapter introduces the basic geometric concepts of multiple-view computer vision. The focus is on geometric models of perspective cameras, and the constraints and properties such models generate when multiple cameras observe the same 3D scene.

Geometric vision is an important and well-studied part of computer vision. A wealth of useful results has been achieved in the last 15 years and reported in comprehensive monographies, e.g., Faugeras (1993); Faugeras and Luong (2001); Hartley and Zisserman (2003), a sign of maturity for a research subject.

Geometric vision plays an essential role in image-based 3D communications. An estimate of the 3D structure of objects in view enables shape-specific, hence potentially very efficient coding (e.g., MPEG4-3DMC or 3D mesh coding). If a 3D model is available, as for instance in the case of the human figure in newscasting or videoconferencing sequences, it is possible to match it to the images, transmit only the model parameters, and animate a CAD figure (avatar) at the receiver. Transmitting rich 3D information, such as that computed by stereo algorithms, enables the receiver to support user-driven 3D effects, e.g., changing the viewpoint interactively. Indeed several chapters of this book draw from the geometric vision repertoire, which motivates this introduction.

It is worth reminding the reader that geometry is an important, but not the only important aspect of computer vision, and in particular of multiple-view vision. The information brought by each image pixel is two-fold: its *position* and its *colour* (or brightness, for a monochrome image). Ultimately, each computer vision system must start with brightness values, and, to smaller or greater depth, link such values to the 3D world.

3D Videocommunication — Algorithms, concepts and real-time systems in human centred communication
Edited by O. Schreer, P. Kauff and T. Sikora © 2005 John Wiley & Sons, Ltd

This chapter is organized as follows. Section 6.2 introduces the basic geometric model of perspective projections, the celebrated *pinhole camera*. Section 6.3 moves on to the case of two cameras, the classic stereo system. We discuss how the pinhole model can be used to derive useful geometric constraints on the position of corresponding points, that is, projections of the same scene point in the two images. The concept of correspondence is a cornerstone of multiple-view vision. In this chapter we assume *known correspondences*, and explore their use in multiple-view vision. Algorithms for computing correspondences are presented in Chapter 7. Section 6.3 also addresses the important, practical problem of *reconstruction*: under what conditions can we estimate the position and shape of objects from two views? Section 6.4 extends our investigation to the general case of multiple (more than two) views. A few concluding remarks are given in Section 6.5.

6.2 PINHOLE CAMERA GEOMETRY

The pinhole camera is at the heart of the geometric model of imaging. It is described by its *optical centre* **C** (also known as the *camera projection centre*) and the *image plane*. The distance of the image plane from **C** is the *focal length f*. The line from the camera centre perpendicular to the image plane is called the *principal axis* or *optical axis* of the camera. The plane parallel to the image plane containing the optical centre is called the *principal plane* or *focal plane* of the camera.

A 3D point is projected onto the image plane with the line containing the point and the optical centre (Figure 6.1). Let the centre of projection be the origin of a Euclidean coordinate system wherein the z-axis is the principal axis. The relationship between the 3D coordinates of a scene point and the coordinates of its projection onto the image plane is described by the *central* or *perspective projection*.

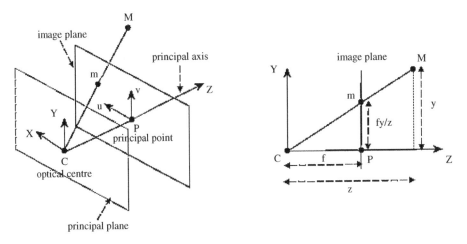

Figure 6.1 Pinhole camera geometry. The left figure illustrates the projection of the point **M** on the image plane by drawing the line through the camera centre **C** and the point to be projected. The right figure illustrates the same situation in the *YZ* plane, showing the similar triangles used to compute the position of the projected point **m** in the image plane

By similar triangles it is readily seen that the 3D point $(x, y, z)^T$ is mapped to the point $(fx/z, fy/z, f)^T$ on the image plane. If the world and image points are represented by homogeneous vectors, then perspective projection can be expressed in terms of matrix multiplication as

$$\begin{pmatrix} fx \\ fy \\ z \end{pmatrix} = \begin{bmatrix} f & 0 & 0 & 0 \\ 0 & f & 0 & 0 \\ 0 & 0 & 1 & 0 \end{bmatrix} \begin{pmatrix} x \\ y \\ z \\ 1 \end{pmatrix} \tag{6.1}$$

The matrix describing the linear mapping is called the *camera projection matrix* P and Equation (6.1) can be written simply as:

$$z\mathbf{m} = P\mathbf{M} \tag{6.2}$$

where $\mathbf{M} = (x, y, z, 1)^T$ are the homogeneous coordinates of the 3D point and $\mathbf{m} = (fx/z, fy/z, 1)^T$ are the homogeneous coordinates of the image point.

The projection matrix P in Equation (6.1) represents the simplest possible case, as it contains only information about the focal distance f. In general, the camera projection matrix is a 3×4 full-rank matrix and, being homogeneous, it has 11 degrees of freedom. Using QR factorization, it can be shown that any 3×4 full rank matrix P can be factorized as

$$P = K[R|\mathbf{t}] \tag{6.3}$$

where K is upper triangular (nonsingular), R is a rotation matrix, and $\mathbf{t}$ is a translation vector.

K is the *camera calibration matrix*; it encodes the transformation from camera coordinates to pixel coordinates. It depends on the so-called *intrinsic* parameters, i.e., focal distance f, image centre coordinates in pixels o_x, o_y, and pixel size in mm s_x, s_y along the two axes of the camera photosensor (Trucco and Verri 1998):

$$K = \begin{bmatrix} f/s_x & 0 & o_x \\ 0 & f/s_y & o_y \\ 0 & 0 & 1 \end{bmatrix} \tag{6.4}$$

The *extrinsic* parameters describe the position and orientation of the camera with respect to an external (world) coordinate system and are stored in the rotation matrix R and the translation vector $\mathbf{t}$.

The camera projection centre $\mathbf{C}$ is the only point for which the projection is not defined, i.e.:

$$P \begin{pmatrix} \mathbf{C} \\ 1 \end{pmatrix} = \mathbf{0} \tag{6.5}$$

After solving for $\mathbf{C}$ we obtain:

$$\mathbf{C} = -P_{3\times3}^{-1} P_{.,4} \tag{6.6}$$

where $P_{3\times3}$ is the matrix composed of the first three rows and first three columns of P, and $P_{.,4}$ is the fourth column of P.

The projection can be geometrically modelled by rays through the optical centre and the point in space that is being projected onto the image plane (Figure 6.1). The *optical ray* of an image point $\mathbf{m} = (u, v, 1)^{\mathrm{T}}$ is the locus of points in space that projects onto $\mathbf{m}$. It can be described as a parametric line passing through the camera projection centre $\mathbf{C}$ and the point at infinity that projects onto $\mathbf{m}$:

$$\mathbf{M} = \begin{pmatrix} \mathbf{C} \\ 1 \end{pmatrix} + \lambda \begin{pmatrix} P_{3\times3}^{-1}\mathbf{m} \\ 0 \end{pmatrix}, \qquad \lambda \in \mathbb{R} \tag{6.7}$$

Notice that $\mathbf{C}$ is expressed in non-homogeneous coordinates, i.e., it is a three-vector. In general, the projection equation writes:

$$\zeta\mathbf{m} = P\mathbf{M} \tag{6.8}$$

where ζ is the distance of $\mathbf{M}$ from the focal plane of the camera (usually referred to as *depth*). Note that, except for a very special choice of the world reference frame, *this 'depth' does not coincide with the third coordinate of* $\mathbf{M}$. Moreover, P is a homogeneous quantity and unless it is properly normalized, ζ contains an arbitrary scale factor.

6.3 TWO-VIEW GEOMETRY

6.3.1 Introduction

The two-view geometry is the intrinsic geometry of two different perspective views of the same 3D scene (Figure 6.2). It is usually referred to as *epipolar geometry*. The two perspective views may be acquired simultaneously, for example in a stereo rig, or sequentially,

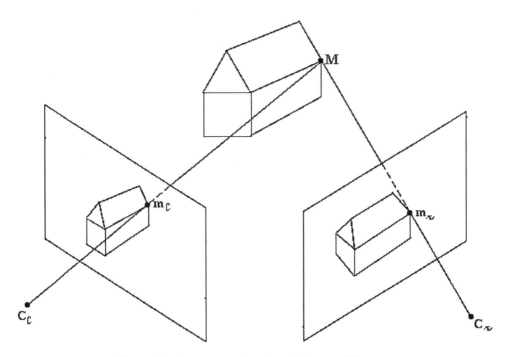

Figure 6.2 Two perspective views of the same 3D scene

for example by a moving camera. From the geometric viewpoint, the two situations are equivalent, but notice that the scene might change between successive snapshots.

Most 3D scene points must be visible in both views simultaneously. This is not true in the case of occlusions, i.e., points visible in only one camera. Any unoccluded 3D scene point $\mathbf{M} = (x, y, z, 1)^T$ is projected to the left and right view as $\mathbf{m}_l = (u_l, v_l, 1)^T$ and $\mathbf{m}_r = (u_r, v_r, 1)^T$, respectively (Figure 6.2). Image points $\mathbf{m}_l$ and $\mathbf{m}_r$ are called *corresponding points* as they represent projections of the same 3D scene point $\mathbf{M}$.

Algebraically, each perspective view has an associated 3×4 camera projection matrix P which represents the mapping between the 3D world and a 2D image. We will refer to the camera projection matrix of the left view as P_l and of the right view as P_r. The 3D point $\mathbf{M}$ is then imaged as Equation (6.9) in the left view, and Equation (6.10) in the right view:

$$\zeta_l \mathbf{m}_l = P_l \mathbf{M} \tag{6.9}$$

$$\zeta_r \mathbf{m}_r = P_r \mathbf{M} \tag{6.10}$$

Geometrically, the position of the image point $\mathbf{m}_l$ in the left image plane I_l can be found by drawing the optical ray through the left camera projection centre $\mathbf{C}_l$ and the scene point $\mathbf{M}$. The ray intersects the left image plane I_l at $\mathbf{m}_l$. Similarly, the optical ray connecting $\mathbf{C}_r$ and $\mathbf{M}$ intersects the right image plane I_r at $\mathbf{m}_r$. The relationship between image points $\mathbf{m}_l$ and $\mathbf{m}_r$ is given by the epipolar geometry, discussed in Section 6.3.2.

The knowledge of image correspondences enables scene reconstruction from images. The correctness of reconstruction crucially depends on the accuracy of image correspondences. What can be reconstructed, in turn, depends on the amount of *a priori* knowledge available about the stereo setup that was used to acquire the images and knowledge about the scene itself. This is discussed in Section 6.3.4.

6.3.2 Epipolar Geometry

The epipolar geometry describes the geometric relationship between two perspective views of the same 3D scene. The key finding, discussed below, is that *corresponding image points must lie on particular image lines*, which can be computed without the information on the camera calibration. This implies that, given a point in one image, one can search for the corresponding point in the other image along a line and not in a 2D region, a significant reduction in complexity.

Figure 6.3 illustrates the rules of the epipolar geometry. Any 3D point $\mathbf{M}$ and the camera projection centres $\mathbf{C}_l$ and $\mathbf{C}_r$ define a plane that is called the *epipolar plane*. The projections of the point $\mathbf{M}$, image points $\mathbf{m}_l$ and $\mathbf{m}_r$, also lie in the epipolar plane since they lie on the rays connecting the corresponding camera projection centre and point $\mathbf{M}$. The corresponding epipolar lines, $\mathbf{l}_l$ and $\mathbf{l}_r$, are the intersections of the epipolar plane with the image planes. The line connecting the camera projection centres $(\mathbf{C}_l, \mathbf{C}_r)$ is called the *baseline*. The baseline intersects each image plane in a point called *epipole*. By construction, the left epipole $\mathbf{e}_l$ is the image of the right camera projection centre $\mathbf{C}_r$ in the left image plane. Similarly, the right epipole $\mathbf{e}_r$ is the image of the left camera projection centre $\mathbf{C}_l$ in the right image plane. All epipolar lines in the left image go through $\mathbf{e}_l$ and all epipolar lines in the right image go through $\mathbf{e}_r$.

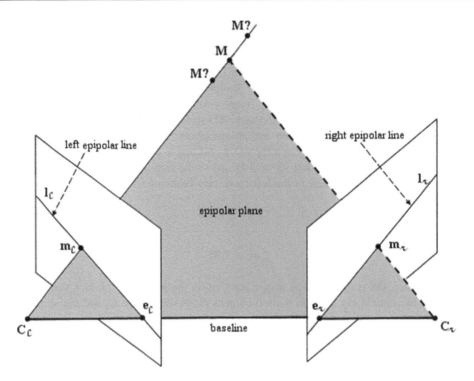

Figure 6.3 The epipolar geometry and epipolar constraint

The epipolar constraint.   It is clear from the previous discussion that an epipolar plane is completely defined by the camera projection centres and one image point, say, in the left image. Therefore, given a point $\mathbf{m}_1$, one can determine the epipolar line in the right image on which the corresponding point $\mathbf{m}_r$ must lie. The logical next step is to determine the equation of the epipolar line in one image given a point in the other image.

The equation of the epipolar line can be derived from the equation describing the optical ray. As we mentioned before, the right epipolar line corresponding to the left image point $\mathbf{m}_1$ geometrically represents the projection (Equation 6.8) of the optical ray through $\mathbf{m}_1$ (Equation 6.7) onto the right image plane:

$$\zeta_r \mathbf{m}_r = P_r \mathbf{M} = P_r \begin{pmatrix} \mathbf{C}_1 \\ 1 \end{pmatrix} + \lambda_1 P_r \begin{pmatrix} P_{3\times3,1}^{-1} \mathbf{m}_1 \\ 0 \end{pmatrix} \tag{6.11}$$

If we now simplify Equation (6.11) we obtain the description of the right epipolar line:

$$\zeta_r \mathbf{m}_r = \mathbf{e}_r + \lambda_1 P_{3\times3,r} P_{3\times3,1}^{-1} \mathbf{m}_1 \tag{6.12}$$

This is the equation of a line through the right epipole $\mathbf{e}_r$ and the image point $\mathbf{m}'_1 = P_{3\times3,r} P_{3\times3,1}^{-1} \mathbf{m}_1$ which represents the projection of the point at infinity, lying on the optical ray of $\mathbf{m}_1$, onto the right image plane. The equation for the left epipolar line is obtained in a similar way.

As we mentioned, the epipolar constraint facilitates the search for corresponding points in two images. The correspondence search can be further simplified by *rectification*. Rectification determines a transformation of each image such that pairs of corresponding epipolar lines become collinear and parallel to one of the image axes, usually the horizontal one. The correspondence search is then reduced to a 1D search along the trivially identified scanline. We will postpone a more detailed description of rectification to Section 6.3.3.

The epipolar geometry can be described analytically in several ways, depending on the amount of the *a priori* knowledge about the stereo system. We can identify three general cases.

- If both *intrinsic* and *extrinsic* camera parameters are known, we can describe the epipolar geometry in terms of the projection matrices (Equation 6.12).
- If only the *intrinsic* parameters are known, we work in normalized coordinates and the epipolar geometry is described by the *essential matrix*.
- If neither intrinsic nor extrinsic parameters are known, the epipolar geometry is described by the *fundamental matrix*.

The essential matrix E.   If the intrinsic parameters are known, we can switch to *normalized coordinates* (note that this change of notation will hold throughout this section):

$$\mathbf{m} \leftarrow K^{-1}\mathbf{m} \tag{6.13}$$

Consider a pair of normalized cameras. Without loss of generality, we can fix the world reference frame onto the first camera, hence:

$$P_1 = [I|0] \quad \text{and} \quad P_r = [R|\mathbf{t}] \tag{6.14}$$

With this choice, the unknown extrinsic parameters have been made explicit.

If we substitute these two particular instances of the camera projection matrices in Equation (6.12), we get

$$\zeta_r \mathbf{m}_r = \mathbf{t} + \lambda_1 R\mathbf{m}_1 \tag{6.15}$$

in other words, the point $\mathbf{m}_r$ lies on the line through the points $\mathbf{t}$ and $R\mathbf{m}_1$. In homogeneous coordinates, this can be written as follows:

$$\mathbf{m}_r^T \mathbf{t} \times (R\mathbf{m}_1) = 0 \tag{6.16}$$

as the homogeneous line through two points is expressed as their cross product. Similarly, a dot product of a point and a line is zero if the point lies on the line.

The cross product of two vectors can be written as a product of a skew-symmetric matrix and a vector. Equation (6.16) can therefore be equivalently written as

$$\mathbf{m}_r^T [\mathbf{t}]_\times R\mathbf{m}_1 = 0 \tag{6.17}$$

where $[\mathbf{t}]_\times$ is the skew-symmetric matrix of the vector $\mathbf{t}$. If we multiply the matrices in Equation (6.17), we obtain a single matrix which describes the relationship between the corresponding image points $\mathbf{m}_1$ and $\mathbf{m}_r$ in normalized coordinates. This matrix is called the *essential matrix E*:

$$E \triangleq [\mathbf{t}]_\times R \tag{6.18}$$

and the relationship between two corresponding image points in normalized coordinates is expressed by the defining equation for the essential matrix:

$$\mathbf{m}_r^T E \mathbf{m}_l = 0 \tag{6.19}$$

E encodes only information on the extrinsic camera parameters. Its rank is two, since $\det[\mathbf{t}]_\times = 0$. The essential matrix is a homogeneous quantity. It has only five degrees of freedom: a 3D rotation and a 3D translation direction.

The Fundamental Matrix F.   The fundamental matrix can be derived in a similar way to the essential matrix. All camera parameters are assumed unknown; we write therefore a general version of Equation (6.14):

$$P_l = K_l[I|0] \quad \text{and} \quad P_r = K_r[R|\mathbf{t}] \tag{6.20}$$

Inserting these two projection matrices into Equation (6.12), we get

$$\zeta_r \mathbf{m}_r = \mathbf{e}_r + \lambda_l K_r R K_l^{-1} \mathbf{m}_l \quad \text{with} \quad \mathbf{e}_r = K_r \mathbf{t} \tag{6.21}$$

which states that point $\mathbf{m}_r$ lies on the line through $\mathbf{e}_r$ and $K_r R K_l^{-1} \mathbf{m}_l$. (It is easy to see that the parameter λ_l is equal to ζ_l, the depth of the point $\mathbf{M}$ with respect to the left camera.) As in the case of the essential matrix, this can be written in homogeneous coordinates as:

$$\mathbf{m}_r^T [\mathbf{e}_r]_\times K_r R K_l^{-1} \mathbf{m}_l = 0 \tag{6.22}$$

The matrix

$$F = [\mathbf{e}_r]_\times K_r R K_l^{-1} \tag{6.23}$$

is the *fundamental matrix F*, giving the relationship between the corresponding image points in pixel coordinates. The defining equation for the fundamental matrix is therefore

$$\mathbf{m}_r^T F \mathbf{m}_l = 0 \tag{6.24}$$

F is the algebraic representation of the epipolar geometry. It is a 3×3, rank-two homogeneous matrix. It has only seven degrees of freedom since it is defined up to a scale and its determinant is zero. Notice that F is completely defined by pixel correspondences only (the intrinsic parameters are not needed).

For any point $\mathbf{m}_l$ in the left image, the corresponding epipolar line $\mathbf{l}_r$ in the right image can be expressed as

$$\mathbf{l}_r = F \mathbf{m}_l \tag{6.25}$$

Similarly, the epipolar line $\mathbf{l}_l$ in the left image for the point $\mathbf{m}_r$ in the right image can be expressed as

$$\mathbf{l}_l = F^T \mathbf{m}_r \tag{6.26}$$

The left epipole $\mathbf{e}_l$ is the right null-vector of the fundamental matrix and the right epipole is the left null-vector of the fundamental matrix:

$$F\mathbf{e}_l = 0 \tag{6.27}$$

$$\mathbf{e}_r^T F = 0 \tag{6.28}$$

One can see from the derivation that the essential and fundamental matrices are related through the camera calibration matrices K_l and K_r:

$$F = K_r^{-T} E K_l^{-1}. \tag{6.29}$$

Further properties of F are discussed in detail in Loung and Faugeras (1996).

Estimating F: the eight-point algorithm.   Equation (6.24) applies to any pair of corresponding points $\mathbf{m}_l \leftrightarrow \mathbf{m}_r$. Writing Equation (6.24) as a set of linear constraints, for a sufficient number of correspondences, yields a linear system in the entries of F. This is the essence of the eight-point algorithm.

Each point correspondence gives rise to one linear equation in the unknown entries of F. For example, the equation corresponding to a pair of points $\mathbf{m}_l = (u_l, v_l, 1)$ and $\mathbf{m}_r = (u_r, v_r, 1)$ is

$$u_r u_l f_{1,1} + u_r v_l f_{1,2} + u_r f_{1,3} + v_r u_l f_{2,1} + v_r v_l f_{2,2} + v_r f_{2,3} + u_l f_{3,1} + v_l f_{3,2} + f_{3,3} = 0$$

where $f_{1,1}, f_{1,2}, \ldots, f_{3,3}$ are the unknown entries of F. All available point correspondences form a homogeneous set of equations with the entries of F as the unknowns. Noise generally present in the data suggests a least-squares solution.

This method does not explicitly enforce F to be singular, so it must be done *a posteriori*. It can be shown that the closest singular matrix in Frobenius norm to a given matrix F is the one obtained by forcing to zero the smallest singular value of F. Geometrically, the singularity constraint ensures that the epipolar lines meet in a common epipole.

As Hartley (1995) pointed out, it is crucial for this linear algorithm that input data is properly preconditioned, by a procedure called *standardization*: points are translated so that their centroid is at the origin and are scaled so that their average distance from the origin is $\sqrt{2}$.

In summary, the eight-point algorithm for computation of F is as follows:

- Standardize the input data. Apply the standardizing transformations T_l and T_r, consisting of translation and scaling, to the image coordinates: $\hat{\mathbf{m}}_l^i = T_l \mathbf{m}_l^i$ and $\hat{\mathbf{m}}_r^i = T_r \mathbf{m}_r^i$.
- Linear solution. Compute the fundamental matrix $\hat{F}$ by solving the homogeneous system of equations defined by the point matches $\hat{\mathbf{m}}_l^i \leftrightarrow \hat{\mathbf{m}}_r^i$.
- Enforce the singularity constraint. Replace $\hat{F}$ by $\hat{F}'$, such that $\det \hat{F}' = 0$, by setting the smallest singular value to zero.
- De-standardize the result. The resulting fundamental matrix, associated to the original point correspondences, is $F = T_r^T \hat{F}' T_l$.

This simple algorithm provides good results in many situations and can be used to initialize a variety of more accurate, iterative algorithms. Details of these can be found in Hartley and Zisserman (2003), Torr and Murray (1997) and Zhang (1998).

6.3.3 Rectification

Given a pair of stereo images, *epipolar rectification* (or simply *rectification*) determines a transformation of each image plane such that *pairs of corresponding epipolar lines become collinear and parallel to one of the image axes* (usually the horizontal one). The rectified images can be thought of as acquired by a virtual stereo pair, obtained by rotating the original cameras and possibly modifying the intrinsic parameters. The important advantage of rectification is that computing stereo correspondences is made simpler, because search is done along the horizontal lines of the rectified images. We assume here that *the stereo pair is calibrated*, i.e., the cameras' internal parameters, mutual position and orientation are known. This assumption is not strictly necessary, but leads to a simple technique.

Hartley (1999), Isgrò and Trucco (1999) and Loop and Zhang (1999) have introduced algorithms which perform rectification given a *weakly calibrated* stereo pair, i.e., a pair for which only point correspondences between images are given (or, equivalently, for which the fundamental matrix could be computed). Some of them concentrate on minimizing the rectified image distortion. We will not address this problem because distortion is less severe than in the weakly calibrated case.

Specifying virtual cameras. Given the actual camera matrices, P_{or} and P_{ol}, the idea behind rectification is to define two new *virtual* cameras, P_{nr} and P_{nl}, obtained by rotating the actual ones around their optical centres until the focal planes become coplanar, thereby containing the baseline (Figure 6.4). This ensures that epipoles are at infinity, hence epipolar lines are *parallel*. To have *horizontal* epipolar lines, the baseline must be parallel to the x-axis of both virtual cameras. In addition, to have a proper rectification, corresponding points must have the *same vertical coordinate*.

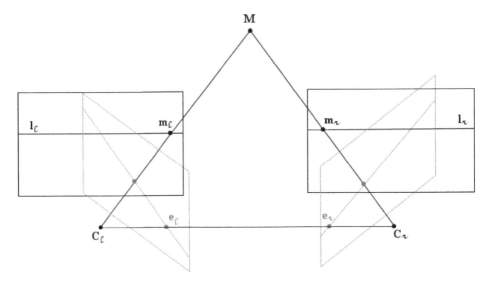

Figure 6.4 Original (light grey) and rectified cameras. The rectified image planes are coplanar and parallel to the baseline

In summary: positions (i.e, optical centers) of the virtual cameras are the same as for the actual cameras, whereas the orientation of both virtual cameras differs from the actual ones by suitable rotations; intrinsic parameters are the same for both cameras. Therefore, the two resulting virtual cameras will differ only in their optical centres, and they can be thought of as a single camera translated along the x-axis of its reference system.

Let us write the virtual cameras, matrices in terms of their factorization:

$$P_{nl} = K[R \mid -R\,C_l] \qquad P_{nr} = K[R \mid -R\,C_r] \tag{6.30}$$

In order to define them, we need to assign K, R, C_l, and C_r. The intrinsic parameters matrix K can be chosen arbitrarily. The optical centres C_l and C_r are the same as for the actual cameras. The matrix R, which gives the camera pose, will be specified by means of its row vectors

$$R = \begin{bmatrix} r_1^T \\ r_2^T \\ r_3^T \end{bmatrix} \tag{6.31}$$

which are the x-, y-, and z-axes, respectively, of the virtual camera reference frame, expressed in world coordinates.

According to the previous comments, we take:

- The x-axis parallel to the baseline: $r_1 = (C_r - C_l)/\|C_r - C_l\|$
- The y-axis orthogonal to x (mandatory) and to an arbitrary unit vector k: $r_2 = k \times r_1$
- The z-axis orthogonal to xy (mandatory) : $r_3 = r_1 \times r_2$

In the second of these cases, k fixes the position of the y-axis in the plane orthogonal to x. In order to ensure that the virtual cameras look in the same direction as the actual ones, k is set equal to the direction of the optical axis of one of the two actual cameras.

We have assumed that both virtual cameras have the same intrinsic parameters. Actually, the horizontal component of the image centre can be different, and this degree of freedom can be exploited to 'centre' the rectified images in the viewport by applying a suitable horizontal translation.

The rectifying transformation. In order to rectify, say, the left image, we need to compute the transformation mapping the image plane of P_{ol} onto the image plane of P_{nl}. It is useful to think of an image as the intersection of the image plane with the cone of rays between points in 3D space and the optical centre. We are moving the image plane while leaving the cone of rays fixed.

According to the equation of the optical ray, for any 3D point M that projects to m_{ol} in the actual image and to m_{nl} in the rectified image, there exist two parameters λ_o and λ_n such that:

$$\begin{cases} M = C_l + \lambda_o P_{3\times3,ol}^{-1} m_{ol} \\ M = C_l + \lambda_n P_{3\times3,n\ell}^{-1} m_{nl} \end{cases} \tag{6.32}$$

hence

$$m_{nl} = \frac{\lambda_o}{\lambda_n} P_{3\times3,nl} P_{3\times3,ol}^{-1} m_{ol} \tag{6.33}$$

The transformation sought is a linear transformation of the projective plane (called *collineation*) given by the 3×3 matrix $\mathbf{H}_1 = P_{3\times3,\mathrm{nl}}P_{3\times3,\mathrm{ol}}^{-1}$. Note that the unknown scale factor $\lambda_\mathrm{o}/\lambda_\mathrm{n}$ can be neglected, as the transformation $\mathbf{H}_1$ is defined up to a scale factor (being homogeneous). The same result applies to the right image.

Reconstruction of 3D points by triangulation can be performed from the rectified images directly, using P_{nr} and P_{nl}. More details on the rectification algorithm can be found in Fusiello *et al.* (2000), from which this section has been adapted.

6.3.4 3D Reconstruction

What can be reconstructed depends on what is known about the scene and the stereo system. We can identify three different cases.

- *If both the intrinsic and extrinsic camera parameters are known*, we can solve the reconstruction problem unambiguously by triangulation.
- *If only the intrinsic parameters are known*, we can estimate the extrinsic parameters and solve the reconstruction problem up to an unknown scale factor. In other words, R can be estimated completely, and $\mathbf{t}$ up to a scale factor.
- *If neither intrinsic nor extrinsic parameters are known*, i.e., the only information available are pixel correspondences, we can still solve the reconstruction problem, but only up to an unknown global projective transformation of the world. This is not really surprising, as the projective transformation is encoded by a matrix, and here the intrinsic parameter matrices are unknown.

Reconstruction by triangulation.   Given the camera matrices P_1 and P_r, let $\mathbf{m}_1$ and $\mathbf{m}_r$ be two corresponding points satisfying the epipolar constraint $\mathbf{m}_r^T F \mathbf{m}_1 = 0$. It follows that $\mathbf{m}_r$ lies on the epipolar line $F\mathbf{m}_1$ and so the two rays back-projected from image points $\mathbf{m}_1$ and $\mathbf{m}_r$ lie in a common epipolar plane. Since they lie in the same plane, they will intersect at some point in the plane. This point is the reconstructed 3D scene point $\mathbf{M}$.

Analytically, the reconstructed 3D point $\mathbf{M}$ can be found using Equation (6.12), by solving for parameters ζ_r and λ_1.

Let us look closely at the epipolar line (Equation 6.12) and let us rewrite it as:

$$\mathbf{m}_r = \frac{1}{\zeta_r}\mathbf{e}_r + \frac{\lambda_1}{\zeta_r}\mathbf{m}_l' \tag{6.34}$$

The unknowns are ζ_r and λ_1. Both encode the position of $\mathbf{M}$ in space, as ζ_r is the depth of $\mathbf{M}$ and λ_1 fixes the position of $\mathbf{M}$ along the optical ray of $\mathbf{m}_1$.

Given that points $\mathbf{m}_r$, $\mathbf{e}_r$ and $\mathbf{m}_l'$ are collinear, there exist only two scalars μ and ν such that

$$\mathbf{m}_r = \mu\mathbf{e}_r + \nu\mathbf{m}_l' \tag{6.35}$$

Knowing all three points, we can calculate the parameters μ and ν using the following closed form expressions:

$$\mu = \frac{(\mathbf{m}_r \times \mathbf{m}_l') \cdot (\mathbf{e}_r \times \mathbf{m}_l')}{||\mathbf{e}_r \times \mathbf{m}_l'||^2}$$

$$\nu = \frac{(\mathbf{m}_r \times \mathbf{e}_r) \cdot (\mathbf{m}_l' \times \mathbf{e}_r)}{||\mathbf{m}_l' \times \mathbf{e}_r||^2} \tag{6.36}$$

The reconstructed point $\mathbf{M}$ can then be calculated by inserting the value λ into Equation (6.7). This reconstruction method was introduced in Faugeras (1992).

In reality, camera parameters and image locations are known only approximately. The back-projected rays therefore do not actually intersect in space. Their intersection can only be estimated as the point of minimum distance from both rays. This problem is addressed in Hartley and Sturm (1997), Hartley and Zisserman (2003) and Trucco and Verri (1998).

Reconstruction up to a Scale Factor.   If only intrinsic parameters are known (plus point correspondences between images), the epipolar geometry is described by the essential matrix (Section 6.3.2). We will see that, starting from the essential matrix, a reconstruction only up to a similarity transformation (rigid + uniform scale) can be achieved.

Unlike the fundamental matrix, the only property of which is to have rank two, the essential matrix is characterised by the following theorem by Huang and Faugeras (1989).

Theorem 6.3.1. *A real 3×3 matrix E can be factorized as a product of a nonzero skew-symmetric matrix and a rotation matrix if and only if E has two identical singular values and a zero singular value.*

The theorem has a constructive proof (Hartley 1992) that describes how E can be factorized into rotation and translation using its singular value decomposition (SVD).

The rotation R and translation $\mathbf{t}$ are then used to instantiate a camera pair as in Equation (6.14), and this camera pair is subsequently used to reconstruct the structure of the scene by triangulation.

The rigid displacement ambiguity arises from the arbitrary choice of the world reference frame. The scale ambiguity results from the fact that, in Equation (6.18), $\mathbf{t}$ can have an arbitrary scale, but the resulting essential matrix is still the same (E is defined up to a scale factor). Translation can thus be recovered from E only up to an unknown scale factor which is inherited by the reconstruction. This is also known as *depth-speed ambiguity*.

Reconstruction up to a Projective Transformation.   Suppose that a set of image correspondences $\mathbf{m}_l^i \leftrightarrow \mathbf{m}_r^i$ are given. It is assumed that these correspondences come from a set of 3D points $\mathbf{M}_i$, which are unknown. Similarly, the position, orientation and calibration of the cameras are not known. This situation is usually referred to as *weak calibration*, and we will see that the scene may be reconstructed up to a projective ambiguity, which may be reduced if additional information is supplied on the cameras or the scene.

The reconstruction task is to find the camera matrices P_1 and P_r, as well as the 3D points $\mathbf{M}_i$ such that

$$\mathbf{m}_l^i = P_l \mathbf{M}^i \qquad \text{and} \qquad \mathbf{m}_r^i = P_r \mathbf{M}^i \qquad \forall i \tag{6.37}$$

In particular, if H is any 4×4 invertible matrix, representing a projective transformation of the 3D space, then replacing points $\mathbf{M}^i$ by $H\mathbf{M}^i$ and matrices P_1 and P_r by $P_1 H^{-1}$ and $P_r H^{-1}$ does not change the image points. This shows that, if nothing is known but the image points, the points $\mathbf{M}^i$ and the cameras can be determined, at best, only up to a projective transformation.

The procedure for reconstruction follows the previous one. Given the weak calibration assumption, the fundamental matrix can be computed (using the algorithm described in Section 6.3.2), and from a (non-unique) factorization of F a pair of camera matrices can be instantiated. The position in space of the points $\mathbf{M}^i$ is then obtained by triangulation. The only difference with the previous case is that F does not admit a unique factorization of the form

$$F = [\mathbf{e_r}]_\times A \qquad (6.38)$$

whence the projective ambiguity follows. Indeed, for any A satisfying Equation (6.38), the pair of camera matrices P_1 and P_r:

$$P_1 = [I|\mathbf{0}] \qquad \text{and} \qquad P_r = [A|\mathbf{e_r}] \qquad (6.39)$$

yields the fundamental matrix F, as can be easily verified. One such matrix A can be obtained as $A = [\mathbf{e_r}]_\times F$.

6.4 *N*-VIEW GEOMETRY

In this section we study the relationship that links three or more views of the same 3D scene, known in the three-view case as *trifocal geometry*. This geometry can be described in terms of fundamental matrices linking pairs of cameras, but in the three-view case a more compact and elegant description is provided by a special algebraic operator, the *trifocal tensor*. We also discover that four views are all we need, in the sense that additional views do not allow us to compute anything we could not compute already (Section 6.4.3).

This section is based on Heyden (2000) and Shashua (2004). The epipolar constraints on three views and the epipolar transfer were introduced in Faugeras and Robert (1994). A starting point to learn more on the trifocal tensor is Shashua (1997). The factorization method for reconstruction is described in Sturm and Triggs (1996). A general reference is Hartley and Zisserman (2003).

6.4.1 Trifocal Geometry

Denoting the cameras by $1, 2, 3$, there are now three fundamental matrices, $F_{1,2}$, $F_{1,3}$, $F_{2,3}$, and six epipoles $\mathbf{e}_{i,j}$, as in Figure 6.5. The three fundamental matrices completely describe the trifocal geometry (as long as the three cameras are not collinear). The plane containing the three optical centres is called the *trifocal plane*. It intersects each image plane along a line which contains the two epipoles. This results in three epipolar constraints:

$$F_{3,1}\mathbf{e}_{3,2} = \mathbf{e}_{1,2} \times \mathbf{e}_{1,3} \qquad F_{1,2}\mathbf{e}_{1,3} = \mathbf{e}_{2,3} \times \mathbf{e}_{2,1} \qquad F_{2,3}\mathbf{e}_{2,1} = \mathbf{e}_{3,1} \times \mathbf{e}_{3,2} \qquad (6.40)$$

Three fundamental matrices include 21 free parameters, less the 3 constraints above; the trifocal geometry is therefore determined by 18 parameters. This description of the trifocal geometry fails when the three cameras are collinear, and the trifocal plane reduces to a line.

If the trifocal geometry is known, given two corresponding points $\mathbf{m}_1$ and $\mathbf{m}_2$ in view 1 and 2 respectively, the position of the corresponding point $\mathbf{m}_3$ in view 3 is completely

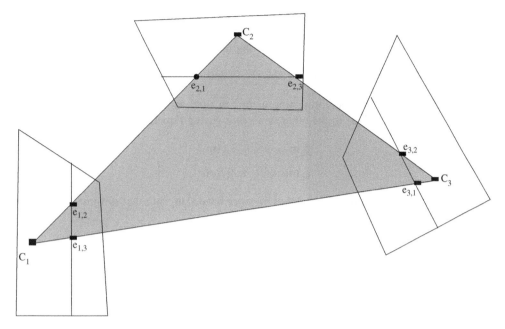

Figure 6.5   Trifocal geometry

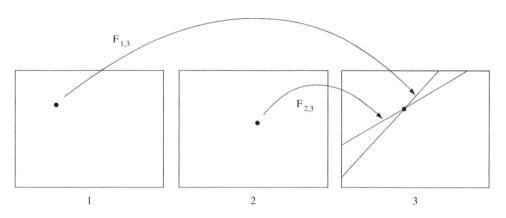

Figure 6.6   Point transfer using epipolar constraints between three views

determined (Figure 6.6). This allows for *point transfer* or prediction. Indeed, $\mathbf{m}_3$ belongs simultaneously to the epipolar line of $\mathbf{m}_1$ and to the epipolar line of $\mathbf{m}_2$, hence:

$$\mathbf{m}_3 = F_{1,3}\mathbf{m}_1 \times F_{2,3}\mathbf{m}_2 \tag{6.41}$$

Point transfer fails for 3D points lying in the trifocal plane, as their epipolar lines are coincident. Even worse, if the three cameras are collinear, the transfer is not possible for any point.

These deficiencies motivate the introduction of the trifocal tensor.

6.4.2 The Trifocal Tensor

Recalling Equation (6.39), consider the following three cameras:

$$P_1 = [I|0] \qquad P_2 = [A|\mathbf{e}_{2,1}] \qquad \text{and} \qquad P_3 = [B|\mathbf{e}_{3,1}] \tag{6.42}$$

Consider a point $\mathbf{M}$ in space projecting to $\mathbf{m}_1$, $\mathbf{m}_2$ and $\mathbf{m}_3$ in the three cameras. Let us write the epipolar line of $\mathbf{m}_1$ in the other two views (using Equation 6.12):

$$\zeta_2 \mathbf{m}_2 = \mathbf{e}_{2,1} + \lambda_1 A \mathbf{m}_1 \tag{6.43}$$

$$\zeta_3 \mathbf{m}_3 = \mathbf{e}_{3,1} + \lambda_1 B \mathbf{m}_1 \tag{6.44}$$

Consider a line through $\mathbf{m}_2$, represented by s; we have $s^T \mathbf{m}_2 = 0$, that substituted in (6.43) gives:

$$0 = s^T \mathbf{e}_{2,1} + \lambda_1 s^T A \mathbf{m}_1 \tag{6.45}$$

Since a point is determined by two lines, we can write a similar independent constraint for a second line $\mathbf{l}$ through $\mathbf{m}_2$:

$$0 = \mathbf{l}^T \mathbf{e}_{2,1} + \lambda_1 \mathbf{l}^T A \mathbf{m}_1 \tag{6.46}$$

To write the last two equations in a more compact way we introduce a 2×3 matrix

$$S = \begin{bmatrix} s^T \\ \mathbf{l}^T \end{bmatrix} \tag{6.47}$$

and switch to tensor notation, where S becomes s_j^μ, and Equations (6.45) and (6.46) become:

$$0 = s_j^\mu e_{2,1}^j + \lambda_1 p_1^i s_j^\mu a_i^j \tag{6.48}$$

By the same token, we can represent two lines through $\mathbf{m}_3$ in tensor notation by means of a 2×3 matrix r_k^ρ, obtaining:

$$0 = r_k^\rho e_{3,1}^k + \lambda_1 p_1^i r_k^\rho b_i^k \tag{6.49}$$

After eliminating λ_1 from Equations (6.48) and (6.49) we obtain

$$(s_j^\mu e_{2,1}^j)(p_1^i r_k^\rho b_i^k) = (r_k^\rho e_{3,1}^k)(p_1^i s_j^\mu a_i^j) \tag{6.50}$$

and after some rewriting:

$$p_1^i s_j^\mu r_k^\rho \mathcal{T}_i^{jk} = 0 \tag{6.51}$$

where

$$\mathcal{T}_i^{jk} \triangleq e_{2,1}^j b_i^k - e_{3,1}^k a_i^j \tag{6.52}$$

is a $3 \times 3 \times 3$ homogeneous tensor, called the *trifocal tensor*. The tensorial Equation (6.51) represents *four trilinear equations*, since $\mu = 1, 2$ and $\rho = 1, 2$ are free indices.

This relationship involves the trifocal tensor and the three corresponding points, as s_j^μ and r_k^ρ define $\mathbf{m}_2$ and $\mathbf{m}_3$ respectively. To make the coordinates of $\mathbf{m}_2$ and $\mathbf{m}_3$ appear explicitly, let us choose the two lines parallel to the coordinate axes, thereby obtaining:

$$s_j^\mu = \begin{bmatrix} -1 & 0 & u_2 \\ 0 & -1 & v_2 \end{bmatrix} \quad \text{and} \quad r_k^\rho = \begin{bmatrix} -1 & 0 & u_3 \\ 0 & -1 & v_3 \end{bmatrix} \tag{6.53}$$

After substituting the expressions for s_j^μ and r_k^ρ, Equation (6.51) becomes:

$$
\begin{aligned}
u_3 \mathcal{T}_i^{13} p_1^i - u_3 u_2 \mathcal{T}_i^{33} p_1^i + u_2 \mathcal{T}_i^{31} p_1^i - \mathcal{T}_i^{11} p_1^i &= 0 \\
v_3 \mathcal{T}_i^{13} p_1^i - v_3 u_2 \mathcal{T}_i^{33} p_1^i + u_2 \mathcal{T}_i^{32} p_1^i - \mathcal{T}_i^{12} p_1^i &= 0 \\
u_3 \mathcal{T}_i^{23} p_1^i - u_3 v_2 \mathcal{T}_i^{33} p_1^i + v_2 \mathcal{T}_i^{31} p_1^i - \mathcal{T}_i^{21} p_1^i &= 0 \\
v_3 \mathcal{T}_i^{23} p_1^i - v_3 v_2 \mathcal{T}_i^{33} p_1^i + v_2 \mathcal{T}_i^{32} p_1^i - \mathcal{T}_i^{22} p_1^i &= 0
\end{aligned}
\tag{6.54}
$$

Every triplet $(\mathbf{m}_1, \mathbf{m}_2, \mathbf{m}_3)$ of corresponding points gives four linear independent equations, hence seven triplets determine the trifocal tensor.

Notice that the trifocal tensor represents the trifocal geometry without singularities: it can be safely used for point transfer in any situation. A transferred point, $\mathbf{m}_3$, can be computed from Equation (6.54) or, in closed form, as:

$$p_3^k = p_1^i s_j \mathcal{T}_i^{jk} \tag{6.55}$$

where s_j represents a line through $\mathbf{m}_2$. In a similar way, the trifocal tensor can be used to transfer lines:

$$q_i = s_j r_k \mathcal{T}_i^{jk} \tag{6.56}$$

where s_j and r_k represent two matching lines in the first two views, and q_i is the transferred line in the third view.

6.4.3 Multiple-view Constraints

We outline here an alternative and elegant way to derive multi-linear constraints, based on determinants. Consider one image point viewed by m cameras:

$$\zeta_i \mathbf{m}_i = P_i \mathbf{M} \qquad i = 1 \ldots m \tag{6.57}$$

By stacking all these equations we obtain:

$$
\begin{bmatrix}
P_1 & \mathbf{m}_1 & 0 & \ldots & 0 \\
P_2 & 0 & \mathbf{m}_2 & \ldots & 0 \\
\vdots & \vdots & \vdots & \vdots & \vdots \\
P_m & 0 & 0 & \ldots & \mathbf{m}_m
\end{bmatrix}
\begin{bmatrix}
\mathbf{M} \\
-\zeta_1 \\
-\zeta_2 \\
\vdots \\
-\zeta_m
\end{bmatrix}
=
\begin{bmatrix}
0 \\
0 \\
\vdots \\
0
\end{bmatrix}
\tag{6.58}
$$

This implies that the $3m \times (m+4)$ matrix (let us call it L) is rank-deficient, i.e., rank $L < m+4$. In other words, all the $(m+4) \times (m+4)$ minors of L are equal to 0.

It has been proven that there are three different types of such minors that translate into meaningful multi-view constraints, depending on the number of rows taken from each view. Since one row has to be taken from each view and the remaining four can be distributed freely, one can choose:

1. Two rows from one view and two rows from another view. This gives a bilinear two-view constraint, expressed by the bifocal tensor, i.e., the fundamental matrix.

2. Two rows from one view, one row from another view and one row from a third view. This gives a trilinear three-view constraint, expressed by the trifocal tensor.

3. One row from each of four different views. This gives a quadrilinear four-view constraint, expressed by the quadrifocal tensor.

All other types of minors can be factorized as products of the two-, three-, or four-views constraints and point coordinates in the other images. This indicates that no interesting constraints can be written for more than four views.

6.4.4 Uncalibrated Reconstruction from *N* views

As in the case of two cameras, given only point correspondences, it is possible to reconstruct scene structure and camera matrices up to a global unknown projective transformation. The reconstruction from *N* views, however, cannot be obtained by simply applying the method of Section 6.3.4 to pairs of views. One would obtain, in general, a set of projective reconstructions linked to each other by an unknown projective transformation (i.e., each camera pair defines its own projective frame). A very elegant method for multi-image reconstruction is described in Sturm and Triggs (1996), based on matrix factorization.

Consider m cameras $P_1 \ldots P_m$ looking at n 3D points $\mathbf{M}_1 \ldots \mathbf{M}_n$. The usual projection equation

$$\zeta_{i,j}\mathbf{m}_{i,j} = P_i\mathbf{M}_j \qquad i = 1 \ldots m, \qquad j = 1 \ldots n \tag{6.59}$$

can be written in matrix form:

$$\underbrace{\begin{bmatrix} \zeta_{1,1}\mathbf{m}_{1,1}, & \zeta_{1,2}\mathbf{m}_{1,2}, & \cdots & \zeta_{1,n}\mathbf{m}_{1,n} \\ \zeta_{2,1}\mathbf{m}_{2,1}, & \zeta_{2,2}\mathbf{m}_{2,2}, & \cdots & \zeta_{2,n}\mathbf{m}_{2,n} \\ \vdots & \vdots & \ddots & \vdots \\ \zeta_{m,1}\mathbf{m}_{m,1}, & \zeta_{m,2}\mathbf{m}_{m,2}, & \cdots & \zeta_{m,n}\mathbf{m}_{m,n} \end{bmatrix}}_{\text{measurements}\, M} = \underbrace{\begin{bmatrix} P_1, \\ P_2, \\ \vdots \\ P_m \end{bmatrix}}_{P} \underbrace{\begin{bmatrix} \mathbf{M}_1, & \mathbf{M}_2, & \cdots & \mathbf{M}_n \end{bmatrix}}_{\text{structure } S} \tag{6.60}$$

In this formula the $\mathbf{m}_{i,j}$ are known, but all the other quantities are unknown, including the projective depths $\zeta_{i,j}$. Equation (6.60) tells us that M can be factored into the product of a $3m \times 4$ matrix P and a $4 \times n$ matrix S. This also means that M has rank four.

If we assume for a moment that the projective depths $\zeta_{i,j}$ are known, then matrix M is known too and we can compute its singular value decomposition:

$$M = UDV. \tag{6.61}$$

In the noise-free case, $D = \text{diag}(\sigma_1, \sigma_2, \sigma_3, \sigma_4, 0, \ldots 0)$ and reconstruction is obtained by setting $P = [U_{.,1}|U_{.,2}|U_{.,3}|U_{.,4}]\text{diag}(\sigma_1, \sigma_2, \sigma_3, \sigma_4)$ (first four columns of UD) and $S = [V_{1,.}|V_{2,.}|V_{3,.}|V_{4,.}]$ (first four rows of V).

Taken individually, the projective depths $\zeta_{i,j}$ are arbitrary (because they depend on arbitrary scale factors), but in a sequence of images they are linked together, and this is the missing constraint that gives a coherent projective reconstruction.

Consider a camera pair. Let $F_{2,1} = F_{1,2}^T$ be the fundamental matrix of the second camera; from Equation (6.21) with $\mathbf{m}_2$ on the left-hand side and Equation (6.23) instantiated for $F_{2,1}$ the following relationship can be obtained:

$$\zeta_2 F_{2,1}\mathbf{m}_2 = \zeta_1(\mathbf{e}_{1,2} \times \mathbf{m}_1) \tag{6.62}$$

This equation relates the projective depths of a single 3D point in two images. From (6.62) one can obtain

$$\zeta_1 = \frac{(\mathbf{e}_{1,2} \times \mathbf{m}_1)F_{2,1}\tilde{\mathbf{m}}_2}{||\mathbf{e}_{1,2} \times \mathbf{m}_1||^2} \zeta_2 \tag{6.63}$$

By estimating a sufficient number of fundamental matrices and epipoles, one can recursively chain together equations like (6.63) to give estimates for the complete set of depths for a given 3D point, starting from $\zeta_1 = 1$.

This technique is fast, requires no initialization, but is non-optimal in a maximum likelihood sense. A global minimization of the reprojection error, called *bundle adjustment* (Triggs *et al.* 2000), is often required to improve the solution.

6.4.5 Autocalibration

A projective reconstruction is what can be achieved from a sequence of images with weakly calibrated cameras, i.e., when point correspondences are the only information available. The aim of *autocalibration* is to compute the intrinsic parameters, or, in general, to recover the Euclidean *stratum*, starting from weakly calibrated cameras. This section explores the constraints available for autocalibration.

Recently, the problems of reconstruction and calibration have been unified within an elegant approach, known as *stratification* (Faugeras 1994). The reader is referred to Fusiello (2000) for a review of autocalibration, and to Hartley *et al.* (1999), Heyden and Aström (1998), Maybank and Faugeras (1992), Mendonça and Cipolla (1999), Pollefeys *et al.* (1998) and Triggs (1997) for classical and recent work on the subject.

Two-view constraints. As seen in Section 6.3.2, the epipolar geometry of two views is described by the fundamental matrix, which depends on seven parameters. Of these, the five parameters of the essential matrix are needed to describe the rigid displacement: two independent constraints are available for the computation of the intrinsic parameters from the fundamental matrix.

In particular, these constraints stem from the equality of two singular values of the essential matrix (Theorem 6.3.1) which can be decomposed in two independent polynomial equations.

N-view constraints. In the case of three views, the trifocal geometry is described by 18 parameters. The rigid displacement is described by 11 parameters: 6 for 2 rotations, 4 for 2 directions of translation and 1 ratio of translation norms. Therefore, in this case there are seven constraints available on the intrinsic parameters.

According to the *canonical decomposition* (Luong and Viéville 1996), $11n - 15$ parameters are needed to describe the geometry of n views. The rigid displacement is described by $6n - 7$ parameters: $3(n - 1)$ for rotations, $2(n - 1)$ for translations, and $n - 2$ ratios of translation norms. Thus, there are $5n - 8$ constraints available for computing the intrinsic parameters.

Let us suppose that n_k parameters are known and n_c parameters are constant. Every view but the first introduces $5 - n_k - n_c$ unknowns; the first view introduces $5 - n_k$ unknowns. Therefore, the unknown intrinsic parameters can be computed provided that

$$5n - 8 \geq (n - 1)(5 - n_k - n_c) + 5 - n_k \tag{6.64}$$

For example, if the intrinsic parameters are constant, three views are sufficient to recover them. If one parameter (usually the skew) is known and the other parameters are varying, at least eight views are needed.

6.5 SUMMARY

This chapter has introduced a number of basic geometric concepts of multiple-view geometry in computer vision. Such concepts are at the heart of vision systems commonly deployed in advanced, image-based, communication systems (Isgrò *et al.* 2004).

The key actors in two-view geometry are the fundamental and essential matrices, in the uncalibrated and calibrated case respectively. They capture the epipolar geometry algebraically, and constrain the position of corresponding points to known lines. The corresponding concept for the three-view case is the trifocal tensor. Importantly, three views trigger the idea of point transfer, that is, predicting a third image acquired by a camera given two images of the same scene and a correspondence map. This is the basis of *view synthesis*, an important class of techniques for image based rendering.

Reconstruction of 3D shape and position is possible to varying degree of ambiguity, depending on what is known about the 3D world and the camera parameters, i.e., their calibration (intrinsic and extrinsic parameters). Interestingly, the problems of reconstruction and calibration are unified in the stratified approach to autocalibration.

REFERENCES

Faugeras O 1992 What can be seen in three dimensions with an uncalibrated stereo rig? *Proceedings of the European Conference on Computer Vision*, S. Margherita Ligure, pp. 563–578.

Faugeras O 1993 *Computer Vision: a Geometric Viewpoint.* MIT Press.

Faugeras O 1994 Stratification of 3-D vision: projective, affine, and metric representations. *Journal of the Optical Society of America A* **12**(3), 465–484.

Faugeras O and Luong QT 2001 *The geometry of multiple images.* MIT Press.

Faugeras O and Robert L 1994 What can two images tell us about a third one? *Proceedings of the European Conference on Computer Vision*, Stockholm, pp. 485–492.

Fusiello A 2000 Uncalibrated Euclidean reconstruction: A review. *Image and Vision Computing* **18**(6–7), 555–563.

Fusiello A, Trucco E and Verri A 2000 A compact algorithm for rectification of stereo pairs. *Machine Vision and Applications* **12**(1), 16–22.

Hartley R 1992 Estimation of relative camera position for uncalibrated cameras *Proceedings of the European Conference on Computer Vision,* Santa Margherita Liqure, pp. 579–587.

Hartely R 1995 In defence of the 8-point algorithm *Proceedings of the International Conference on Computer Vision,* Cambridge, MA, pp. 1064–1075.

Hartley R 1999 Theory and practice of projective rectification. *International Journal of Computer Vision* **35**(2), 1–16.

Hartley R and Sturm P 1997 Triangulation. *Computer Vision and Image Understanding* **68**(2), 146–157.

Hartley R and Zisserman A 2003 *Multiple view geometry in computer vision* 2nd edn. Cambridge University Press.

Hartley R, Hayman E, de Agapito L and Reid I 1999 Camera calibration and the search for infinity *Proceedings of the IEEE International Conference on Computer Vision,* Corfu, Greece, pp. 510–517.

Heyden A 2000 Tutorial on multiple view geometry. In conjunction with ICPR00.

Heyden A and Aström K 1998 Minimal conditions on intrinsic parameters for Euclidean reconstruction *Proceedings of the Asian Conference on Computer Vision,* Hong Kong, pp. 169–176.

Huang T and Faugeras O 1989 Some properties of the E matrix in two-view motion estimation. *IEEE Transactions on Pattern Analysis and Machine Intelligence* **11**(12), 1310–1312.

Isgrò F and Trucco E 1999 Projective rectification without epipolar geometry *Proceedings of the IEEE Conference on Computer Vision and Pattern Recognition,* Fort Collins, CO, pp. I:94–99.

Isgrò F, Trucco E, Kauff P and Schreer O 2004 3-D image processing in the future of immersive media. *IEEE Transactions on Circuits and Systems for Video Technology* **14**(3), 288–303.

Loop C and Zhang Z 1999 Computing rectifying homographies for stero vision *Proceedings of the IEEE Conference on Computer Vision and Patern Recognition,* Fort Collins, CO, pp. I:125–131.

Luong QT and Faugeras OD 1996 The fundamental matrix: Theory, algorithms, and stability analysis. *International Journal of Computer Vision* **17**, 43–75.

Luong QT and Viéville T 1996 Canonical representations for the geometries of multiple projective views. *Computer Vision and Image Understanding* **64**(2), 193–229.

Maybank SJ and Faugeras O 1992 A theory of self-caliberation of a moving camera. *International Journal of Computer Vision* **8**(2), 123–151.

Mendonça P and Cipolla R 1999 A simple techinique for self-calibration *Proceedings of the IEEE Conference on Computer Vision and Pattern Recognition,* pp. I:500–505.

Pollefeys M, Koch R and Van Gool L 1998 Self-calibration and metric reconstruction in spite of varying and unknown internal camera parameters *Proceedings of the IEEE International Conference on Computer Vision,* Bombay, pp. 90–95.

Shashua A 1997 Trilinear tensor: The fundamental construct of multiple-view geometry and its applications. *International Workshop on Algebraic Frames For The Perception Action Cycle,* Kiel, Germany, pp. 190–206.

Shashua A 2004 Multi-view geometry from a stationary scene Lecture given at U. of Milano.

Sturm P and Triggs B 1996 A factorization based algorithm for multi-image projective structure and motion *Proceedings of the European Conference on Computer Vision,* Cambridge, UK, pp. 709–720.

Torr P and Murray DW 1997 The development and comparison of robust methods for estimating the fundamental matrix. *International Journal of Computer Vision* **24**(3), 271–300.

Triggs B 1997 Autocalibration and the absolute quadric *Proceedings of the IEEE Conference on Computer Vision and Pattern Recognition,* Puerto Rico, pp. 609–614.

Triggs B, McLauchlan PF, Hartley RI and Fitzgibbon AW 2000 Bundle adjustment — a modern synthesis *Proceedings of the International Workshop on Vision Algorithms,* Springer–Verlag, pp. 298–372.

Trucco E and Verri A 1998 *Introductory Techniques for 3-D Computer Vision.* Prentice-Hall.

Zhang Z 1998 Determining the epipolar geometry and its uncertainty: A review. *International Journal of Computer Vision* **27**(2), 161–195.

7

Stereo Analysis

Nicole Atzpadin[1] and Jane Mulligan[2]

[1] *Fraunhofer Institute for Telecommunications/Heinrich-Hertz-Institut, Berlin, Germany*
[2] *University of Colorado, Boulder, CO, USA*

In the preceding chapter we examined the geometry of multiple-camera systems. Many of the ideas presented there depend on knowing or determining the points in two or more views to which a particular world point is projected. In real-time 3D applications, where we are presented with two or more images of natural scenes, the problem of determining corresponding image points for all views is both challenging and critical to the quality of our 3D models.

In this chapter we will present the basic approaches used for determining these image correspondences. Section 7.1 addresses the problem for two views and introduces the main definitions, metrics and search and refinement techniques. Section 7.2 discusses the advantages of using three or more views to establish correspondences, and describes the basic approaches for searching with more views. Our emphasis is on techniques and systems which achieve real-time performance suitable for 3D communication applications.

7.1 STEREO ANALYSIS USING TWO CAMERAS

Finding correspondences in natural stereo image pairs plays an important role in a large number of applications such as robot navigation, augmented reality applications and telecommunications. The main goal of stereo correspondence search is the matching of points in two images (I_l, I_r) such that the two corresponding pixels $\mathbf{m}_l = (u_l, v_l)$ and $\mathbf{m}_r = (u_r, v_r)$ are projections of the same point $\mathbf{P}$ in the 3D scene. The resulting disparity vector associates the matching pair by giving a difference or disparity in image positions:

$$\mathbf{d} = \mathbf{m}_l - \mathbf{m}_r = (d_u d_v) \tag{7.1}$$

3D Videocommunication — Algorithms, concepts and real-time systems in human centred communication
Edited by O. Schreer, P. Kauff and T. Sikora © 2005 John Wiley & Sons, Ltd

As we learned in Chapter 6, the search for correspondences can be constrained to a 1D search along the epipolar line. If we exploit rectification techniques (Section 6.3.2) the epipolar lines will lie along scanlines or along columns of the transformed image pair. For the first widely used case we have the simplified expression:

$$\mathbf{d} = (u_1 - u_r, 0) \tag{7.2}$$

A disparity map is the set of all such pixelwise disparities for an image pair.

The guiding principle to solve the correspondence problem is similarity. Single-colour samples do not provide sufficient and distinct information for correspondence search, hence similar areas or features are required in the two stereo images.

Several approaches for correspondence search are mentioned in the literature :

- Area-based algorithms consider an area or window around the pixel in order to handle ambiguities in matching. They result in dense disparity maps, but fail in occluded or low-structure regions. Due to the classical depth reliability versus depth accuracy trade-off the definition of the window size is a central problem in area-based stereo.
- Feature-based algorithms extract and match local features like edges, corners or lines (Marr and Poggio 1979). They provide sparse, but robust disparity maps.
- Pixel-recursive approaches were originally developed for image sequence coding (Böröczky 1991). They obtain a dense field by scanning the whole image, predicting disparities from the previous one.
- Optical-flow-based algorithms rely on the relation between photometric correspondence vectors and spatiotemporal derivatives of luminance (Barron *et al.* 1994; Horn and Schunck 1981).

This section focuses on area-based algorithms, which will be described in detail in Section 7.1.1. These have proven effective in real-time generation of dense disparity maps. To detect confident correspondences, additional constraints are usually exploited for several fundamental reasons. Projective distortions, especially in the case of a convergent camera setup, result in different projections of an object in the two views. Unstructured regions (e.g., smooth white walls) or regions with periodic structures (e.g., regular stripes) lead to ambiguity and wrong correspondences because the same region in the left image will match equally well with many windows in the right image. Areas of the scene visible in one image, but occluded in the other have no valid match.

The commonly used constraints to enhance the quality of stereo analysis are the epipolar constraint, the uniqueness constraint and the continuity constraint.

- *Epipolar constraint*: as mentioned in the previous section, the epipolar constraint restricts the search range for a corresponding point in one image to the epipolar line in the other image. The search area is therefore reduced to one dimension.
- *Uniqueness constraint*: the uniqueness constraint indicates that for opaque objects a point in one image should have at most one corresponding point in the other image. This constraint is useful to detect outliers and occlusions.
- *Continuity constraint*: the basic idea of the continuity constraint is that disparity tends to vary slowly across a surface. This works for smooth objects, but is violated at object boundaries. With the help of this constraint the search range can be further restricted.

Although these constraints help to enhance the quality of disparities, a post-processing of disparity fields is unavoidable to generate reliable disparities for all parts of the image. A detailed description of post-processing can be found in Section 7.1.3.

7.1.1 Standard Area-based Stereo Analysis

Area-based stereo approaches are some of the oldest methods used in computer vision. The general principle of area-based stereo approaches is based on the photometric compatibility constraint which is the strongest constraint that relates correspondence and luminance (Nagel and Enkelmann 1986). Because of ambiguities this constraint is not sufficient for a unique correspondence. Instead of comparing individual pixels, several neighbouring pixels are grouped in a window and their intensities are compared with those of the pixels in another window. Figure 7.1 points up this principle. To find the corresponding pixel to u, v in the original image a rectangular window is centred at u, v and compared with a window in the reference image, which is shifted along the epipolar line.

Numerous matching measures for the comparison of two windows have been introduced in the literature. Their performance and computational complexity vary. The simplest matching measure which is used by many real-time algorithms is the sum of absolute differences (SAD). The disparity which minimizes the SAD score for each pixel is chosen.

$$\text{SAD}(u, v, \mathbf{d}) = \sum_{(i,j)} | I_1(u+j, v+i) - I_r(u+d_u+j, v+d_v+i) | \tag{7.3}$$

Here $i \in [-n, n]$ and $j \in [-m, m]$. A slightly more complex matching measure is the sum of squared differences (SSD) which weights big differences more than small ones.

$$\text{SSD}(u, v, \mathbf{d}) = \sum_{(i,j)} \{I_1(u+j, v+i) - I_r(u+d_u+j, v+d_v+i)\}^2 \tag{7.4}$$

A version of SSD with better tolerance to camera differences is:

$$\text{ZSSD}(u, v, \mathbf{d}) = \sum_{(i,j)} \{(I_1(u+j, v+i) - \mu_l) - (I_r(u+d_u+j, v+d_v+i) - \mu_r)\}^2 \tag{7.5}$$

with

$$\mu_k(u, v) = \frac{1}{W} \sum_{(i,j)} \{I_k(u+j, v+i)\}.$$

Here $W = (2n+1)(2m+1)$. The idea of subtracting the mean of the matching window from each intensity value is also applicable to other matching measures. A further improvement of the quality of disparity maps can be obtained with the use of correlation based

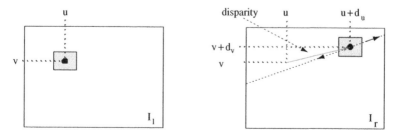

Figure 7.1 Principle of area-based stereo analysis

measures. The standard cross-correlation (CC) is sensitive to noise and therefore not often used. A more complex, but even better matching measure is the normalized cross-correlation (NCC). Especially for two stereo cameras with photometric differences NCC produces disparity maps of higher quality than distance-based measures.

$$\text{NCC}(u, v, \mathbf{d}) = \frac{\sigma_{\text{lr}}^2(u, v, \mathbf{d})}{\sqrt{\sigma_{\text{l}}^2(u, v)\ \sigma_{\text{r}}^2(u + d_u, v + d_v)}} \tag{7.6}$$

with

$$\sigma_k^2(u, v) = \tfrac{1}{w}\sum_{i,j}\{I_k(u + j, v + i) - \mu_k\}^2$$
$$\sigma_{\text{lr}}^2(u, v, \mathbf{d}) = \tfrac{1}{w}\sum_{i,j}\{I_l(u + j, v + i) - \mu_l\}\{I_r(u + d_u + j, v + d_v + i) - \mu_r\}.$$

A variant of NCC proposed by Moravec (1980/81) normalizes using the sum of variances from the left and right window rather than the product. This modified normalized cross-correlation (MNCC) yields a metric which varies from -1 to 1, which is somewhat more stable.

$$\text{MNCC}(u, v, \mathbf{d}) = \frac{2\sigma_{\text{lr}}^2(u, v, \mathbf{d})}{\sigma_{\text{l}}^2(u, v) + \sigma_{\text{r}}^2(u + d_u, v + d_v)} \tag{7.7}$$

In an attempt to reduce the effects of radiometric gain and bias, Zabih and Woodfill (1994) have proposed two more qualitative neighbourhood transforms based on the ordering of image intensities. The *rank transform* for a window is the number of pixels with intensity less than the central pixel. Rank transformed images are compared using an SAD metric. To combat the loss in discriminatory power caused by this transform, the authors further proposed the *census transform* which encodes the window as a bitstring with ones for pixels less than the centre. Matching is performed using the Hamming distance (number of differing bits) between strings/windows.

The basic problem when using any area-based algorithm for stereo analysis is to provide sufficient intensity variation within the measurement window considered, while still localizing the correct disparities accurately (Kanade and Okutome 1994). A large window increases the probability of sufficient intensity variation and therefore avoids ambiguities and mismatches, but it also reduces the selectivity of finding correct disparities, especially in areas with depth discontinuities. Thus, the larger the window considered the greater the robustness, and the smaller the window the better the accuracy. Various approaches have been proposed in the literature to try to solve this depth reliability versus depth accuracy trade-off.

Kanade and Okutomi (1994) solve the reliability – accuracy problem by the definition of adaptive windows. They attempt to find an ideal window in size and shape for each pixel in an image. An ideal window should not include a depth discontinuity, and the intensity variation should be sufficient to produce a correct correspondence. The intensity variation can be directly measured in the images, but depth discontinuity can be determined only after disparity estimation and is therefore unknown. Levine *et al.* (1973) handle this problem by exclusively considering the intensity variation. Kanade and Okutomi go a step further and introduce a statistical model of the disparities. Starting with an initial estimate of the disparity map, the algorithm iteratively updates the disparities. The variance of disparities and intensities define a so-called uncertainty which influences not only the size, but also the shape of the windows. The algorithm starts with a window of size 3×3 for each pixel, which is expanded in each direction until the uncertainty increases.

A less computational approach to adaptive windows was introduced by Bobick and Intille (1999). In contrast to the approach of Kanade and Okutomi the windows are not adapted to the image contents. For each pixel nine asymmetric windows of the same size with shifted window centers are tested. The idea behind this approach is that a window minimizing the matching criterion is more likely to cover a region of constant depth. Windows containing depth discontinuities result in high matching scores due to projective distortions.

Figure 7.2 shows examples for the two adaptive window approaches for some selected pixels. Areas of different colour are areas of different depth in this example. Windows of arbitrary shape result from the approach introduced by Kanade and Okutomi, whereas windows are of constant size in the approach of Bobick and Intille. All chosen windows for both approaches exclude depth discontinuities.

Another group of well known methods for disparity estimation are hierarchical approaches (Anandan 1989). The basic idea in hierarchical systems is that robustness is achieved by starting with large windows to estimate reliable, but inaccurate disparities. These results are refined by using smaller windows and a small search area defined by the results obtained by the large windows. All hierarchical approaches can be divided into two groups, fine-to-fine and coarse-to-fine approaches. Fine-to-fine approaches start with large matching windows which become smaller from one layer to the next. In coarse-to-fine approaches the window size is fixed, but a resolution pyramid of the images is generated and the correspondence search starts in the lowest resolution. In both approaches the relation between window size and image size changes and the localization of the disparity is improved from layer to layer (Park and Inoue 1997). Figure 7.3 shows the difference between the two methods.

Coarse-to-fine approaches are suitable candidates for real-time applications. Due to the fact that the search ranges are defined by the previous layer and are strongly minimized from layer to layer computational costs can be minimized. However minor additional computations are necessary to build up the multi-resolution image pyramid and to calculate the epipolar lines for all layers if no rectification is used.

A great disadvantage of all hierarchical approaches is the error propagation from one layer to the next. This error propagation is excluded in the coarse-to-fine approach introduced by Faugeras *et al.* (1993). The stereo analysis is performed independently on all layers of the

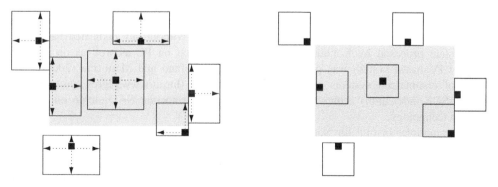

Figure 7.2 Examples of chosen windows for the adaptive window approach introduced by Kanade and Okutomi (left) and the less complex approach introduced by Bobick and Intille (right)

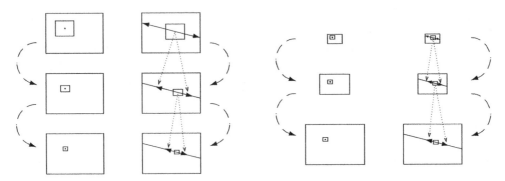

Figure 7.3 Hierarchical stereo matching:fine-to-fine (left) and coarse-to-fine approach (right)

resolution pyramid, which increases the computational complexity dramatically, especially for large disparity ranges. Reliability is calculated for each disparity in each layer to merge the disparity maps by selecting the highest level of resolution for which a valid disparity is found.

7.1.2 Fast Real-time Approaches

The focus of research in stereo reconstruction has increasingly been concentrated on the development of algorithms applicable for real-time systems. This is based on the fact that real-time algorithms are more and more feasible on standard personal computers. This is much cheaper than using expensive dedicated hardware.

Faugeras *et al.* (1993) developed a system based on dedicated DSP systems in 1993. A full search is performed using the normalized cross-correlation (NCC) as matching measure followed by a post-processing step. Later Kanade introduced an implementation on a custom DSP machine (Kanade *et al.* 1996). Up to six cameras are used in this system. A dedicated hardware board which fits on a standard PCI card in a standard PC or workstation was presented by Woodfill and Von Herzen (1997). In the field of videoconferencing a real-time system was developed by the European ACTS project PANORAMA in 1997 (Ohm *et al.* 1997). FPGAs and DSPs on a dedicated hardware board are used to achieve real-time performance for the whole system. Sarnoff Labs have developed a series of high-speed hierarchical stereo systems including the Acadia Vision Processor board (van der Wal *et al.* 2000).

Only a few algorithms achieve real-time performance without any dedicated hardware. The system proposed by Konolige (1997) is implemented on PC as well as on a DSP. Another PC-based system was presented in 2000 by Stefano and Mattoccia (2000). The series of commercial systems by Point Grey Research (http://www.ptgrey.com/products /triclopsSDK/index.html) achieve video frame rates for trinocular SAD stereo on current desktop technology.

The Fraunhofer Institut für Nachrichtentechnik, Heinrich-Hertz-Institut, developed a fast algorithm, which circumvents local search by using a hybrid recursive matching strategy based on a small number of candidate vectors (Atzpadin *et al.* 2004). It is a pure software solution running on standard PCs, which processes full-resolution CCIR 601 video (720×576 pixel) at 25 frames per second with a disparity range of around 150 pixels. The main idea of the hybrid recursive stereo matching algorithm is to unite the advantages of block-recursive disparity matching and pixel-recursive optical flow estimation in one common scheme. The block-recursive part assumes that depth does not change significantly from one image to the next and is nearly the same in the local neighbourhood. Obviously this assumption is not valid in all image areas — especially not in areas with high motion, nor at depth discontinuities. To update the results of the block-recursive stage in these areas, the pixel-recursive stage calculates the optical flow by analysing gradients and grey value differences.

In more detail, the structure of the whole algorithm can be outlined in three processing steps :

1. Three candidate vectors are evaluated for the current block position by recursive block matching.

2. The candidate vector with the best result is chosen as the start vector for the pixel-recursive algorithm, which yields an update vector.

3. The final vector is obtained by testing whether the update vector from the pixel recursive stage is of higher quality than the start vector from block-recursion.

Figure 7.4 shows this concept. Compared with former approaches, this new hybrid recursive matching has two main advantages. The recursive structure speeds up the analysis dramatically and allows the computation of disparity vectors for a full-size CCIR 601 stereo image pair on a common Pentium IV processor. The combined choice of spatial and temporal candidates yields spatially and temporally consistent disparity fields due to an efficient strategy of testing particular vector candidates. The latter aspect is very important to avoid temporal inconsistencies in disparity sequences, which may cause strongly visible and very annoying artifacts in virtual views synthesized on the basis of these disparities.

The idea of the block recursion is to use information from both the previous image and the spatial neighbourhood. This kind of recursion forces temporal and spatial consistency and

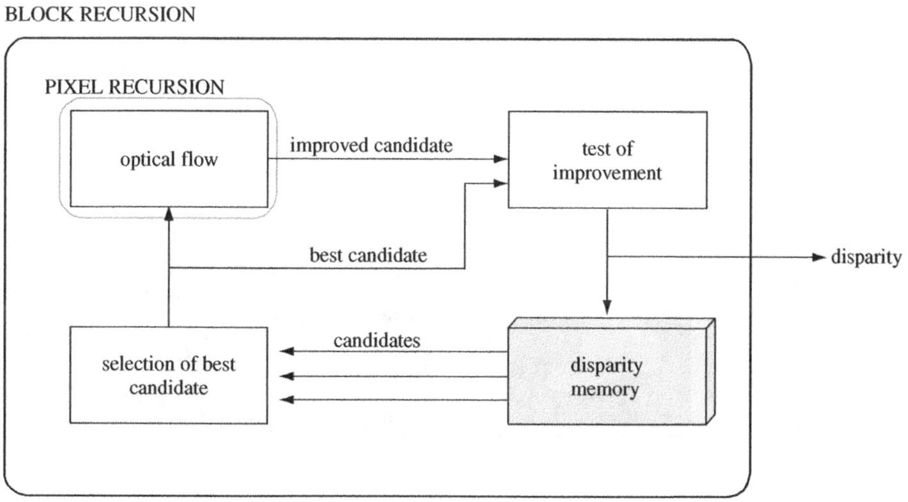

Figure 7.4 Concept of hybrid recursive matching

it additionally reduces the local search range to a few pixels. Usually, calculation of three matching scores per pixel considered is fully sufficient to achieve results comparable to a full search method. A further feature which is very important for real-time performance, is that the HRM algorithm does not use any local search around the candidate vector in the block-recursive stage. However, this feature also requires an update mechanism. Obviously, without any update the block recursive stage is not able to follow spatiotemporal deviations in the disparity field. It is therefore the task of the pixel recursion to act as a permanent update process. For this purpose it injects one new candidate vector per pixel position into the block recursion.

Pixel-recursive disparity estimation is a low-complexity method, which calculates dense displacement fields using a simplified optical flow approach. Following the principal of optical flow, an update vector $\mathbf{d}$ is calculated on the basis of spatial gradients and gradients between the two frames. The gradient between the frames is approximated by the displaced pixel difference (DPD) given by corresponding points in the left and right images.

$$\mathbf{d}(u, v) = \mathbf{d}_i(u, v) - \text{DPD}(\mathbf{d}, u, v) \frac{\text{grad } I_1(u, v)}{\|\text{grad } I_1(u, v)\|^2} \tag{7.8}$$

with

$$\text{DPD}(\mathbf{d}, u, v) = |I_1(u, v) - I_r(u + d_u, v + d_v)|.$$

To meet real-time requirements, the gradient calculation is approximated. Multiple pixel-recursive processes are started. Finally, after processing all paths of multiple pixel-recursions, the vector with the smallest DPD among all pixel-recursive processes is taken as the final update vector. Its SAD is calculated and compared with the SAD of the start vector from block-recursion. If the SAD of the update vector is smaller than the one of the start vector, the update vector is chosen as the final output vector, otherwise the start vector is retained.

Figure 7.5 shows results of the hybrid recursive matching. Smaller disparities are labelled darker than larger disparities. Block recursion is mainly responsible for spatial and temporal smoothing of disparities in areas of homogeneous depth and moderate motion. Pixel recursion considerably improves the transient spatial and temporal behaviour of the whole algorithm. However the results clearly show that post-processing for scenarios with large baselines is unavoidable due to the large error regions caused by occlusions.

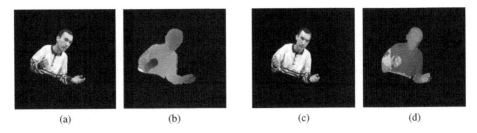

(a)   (b)   (c)   (d)

Figure 7.5 Matching results with hybrid recursive matching: (a) left; (b) left disparity; (c) right; (d) right disparity

7.1.3 Post-processing

Mismatches often occur in regions of homogeneous or periodic texture as well as in regions occluded in one view, where it is not possible to estimate disparities because there is no matching region in the other view. In recent years many techniques have been proposed to detect mismatches (Banks *et al.* 1998; Egnal *et al.* 2002):

- *Value of matching measure*: the value of the matching measure indicates the similarity of the two matching windows and is therefore an indicator of match reliability. The more similar the matching windows the more probable is a correct match.
- *Identification of homogeneous regions*: homogeneous areas do not contain enough pixel variation to be successfully matched. The calculation of the variance within a small window can be used to identify these regions. If the variance is below a certain threshold the disparity is marked as unreliable. Another possibility is the calculation of the entropy within a small window. The entropy measures the randomness of a random variable. If image texture is interpreted as random variable the entropy measures image texture (Leclerc 1989). High-texture regions have a higher entropy than low-texture regions, which means that disparities in areas with low entropy can be discarded.
- *Curvature of matching measure*: a small value of the curvature of the matching measure (no peak) around the chosen disparity identifies a bad match. Especially mismatches in homogeneous regions are detected by this criterion.
- *Peak ratio*: ideally a match should be unique. If the matching measure for the optimum and the second best optimum are close in value the match is ambiguous and it is unclear which is the better match. The ratio between the best and the second best match defines the reliability. If it is close to one a false match is detected. This is the case in periodic or low-texture regions.
- *Neighbourhood consideration*: the deviation of a disparity from the average of its neighbours defines the reliability (Cochran and Medioni 1992; Fua 1993). Small islands of bad matches can be discarded. The difficulty with this criterion is the selection of an adequate threshold which defines where disparities are discarded. Especially at depth discontinuities correct disparities can be discarded with a poorly chosen threshold.
- *Left–right consistency check*: the consistency check concatenates corresponding vectors of left-to-right and right-to-left disparity fields, taking into account that the difference between starting and end point should be close to zero or nearby (Figure 7.6). If different matches are found in each direction, the estimated correspondence is obviously wrong due to occlusions, homogeneous regions or other reasons for mismatches (e.g., periodic structures). Therefore, if the difference is greater than a predefined threshold Δ, the disparity is discarded.

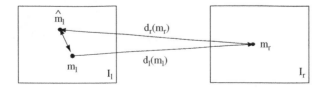

Figure 7.6 Left–right consistency check

The most promising reliability criterion for real-time applications seems to be the value of the matching measure because no additional computation is required. Unfortunately this criterion is insufficient to detect all mismatches, especially in homogeneous regions. In these regions the matching windows are quite similar in spite of the high probability for a mismatch. An additional criterion such as entropy or the peak ratio is required. However this increases the computational complexity. The curvature of the matching measure is a more promising reliability criterion for stereo analysis approaches using a full search, because few additional calculations are necessary. This criterion fails in detecting occluded regions. The left-right consistency check is used by many real-time algorithms (Faugeras *et al.* 1993; Konolige 1997; Stefano and Mattoccia 2000) due to the fact that only a few additional computations are required to detect mismatches in occluded and homogeneous regions. Figure 7.7 shows results for the left-right consistency check for small occlusions on the left (a,b) and large occlusions on the right (c,d). The threshold has been set to five pixels in this example and discarded disparities are marked in black. The effectiveness of the consistency check can clearly be seen. All the occluded regions are detected. Only reliable disparities remain in the disparity map.

The quality of disparity maps is greatly influenced by the substitution of mismatches, especially for images containing large occluded or low-structure regions. In a lot of applications a simple horizontal interpolation is sufficient to close the holes in post-processed disparity maps (Ohm *et al.* 1997). However, simple interpolation works only in regions with homogeneous depth, whereas it can produce crucial artifacts at depth discontinuities as can be seen in the right-to-left disparity map in Figure 7.8(c). Occlusion handling helps to avoid the annoying artifacts in the synthesized images (Birchfield 1999). The basic idea of this occlusion handling is to transform the positions of the disparities considered for interpolation into

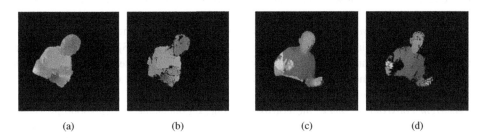

(a)   (b)   (c)   (d)

Figure 7.7 Matching result and detected mismatches for small (left a,b) and large occluded regions (right c,d)

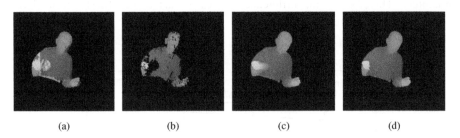

(a)   (b)   (c)   (d)

Figure 7.8 Linear interpolation and occlusion handling: (a) matching result; (b) mismatches; (c) linear interpolation; (d) occlusion handling

the other image and to compare the distances between them. If the transformed positions are significantly closer together than the positions in the starting image an occlusion is detected. The pixels inside the occluded regions are always substituted by the surrounding disparity with higher depth. Figure 7.8 (d) shows the convincing results of this simple approach.

The detection and substitution of mismatches is followed by filtering the whole disparity map in order to eliminate noise which produces undesirable flicker in calculated synthetic sequences. Using a moving average filter is only recommended for disparity maps which do not contain depth discontinuities. Many applications use a median filter to eliminate noise without smoothing depth discontinuities. A size of 3×3 seems to be appropriate for many applications due to the fact that large filter sizes eliminate details.

7.2 DISPARITY FROM THREE OR MORE CAMERAS

Trinocular stereo approaches were introduced in the mid-1980s as a more physical approach to choosing correct correspondences by adding a third camera. The trinocular epipolar constraint developed in Chapter 6 tells us that a matching pair of points in two images determines a position in the third where a corresponding feature must appear. This added constraint allows us to make better selections among possibly ambiguous peaks in two-view matching, by adding more evidence.

A typical image triple appears in Figure 7.9. We will refer to the images and their associated points and matrices using the subscripts 1, 2 and 3. Generally, one of the images is designated the reference image and disparities are calculated with respect to it. The images in the figure are from converged cameras which are mounted in a horizontal line, but many systems arrange cameras in an 'L'-shape with a right-left pair and the third camera above (or below) the left reference image. This configuration is convenient for systems using rectification because epipolar lines can be row-aligned for the horizontal pair and column-aligned for the vertical pair. Camera triples can take on a wide variety of configurations as long as the views provide sufficient overlap for matching.

If we know the trifocal tensor $\mathcal{T}_i^{jk}$ (from Chapter 6, Equation 6.52) for the trinocular camera triple, we can use a matched pair of points $(\mathbf{m}_1, \mathbf{m}_2)$ in two views to determine the corresponding point $\mathbf{m}_3$ in the third view (Figure 6.7). This is the *point transfer problem* (Hartley and Zisserman 2003). Given a line $\mathbf{l}_2$ passing through $\mathbf{m}_2$, the transferred point is:

$$p_3^k = p_1^i l_{2j} \mathcal{T}_i^{jk} \tag{7.9}$$

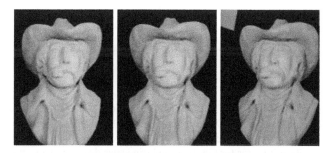

Figure 7.9 Trinocular image triple (C_1, C_2, C_3)

(See Equations 6.54 and 6.55 in Section 6.4.2). Equivalently we could triangulate the 3D point represented by the two-view correspondence and project it into the third view using the camera projection matrix P_3.

Multiview geometry can be exploited in a number of ways. For correspondence, the main issue is how to compute and optimize similarity among three or more images. There are two basic approaches to this matching problem: image-based and scene-based. The image-based approach starts with a pixel in a reference image and searches for correspondences in each of the other views. The scene-based approach refers similarity to points in the world, rather than an image. Particularly for systems with many cameras, referring to a scene model makes it easy to consider points not visible in a particular reference image. Space-carving and level-set methods, which search in the space of world points, varying surfaces until they explain the appearance in all views, can be thought of as scene-based correspondence techniques.

Many trinocular systems combine pairwise similarity metrics for the reference view paired with all other views. Some systems simply treat the triple as two independent pairs and choose the best match from all pairs for each reference pixel (Faugeras *et al.* 1993). Bhat and Nayar (1995) select the correspondence according to which pair is positioned to avoid specularities for that pixel. Another common approach is to find a set of candidate matches from a base pair, and select the best match using scores from the third image (Dhond and Aggarwal 1991; Mulligan and Daniilidis 2000). Perhaps the most robust approach uses standard area-based cost functions augmented to combine scores from the reference image paired with all other views into a single metric. Figure 7.10(a) illustrates the SSD metric for a face pixel from the centre image in Figure 7.9(b). Neither the left pair (cameras 1 and 2) nor the right pair (cameras 2 and 3) SSD profile shows a distinct minimum, however, the sum of SSD profile combining the two has a well defined minimum.

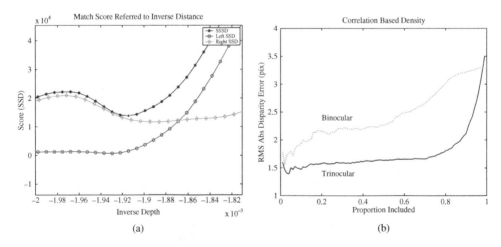

Figure 7.10 Comparison of binocular and trinocular disparity results. In (a) the SSD metric for a pixel in the reference image in Figure 7.9(b) is plotted against $1/z$. SSD for left and right views do not show distinct minima, while the combined Sum of SSD has a clear minimum; (b) illustrates a density plot of RMS absolute disparity error for binocular and trinocular disparity maps, versus proportion of pixels ordered by best correlation score

Of course computing similarity metrics, and potentially searching in a third image, adds considerable overhead, particularly for the real-time systems we are addressing. Nonetheless, high-speed commercial systems such as Point Grey Research's DigiClops (http://www.ptgrey.com/products/triclopsSDK/index.html) compute dense SAD trinocular disparity at rates of 30 frames per second at 320×240 pixel resolution for 48 disparities on current desktop systems.

7.2.1 Two-camera versus Three-camera Disparity

Chapter 6 has described some of the mathematical advantages of a third camera. Other researchers have addressed the trade-off between increased cost and improved accuracy empirically.

Dhond and Aggarwal (1991) discuss costs and benefits of a third camera for stereo. For edge-pixels in the reference image they first find candidates on the epipolar lines based on similar sign of zero crossing of $\nabla^2 G$, and edge orientation difference (eod). The candidate with the most similar edge orientation in two or three views is selected as the correspondence. The ground truth for the evaluation consists of aerial images constructed from digital elevation models. On the basis of their experiments, the authors conclude that the implemented trinocular stereo method reduces outliers (pixels with large disparity error) by one half, while increasing computational cost by one fourth. Binocular and trinocular disparity maps for the images in Figure 7.9 appear in Figure 7.11.

Mulligan *et al.* (2001) have presented ground truth analysis, comparing binocular and trinocular reconstructions to registered laser datasets. Binocular–trinocular disparity comparison is illustrated by the density plot in Figure 7.10(b). The figure plots the RMS absolute disparity error for proportions of the disparity data sorted according to a modified NCC metric. By $n\%$ disparity density we denote the highest $n\%$ of image points sorted according to their MNCC score. RMS absolute disparity difference for corresponding percentiles of points included is consistently lower for the trinocular system.

 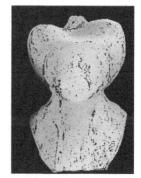

Binocular Disparity   Trinocular Disparity

Figure 7.11 Binocular versus trinocular disparity maps: the binocular map (left) has more missing disparities, indicating outliers suppressed by median filtering compared with the trinocular map (right). Maps computed using normalized cross-correlation and post-processed with median filtering

7.2.2 Correspondence Search with Three Views

As with the binocular systems described in the previous sections the common metrics for dense trinocular stereo are SAD, SSD, and NCC. The key difference involves combining these scores to exploit information from the third image. It is at the level of correspondence search that the third (or more) camera actually makes its contribution.

Many early trinocular systems were edge-based (Ayache 1991; Dhond and Aggarwal 1991). Match criterion included similarity in edge gradient, orientation, segment length and zero crossing direction. Generally a set of candidate edges or edge pixels is selected in the reference pair, and these are transferred and verified in the third image.

Fua (1993) presented an early dense correlation-based trinocular stereo algorithm. As previously described, a reference image is correlated with two or more images of the scene at several resolutions in parallel, with a fixed window size. A left–right consistency check is used to validate matches for each pair. For each pair the disparity with the highest score over all levels of detail is chosen. The pairwise disparity maps are all relative to a single reference frame so they in turn are merged by choosing the valid disparity with the highest score for each pixel. Presumably the fact that disparities are from different pairs is accounted for during reconstruction. The transfer of matches to new views is not explicitly computed, only the relative match score for a reference pixel over multiple disparity maps is used for match selection. Faugeras *et al.* (1993) present a real-time version of this system. They exploit the epipolar geometry further by using a triple of cameras mounted in an L-shape, which allows aligning the columns of the up–down pair and the rows of the left–right pair to optimize search. On special purpose hardware of the era, the system achieves four frames per second at 256×256 pixel resolution and 32 disparities.

Another early dense multiview stereo system was the multiple-baseline stereo of Okutomi and Kanade (1993). This was the first system to combine SSD scores from multiple pairs of images (with a variety of baselines) using a sum of SSD (SSSD) metric referred to $1/z$ which is called the SSSD-in-inverse-distance metric. The authors point out that correlation scores cannot be combined based on raw disparities for independent pairs. Given the SSSD-in-inverse-distance metric, matching proceeds as usual. The advantage over methods which select candidate matches from one or more stereo pairs and cross-validate them, is that the combined metric includes all of the available information and false intermediate matches will not influence the result. For example we can see in Figure 7.10(a) that neither pairwise SSD score has a clear minimum at the point indicated by SSSD. This basic method has been implemented in special purpose hardware as a video-rate stereo machine (Kanade *et al.* 1996).

The National Tele-Immersion Initiative (Lanier 2001; Mulligan *et al.* 2004) spawned a sequence of high-speed stereo systems, including a number of trinocular stereo approaches. The main trinocular system (Mulligan *et al.* 2001) uses the modified NCC (MNCC) metric from Equation (7.7) to generate five depth images, from a set of seven cameras used as overlapped triples. Each triple is rectified as two separate pairs, the (right) reference pair (C_2, C_3) and the left pair (C_2, C_1). Transfer to C_1 is precomputed for a range of right pair disparities for each pixel in the reference image C_2. The resulting left disparities are used to estimate a linear model for the disparity $\widehat{d}_L$ in the left pair parameterized by the pixel location (u_2, v_2) and disparity d_R in the right pair. As the search proceeds in the reference image, for each (u_2, v_2, d_R) the MNCC score is computed, then the left disparity is estimated $\widehat{d}_L = M(u_2, v_2)d_R + b(u_2, v_2)$. The sum of MNCC profile is computed as

SMNCC $= MNCC(u_2, v_2, d_R) + MNCC(u_2, v_2, \widehat{d_L})$. As in Okutomi and Kanade (1993), the selected disparity is determined using the extremum of the SMNCC. This system computes disparity frames for one camera triple at 15–20 frames per second at 320×240 pixel resolution for 64 disparities on current desktop technology.

A group of multiview stereo techniques which are typically not real-time are those which exploit iterative optimization methods. Level set methods refine an initial world surface by iterating partial differential equations derived from the constraints of the problem (Zhao *et al.* 2001). They automatically accommodate changes in topology as surfaces evolve and allow authors to relax some of the assumptions inherent in correlation-based approaches. The key for multiview stereo is to formulate equations which capture the imaging constraints and deform the estimated world surface to match its appearance in a set of calibrated views (correspondence). Faugeras and Keriven (1998) incorporate the normalized cross-correlation metric, but account for the fact that surface patches are unlikely to be frontoparallel to all cameras. Jin *et al.* (2003) describe their approach as correlating images to models rather than images to images. They attempt to eliminate the Lambertian assumption inherent in correlation by adopting a diffuse plus specular model of scene radiance. Under this model a matrix R, composed of intensities from the projection of a tessellated surface patch into each view, should have rank two. The differential equations used for iteration are based on the difference between the observed R and the current specular and diffuse components for the estimated surface.

The topic of volumetric reconstruction from multiple views will be addressed in Chapter 8. We will only mention here that space-carving approaches (Kutulakos and Seitz 2000; Seitz and Dyer 1999) exploit similar constraints on multi-camera geometry to those employed in multiview stereo. Effectively they search along the ray from a reference image pixel, projecting world points represented by voxels to a set of image views and keeping those voxels which produce the same appearance in all views (correspondences). The similarity or consistency metric is computed on the set of pixels from all views about the current projected point. Measures such as standard deviation or a likelihood ratio test are used to determine consistency.

7.2.3 Post-processing

Many of the post-processing techniques described in Section 7.1.3 are used in trinocular systems as well. One special claim for systems which use more than two views is that occlusion problems are reduced when regions are occluded in one view, but visible in several others. The truth of this claim depends on the approach the system takes to correspondence. For example, if it depends on hypotheses from a reference pair, then any regions semioccluded in that pair, are unlikely to be matched correctly in spite of being visible in the third image.

Kang *et al.* (2001) address this question of pixels visible in some, but not all images in a multiview stereo system. They propose using shiftable windows similar to the approach of Bobick and Intille (1999), combined with temporal selection. Temporal selection means computing a similarity metric such as SSD for all views, then choosing a selection (50%) of views with the lowest scores and summing the scores. Presumably a pixel's score will be poor for pairs in which it is semioccluded, and better for pairs where it is fully visible. Effectively this is an SSSD approach, which discards scores which are too large.

7.3 CONCLUSION

Much of the 3D information which can be derived from multiview systems depends on identifying the projections of world points in two or more views. The critical question is how to measure similarity between image features or regions and choose the best match. Determining these corresponding image points is a challenging and active research problem. This chapter has reviewed the basic definitions and approaches for stereo analysis for two or more cameras, as well as describing many of the seminal systems in the area.

Computing robust disparity maps in real-time requires exploiting constraints such as restricting correspondence search to 1D along the epipolar line, enforcing unique matches and assuming the scene contains smooth surfaces. Area-based techniques have proven effective in generating dense disparity maps at high frame rates, even on common desktop computers. Unfortunately they are unreliable in image regions with low or repeated texture, and at occlusion boundaries and half-occluded regions. Bad matches can be reduced by post-processing using techniques such as right–left checking, region entropy and peak sharpness in the similarity metric.

For systems with three or more cameras similarity metrics must combine evidence from all views. For calibrated systems we can use the trinocular epipolar constraint to transfer a match in one image pair to a third image. Knowledge of camera geometry allows us to refer the metric to a reference image or to world scene points. We can think of these systems as searching in space of possible surfaces which have projections matching the images (scene-based), or as a straightforward comparison among images (image-based). Post-processing particular to multi-camera systems consists of discarding evidence from images for which the considered scene point is occluded.

REFERENCES

Anandan P 1989 A computational framework and an algorithm for the measurement of visual motion. *International Journal of Computer Vision* **2**, 283–310.

Atzpadin N, Kauff P and Schreer O 2004 Stereo analysis by hybrid recursive matching for real-time immersive video conferencing. *IEEE Transactions on Circuits and Systems for Video Technology* **14**(3), 321–334.

Ayache N 1991 *Artificial Vision for Mobile Robots: Stereo Vision and Multisensory Perception.* MIT Press, Cambridge, MA.

Banks J, Bennamoun M, Kubik K and Corke P 1998 Evaluation of new and existing confidence measures for stereo matching. *Proceedings of the Image and Vision Computing NZ conference (IVCNZ'98)*, Auckland, NZ, pp. 252–261.

Barron J, Fleet DJ and Beauchemin S 1994 Performance of optical flow techniques. *International Journal of Computer Vision* **12**(1), 43–77.

Bhat DN and Nayar SK 1995 Stereo in the presence of specular reflection. *Proceedings of the 5th International Conference on Computer Vision,* pp. 1086–1092.

Birchfield S 1999 *Depth and Motion Discontinuities* Department of Electrical Engineering, Stanford University.

Bobick AF and Intille SS 1999 Large occlusion stereo. *International Journal of Computer Vision* **33**(3), 181–200.

Böröczky L 1991 *Pel-recursive Motion Estimation.* Department of Electrical Engineering, Delft University of Technol.

Cochran S and Medioni G 1992 3-d surface description from binocular stereo. *IEEE Transactions on pattern analysis and machine intelligence* **14**(10), 981–994.

Dhond UR and Aggarwal JK 1991 A cost-benefit analysis of a third camera for stereo correspondence. *International Journal of Computer Vision* **6**(1), 39–58.

Egnal G, Mintz M and Wildes R 2002 A stereo confidence metric using single view imagery. *Proceedings of Vision Interface,* pp. 162–170.

Faugeras O and Keriven R 1998 Variational principles, surface evolution, pde's, level set methods, and the stereo problem. *IEEE Transactions on Image Processing* **7**(3), 336–344.

Faugeras O, Vieville T, Theron E, Vuillemin J, Hotz B, Zhang Z, Moll L, Bertin P, Mathieu H, Fua P, Berry G and Proy C 1993 Real-time correlation-based stereo: algorithm, implementations and applications. *Technical Report RR-2013,* INRIA-Sophia Antipolis.

Fau P 1993 A parallel stereo algorithm that produces dense depth maps and preserves image features. *Machine Vision and Applications* **6**, 35–49.

Hartley R and Zisserman A 2003 *Multiple View Geometry in Computer Vision,* 2nd edn. Cambridge University Press, Cambridge, UK.

Horn BKP and Schunck BG 1981 Determining optical flow. *Artificial Intelligence* **17**, 185–204.

Jin H, Soatto S and Yezzi A 2003 Multi-view stereo beyond lambert. *Proceedings of 2003 IEEE Computer Society Conference on Computer Vision and Pattern Recognition,* vol. 1, pp. I-171–I-178.

Kanade T and Okutomi M 1994 A stereo matching algorithm with an adaptive window: Theory and experiment. *IEEE Transactions on Pattern Analysis and Machine Intelligence (PAMI)* **16**(9), 920–932.

Kanade T, Yoshida A, Oda K, Kano H and Tanaka M 1996 A stereo machine for video-rate dense depth mapping and its new applications. *Proceedings of the 15th Computer Vision and Pattern Recognition Conference (CVPR '96),* pp. 196–202.

Kang SB, Szeliski R and Chai J 2001 Handling occlusions in dense multi-view stereo. *Proceedings IEEE Computer Society Conference on Computer Vision and Pattern Recognition (CVPR'01),* Kauai, HI, vol. I, pp. 103–110.

Konolige K 1997 Small vision systems: Hardware and implementation. *Proceedings of the Eigth International Symposium on Robotics Research (Robotics Research 8),* Hayama, Japan, pp. 203–212.

Kutulakos KN and Seitz SM 2000 A theory of shape by space carving. *International Journal of Computer Vision* **38**(3), 199–218.

Lanier J 2001 Virtually there. *Scientific American,* April, pp. 66–75.

Leclerc YG 1989 Constructing simple stable descriptions for image partitioning. *International Journal of Computer Vision* **3**(1), 73–102.

Levine MD, O'Handley DA and Yagi GM 1973 Computer determination of depth maps. *Computer Graphics and Image Processing* **2**, 131–150.

Marr D and Poggio T 1979 A theory of human stereo vision. *Proceedings of the Royal Society London* **B204**, 301–328.

Moravec H 1980/81 Robot rover visual navigation. *Computer Science: Artificial Intelligence,* No. 3, pp. 105–108.

Mulligan J and Daniilidis K 2000 Trinocular stereo for non-parallel configurations. *Proceedings of the 15th International Conference on Pattern Recognition,* vol. 1, Barcelona, Spain. pp. 567–570.

Mulligan J, Isler V and Daniilidis K 2001 Performance evaluation of stereo for tele-presence. *Proceedings of the 8th IEEE International Conference on Computer Vision (ICCV'01),* vol. 2, Vancouver, BC, Canada, pp. 558–565.

Mulligan J, Zampoulis X, Kelshikar N and Daniilidis K 2004 Stereo-based environment scanning for immersive telepresence. *IEEE Transactions on Circuits and Systems for Video Technology* **14**(3), 304–320.

Negal HH and Enkelmann W 1986 An investigation of smoothness constraints for the estimation of displacement vector fields from image sequences. *IEEE Transaction on Pattern Analysis and Machine Intelligence (PAMI)* **8**, 565–593.

Ohm J, Grüneberg K, Izquierdo E, Hendriks MKE, Redert A, Kalivas D and Papadimatos D 1997 A realtime hardware system for stereoscopic videoconferencing with viewpoint adaptation. *Proceedings of International Workshop on Synthetic – Natural Hybrid Coding and Three Dimentional Imaging,* Rhodes, pp. 147–150.

Okutomi M and Kanade T 1993 A multiple-baseline stereo. *IEEE Transactions on Pattern Analysis and Machine Intelligence (PAMI)* **15**(4), 353–363.

Park JI and Inoue S 1997 Hierarchical depth mapping from multiple cameras. *Proceedings of International Conference on Image Analysis and Processing ICIAP,* pp. 685–692.

Seitz SM and Dyer CR 1999 Photorealistic scene reconstruction by voxel coloring. *International Journal of Computer Vision* **35**(2), 151–173.

Stefano LD and Mattoccia S 2000 Real-time stereo for a mobility aid dedicated to the visually impaired. *Proceedings of the 6th International Conference on Control, Automation, Robotics and Computer vision (ICARCV 2000),* Singapore.

van der Wal G, Hansen M, and Piacentino M 2000 The acadia vision processor. *IEEE Proceedings of International Workshop on Computer Architecture for Machine Perception,* Padua, Italy, pp. 31–40.

Woodfill J and Herzen BV 1997 Real-time stereo vision on the parts reconfigurable computer. *Proceedings IEEE Symposium on Field-Programmable Custom Computing Machines,* Napa, CA, pp. 201–210.

Zabih R and Woodfill J 1994 Non-parametric local transforms for computing visual correspondence. *Proceedings of the 3rd European Conference on Computer Vision,* Stockholm, pp. 151–158.

Zhao HK, Osher S and Fedkiw R 2001 Fast surface reconstruction and deformation using the level set method. *Proceedings of the 1st Workshop on Variational and Level Set Methods in Computer Vision (VLSM'01),* pp. 194–201.

8

Reconstruction of Volumetric 3D Models

Peter Eisert

Fraunhofer Institute for Telecommunications/Heinrich-Hertz-Institut, Berlin, Germany

8.1 INTRODUCTION

Photo-realistic 3D computer models of real objects and scenes are key components in many 3D multimedia systems. The quality of these models often has a large impact on the acceptability of applications such as virtual walk-throughs (e.g., city guides or virtual museums), caves, computer games, product presentations in e-commerce, or other virtual reality systems. Although the rendering of 3D computer scenes can often be performed in real-time even on hand-held devices, the creation and design of 3D models with high quality is still time-consuming and thus expensive. This has motivated the investigation of a large number of methods for the automatic acquisition of textured 3D geometry models from multiple views of an object.

The large body of work devoted to this problem can basically be divided into two different classes of algorithms. The first class of 3D model acquisition techniques computes depth maps from two or more views of the object. Here, depth is estimated from changes in the views caused by altering such properties as position of cameras (shape-from-stereo, shape-from-motion), focus (shape-from-focus, shape-from-defocus), or illumination (shape-from-shading, photometric stereo). For each view, a depth map can be computed that specifies for each pixel the distance of the object to the camera. Since the object is only partially represented by a single depth map, due to occlusions, multiple depth maps must be registered into a single 3D surface model.

This registration process can often be avoided in a second class of reconstruction methods which relies on a volumetric description of the scene. In the simplest case, the space containing the object or scene to be reconstructed is equidistantly divided into small cubical

3D Videocommunication — Algorithms, concepts and real-time systems in human centred communication
Edited by O. Schreer, P. Kauff and T. Sikora   © 2005 John Wiley & Sons, Ltd

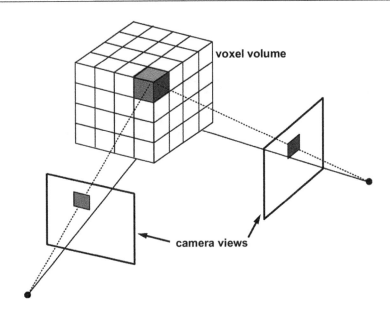

Figure 8.1 Representation of a scene by a 3D array of volume elements (voxels)

volume elements called *voxels* (Figure 8.1). If a particular voxel does not belong to the object it is set transparent, whereas voxels within the object remain opaque and can additionally be coloured. Thus, the entire scene is composed of small cubes approximating the volume of the objects. The finer the discretization of the 3D space, the more accurate the shape.

One advantage of volumetric representations for reconstruction purposes is the simple joint consideration of all available views. Once the cameras are calibrated, each voxel can be projected into all views as shown in Figure 8.1 and the information of the corresponding pixels can determine whether the voxel belongs to the object or not. Instead of fusing multiple depth maps computed, e.g., from pairs of images, the joint computation leads to 3D computer models that are consistent with all views. Since fewer smoothness constraints are incorporated, volumetric methods can often be well exploited to reconstruct fine structures with multiple occlusions.

This chapter focuses on methods for the reconstruction of volumetric computer models from multiple camera views. The emphasis is on vision-based applications where the 3D representation usually describes a coloured sampled surface. The interior of the object is here of less importance. Other applications for volumetric data that exploit the entire volume for visualization (e.g., computer graphics or medical applications such as computer tomography) are not considered in the following. The chapter is organized as follows. First, the important class of *shape-from-silhouette* algorithms is reviewed. Due to its simplicity, robustness, and efficiency, this method is very popular in many multimedia applications (Goldlücke and Magnor 2003; Grau 2003; Gross 2003; Wu and Matsuyama 2003). One drawback of shape-from-silhouette methods is their inability to recover concavities in surfaces. If colour information is additionally considered it is possible to recover such surfaces as well. Volumetric algorithms that exploit colour information are, e.g., space-carving algorithms described in Section 8.3 and the image cube trajectory analysis presented in Section 8.4.2.

8.2 SHAPE-FROM-SILHOUETTE

Shape-from-silhouette or *shape-from-contour* is a class of algorithms for 3D scene reconstruction that uses the outline of the objects to recover their shape. Especially under controlled
situations, e.g., in studio scenarios, the silhouette of an object can be determined very reliably.
Therefore, these methods are very robust against lighting changes or photometric variations
between cameras which are critical for other algorithms based on the brightness constancy
assumptions (Horn 1986).

The concept of shape-from-silhouette dates back to the early seventies (Baker 1977; Greenleaf *et al.* 1970; Martin and Aggarwal 1983) and was initially used for medical applications.
The basic idea of these approaches is that any object must be entirely located somewhere
within its contour. If an object is viewed from a particular known position under perspective
projection the rays from the focal point through the silhouette contour form the hull of a
viewing cone. This viewing cone, illustrated on the left-hand side of Figure 8.2, defines
an upper bound for the object shape — the correct object volume is definitely less or equal
than this rough approximation. In this consideration, no assumption about the viewing position is made except for the knowledge of this calibration data. For any viewing position,
the silhouette defines a viewing cone which entirely contains the object. Since the object
volume is bounded by all particular cones, it must also reside within the intersection of these
viewing cones. Only points in the 3D space that are inside all viewing cones may belong to
the object to be reconstructed. This is shown on the right-hand side of Figure 8.2.

The algorithm starts without any restrictions on the object shape. A large bounding
volume enclosing the entire scene can be used as initialization. Multiple views from different
viewing positions are captured. Each view provides a viewing cone which is intersected
with the current object shape approximation, leading to a volume with increasing accuracy.
However, not all possible shapes can be reconstructed with shape-from-silhouette methods.
The concavity, e.g., on the right-hand side of Figure 8.2 is never visible in the silhouette
(independent of the viewing direction) and cannot be recovered. Instead of the true shape, only
the *visual hull* (Laurentini 1994) can be estimated. However, for many practical applications
this leads to sufficiently accurate representations. In order to model, e.g., small bumps
in the surface, additional information such as colour has to be evaluated as it is done in

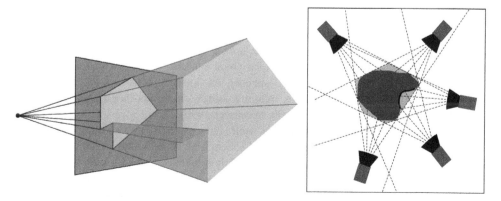

Figure 8.2 *Left*: viewing cone through the image silhouette containing the object. *Right*: intersection
of multiple viewing cones

space-carving or *voxel coloring* approaches described in Section 8.3. The entire procedure of shape-from-silhouette consists of the following steps:

- Calibration of the cameras to determine position, orientation, and intrinsic camera parameters
- Segmentation of the object from the background in the captured images to derive the object contour
- Intersection of all viewing cones

The intersection of the viewing cones can be computed exactly with polyhedral representations (Matusik *et al.* 2001). But most often, a volumetric discretization of the space is used, as shown in Figure 8.1. For a particular voxel with given 3D position, its projection into the camera frames is computed. If the voxel falls outside the silhouette in at least one view, it is discarded from the volume. After all voxels are processed, the remaining volume elements approximate the visual hull. Although some aliasing effects may occur, this discrete volume intersection is very popular due to its low complexity. With current technology, even real-time reconstruction is possible (Goldlücke and Magnor 2003; Wu and Matsuyama 2003), enabling new interactive 3D applications in computer vision and graphics.

8.2.1 Rendering of Volumetric Models

Once a 3D volume is reconstructed, it can be viewed from different directions. For rendering, a colour can also be assigned to each voxel. This colour information is extracted from those camera views, where the corresponding voxel is visible. In order to consider occlusions of multiple voxels, a z-buffer may be added for rendering. Each voxel is then projected into the image plane and modifies the pixel colour if it is closer to the virtual camera than the current value stored in the z-buffer. Different approaches can be classified by the direction of volume traversal (front-to-back, back-to-front, or arbitrary traversal).

The simplest approach in voxel rendering is to assume a voxel to be infinitesimally small and to use it to exactly colour one pixel. However, dependent on the viewing distance and the discretization of the image and volume space, this method can lead to aliasing or holes in the surface. Better results can be obtained if the finite size of the projected voxel is considered. Figure 8.3 shows the projection of a cube into the image plane that results in a 2D polygon of up to six corners that may cover multiple pixels.

Due to the discrete nature of pixels, special attention must be paid to the boundary of the polygon in order to avoid aliasing. Since the use of six-sided polygons for each voxel is computationally demanding, the footprint of a voxel can also be approximated by simpler shapes. For example, the bounding box, spheres, or Gaussian functions can be used for projection leading to techniques such as splatting, surfels, and point-based rendering (Pauly *et al.* 2003; Rusinkiewicz and Levoy 2000; Westover 1990).

Often, the voxel volume is not rendered itself, but converted into a triangle mesh that can efficiently be handled by a graphics card (Niem and Wingbermühle 1997). Modeling the outer surface of the cubical structures with polygons may lead to rough surfaces with discrete steps. Better results can be obtained using *marching cubes* (Lorensen and Cline 1987) or *marching intersections* (Tarini *et al.* 2002) methods that use a local neighborhood of voxels to derive oriented surface patches. After smoothing and mesh reduction, efficient representations of the 3D objects can be created. Figure 8.4 shows an example of a vase

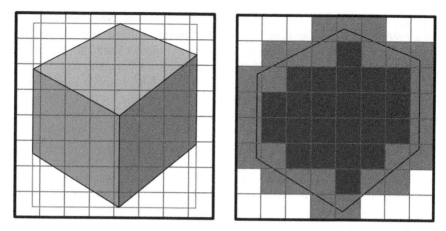

Figure 8.3 *Left*: voxel projection onto image grid. *Right*: voxel footprint and pixel contribution

Figure 8.4 *Left*: voxel volume of a Peruvian vase reconstructed from 72 images. *Middle*: triangle mesh derived from the voxel volume. *Right*: textured 3D model

reconstructed with shape-from-silhouette techniques and rendered as a voxel volume (left), a reduced triangle mesh (middle), and as a textured 3D model.

8.2.2 Octree Representation of Voxel Volumes

Volumetric descriptions tend to impose high demands on memory due to their three dimensions. Therefore, efficient representations of large voxel volumes are essential. Mostly, *octrees* are used in order to represent the 2D surface with a 3D voxel grid structure (Chien and Aggarwal 1986; Potmesil 1987; Srivastava and Ahuja 1990; Szeliski 1993; Veenstra and Ahuja 1986). In this case, an octree is a tree that hierarchically defines the object shape starting with a root node that covers the entire bounding volume. Each cube represented by a node in the tree is then successively subdivided into eight smaller cubes (dependent on the contour information) with the cube's edge lengths halved. Further subdivision of a cube is stopped if it either lies completely inside or outside the object.

Figure 8.5 shows an example for such an octree subdivision. The volumetric object on the left is defined by the tree on the right. Each node in this tree specifies if one of its sub-cubes lies completely inside (black), completely outside (white), or on the object surface (grey). In the latter case, the corresponding voxel is further sub-divided until a maximum layer defining the highest resolution is reached. In this way, it is assured that the object surface is accurately modeled while regions outside or inside the object can be efficiently described by large cubes. Figure 8.6 gives an example of a 3D object represented with different maximum octree levels. For practical scenarios, drastic memory savings can be achieved compared to the full 3D voxel volume with the same resolution.

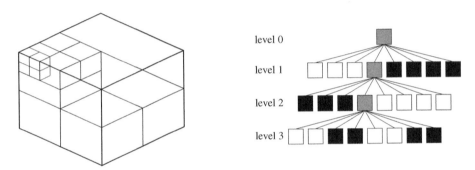

Figure 8.5   *Left*: octree representation of an object. *Right*: tree structure defining subdivision

Figure 8.6   voxel volume of the vase in Figure 8.4 rendered up to octree level 5, 6, and 7, respectively

For hierarchical shape-from-silhouette, an octree can be constructed bottom-up or top-down. If reconstruction is started from level 0 with recursively refining the object, the spatial extension of the cube must be considered for projection and the footprint of the voxel, as illustrated in Figure 8.3, need to be computed. The intersection of all viewing cones in the shape-from-silhouette procedure can be performed as follows (Smolic *et al.* 2004). A voxel of a particular resolution level that is completely inside the silhouettes of all views belongs to the object and is not further subdivided. Similarly, a cube whose footprint is completely outside the silhouette of at least one view is marked as outside and is not further refined. All other voxels are subdivided into eight sub-cubes and the process is continued on all sub-cubes until a predefined resolution is reached.

8.2.3 Camera Calibration from Silhouettes

For shape-from-silhouettes algorithms, the accuracy of the reconstructed geometry is considerably affected by the knowledge of the true camera parameters, which requires an accurate camera calibration. Deviations from the correct values can lead to incorrect 3D models, since valid object parts might be cut off during viewing cone intersection. As a result, the silhouette of the reconstructed object is always smaller than or equal to the true contour. However, this unwanted effect can be used to refine the camera parameters again. In Grattarola (1992), pairs of images are used and the viewing cone of one image contour is projected into the other image plane, computing silhouette mismatches. Minimization of these mismatches over all camera parameters optimizes the calibration. Similarly, Niem (1998) minimizes the deviation of the back-projected silhouette of the reconstructed object to the true silhouette in the camera images. Both approaches have in common, that a nonlinear optimization in a high dimensional space has to be performed.

For the particular case of turntable scenarios where shape-from-silhouette is often used, additional constraints apply that simplify the optimization of camera parameters. Instead of placing multiple cameras around the object, a single camera is used capturing images while the object slowly rotates. The rotation angle between two shots can usually be accurately controlled, whereas the position of the camera relative to the turntable is in general unknown and requires camera calibration. The knowledge of circular motion adds severe constraints which increase robustness and result in a very low-dimensional parameter space that has to be searched. If the size of the reconstructed object need not be recovered three parameters are sufficient to determine the entire extrinsic camera geometry (Eisert 2004). Similar constraints are also utilized in Fitzgibbon *et al.* (1998), where extrinsic and intrinsic camera parameters are derived from feature point correspondences instead of silhouette.

Figure 8.7 shows the result from a camera refinement exploiting silhouette mismatches. The left hand side of this figure is an original camera frame from a turntable sequence of a small tree. Mismatches (white pixels) between the silhouette of the camera frame and the silhouette of the reconstructed object with only inaccurate initial calibration are depicted in the middle. These deviations can be exploited in order to optimize the camera parameters leading to the smaller deviations shown on the right. Due to the improved camera parameters, the accuracy of the reconstructed 3D model is highly increased by this preprocessing step, making initial camera calibration less important or even unnecessary.

Figure 8.7 Geometry deviations for a turntable sequence. *Left*: original camera frame. *Middle*: silhouette mismatch between original and synthetic view. *Right*: silhouette error after refinement

8.3 SPACE-CARVING

In shape-from-silhouette methods, binary masks specifying the 2D object contours in the images are used as input in order to estimate the 3D object shape. Colour information is not exploited. This leads to very fast and robust algorithms, but not all object shapes can be reconstructed accurately. The recovered object is bounded by the visual hull — small dents and concavities cannot be determined. In the extreme case, for example the reconstruction of rooms in indoor environments with centred, outward-facing cameras, no silhouette information is available and the reconstruction fails. This limitation can be overcome if colour information from the images is used in addition to the binary contour masks. *Voxel colouring* (Seitz and Dyer 1997), *multi-hypothesis volumetric reconstruction* (Eisert *et al.* 1999), and *space-carving* (Kutulakos and Seitz 2000) belong to this group of algorithms that exploit colour in order to improve the reconstruction accuracy.

The benefit of exploiting surface colour information is exemplarily illustrated in Figure 8.8. The image shows an object with a small concavity which cannot be detected using contour information only. Textured surfaces, however, allow to distinguish between the correct shape and the visual hull. Imagine a point M_1 lying on the visual hull and being projected on particular pixels of cameras A and B, respectively. Since the true shape contains a concavity at that position, the pixels in both cameras are coloured from two different points on the surface M_2 and M_3 instead of M_1. If the surface is textured these points and therefore the pixels may have different colours. Thus, point M_1 cannot belong to the true object volume and can be removed. Space-carving methods use that phenomenon in order to carve out concavities from the visual hull, like a sculptor from a block of stone, until the object surface looks correct in all views.

Shape-from-silhouette approaches reconstruct a binary voxel volume that has the same silhouettes as in the camera images. Similarly, space-carving or voxel colouring methods create a coloured voxel representation, where the volume reprojections into the image planes show the same pixel colours as the original views. Again, this need not be the correct geometry of the object. It is only assured that the rendered views are *photo-consistent*

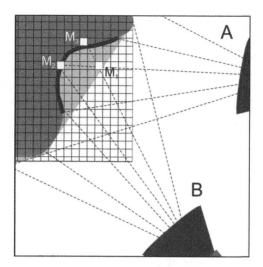

Figure 8.8   Refinement of object shape using surface colour

(Seitz and Dyer 1997) with the camera views. There are multiple possible geometries that all lead to photo-consistent views. Since generally no smoothness assumptions are made, there are different voxel configurations (with mostly local variations at the object surface) that result in the same projections into the finite number of available views. The union of all photo-consistent configurations is called the *photo-hull*, following the concept of the visual hull.

The basic principle of all volumetric colour reconstruction methods is quite similar. All voxels in the volume are usually processed one after the other. For each voxel, its projection into all available views is calculated and, depending on the pixel colours at the corresponding 2D image positions and considering visibility, it is determined whether the voxel belongs to the object surface or not. If it lies on the surface, the voxel colour is computed from the pixel colours in those views, where the voxel is visible. In the other case, the voxel is set transparent and the algorithm proceeds with the remaining volume elements until the reconstructed object looks correct from all available viewing directions. Thus, all available views are considered simultaneously without the need of determining point correspondences or fusing multiple depth maps into a common 3D model.

Although the principle of projecting voxels into all available views is common to a wide class of algorithms, there are many differences in the way the voxel volume is traversed, how visibility is considered, and how photo-consistency is checked. For the consistency test, the voxel is projected into all visible views and the pixel colours are compared. In the ideal case, all these colours should be the same, since they represent the same surface point. In practice, however, camera noise, mismatch in calibration data, or non-Lambertian surface reflections lead to variations between the views. This can be considered by allowing some colour deviations for the test. If this range is too small, holes can occur in the model, if the threshold is too large, the resulting voxel object might be too large.

Dependent on the camera configurations, different voxel access schemes can be applied. If the convex hulls of all cameras' focal points lie completely outside the objects, the voxel volume can be traversed in a single sweep as is done in voxel colouring (Seitz and Dyer

1997). In this case, a voxel access order can be defined that assures that visibility of a voxel can be determined by considering only already processed ones. For general camera configurations, this is no longer possible. Here, camera views are grouped according to their position and an iterative scheme is applied on the voxel volume, that carves away voxels at the surface until photo-consistency is reached (Culbertson *et al.* 1999; Eisert *et al.* 1999; Kutulakos and Seitz 2000). Similar to the shape-from-silhouette methods, many extension can be applied, for example the use of octrees to deal with large environments (Prock and Dyer 1998), or to consider extended voxels for reducing aliasing effects (Steinbach *et al.* 2000).

Figure 8.9 shows examples of objects reconstructed with the multi-hypothesis volumetric reconstruction method described in Eisert *et al.* (1999). The top left image shows one frame of the *Penguin* sequence recorded in CIF resolution. The algorithm was applied on 65 unsegmented frames, leading to the 3D coloured voxel representation shown on the upper right of Figure 8.9. Since the bounding box of the volume does not include the background objects, an inherent segmentation is achieved. In the second example shown on the lower row of Figure 8.9, an indoor scene of an office is reconstructed from 27 views with no available silhouette information. Although shape-from-silhouette methods would fail, the use of colours allows a photo-consistent reconstruction of the room, as shown on the right-hand side of Figure 8.9.

Figure 8.9 *Upper left*: original view of the *Penguin* sequence. *Upper right*: reconstructed view. *Lower left*: View of an indoor sequence. *Lower right*: reconstructed view

8.4 EPIPOLAR IMAGE ANALYSIS

Similar to space-carving, *epipolar image analysis* is a technique for the reconstruction of object shape using colour information from multiple calibrated views of the scene. Instead of starting from the 3D volume by projecting single voxels into the available views, epipolar image analysis makes use of particular structures in 2D images to derive 3D information. For that purpose, the sequence of images is usually collated to form a 3D volume, the so called *image cube* (Criminisi *et al.* 2002). In this volume, shown on the left hand side of Figure 8.10, vertical slices represent the original camera views.

Since the camera moves during acquisition of the slices, object points also change their position in the image cube. For the particular case of a linear and purely horizontal camera movement described in the next section, object point trajectories form lines in the image cube which can be detected easily with image processing techniques. 3D information is then derived from the line parameters. Usually, a large number of views is jointly exploited which allows the inherent consideration of multiple occlusions as well as homogeneous regions. The initial method is restricted to linear equidistant camera movements, but can be extended to other camera configurations as well. Section 8.4.2 describe a generalization to circular setups occurring, e.g., in turntable acquisition or concentric mosaics (Shum and He 1999).

8.4.1 Horizontal Camera Motion

The most important case of epipolar image analysis is the case of an equidistant linear camera movement parallel to the horizontal axis of the image plane (Bolles *et al.* 1987). This constraint forces all *epipolar lines* to be purely horizontal. Moreover, all epipolar lines of a particular object point coincide and remain at the same vertical position. As a result, 3D reconstruction can be performed on horizontal slices through the image cube, since all projections of 3D object points remain in the same slice throughout the entire sequence. These horizontal slices shown on the right hand side of Figure 8.10 are called *epipolar images*.

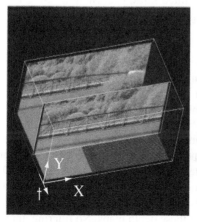 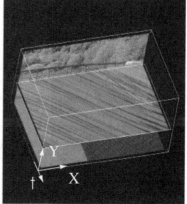

Figure 8.10 Image cube representation of an image sequence. *Left*: time slices represent camera images. *Right*: horizontal slices are epipolar images

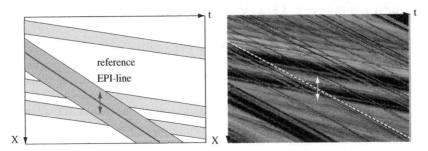

Figure 8.11 Linear camera movement. *Left*: occlusion of EPI lines *Right*: line structure of an epipolar image

These epipolar images represent the trajectories of object points. If the camera is moved equidistantly, the path of an arbitrary 3D point becomes a straight line, called *EPI line* (Figure 8.11, right). For each object point in the image sequence, exactly one corresponding line can be detected in the epipolar image. In order to determine depth of the points, the slope of these lines is analyzed. Object points that are closer to the camera move faster through the image (large disparities), whereas projections of points further away change their position much slower. Therefore, the slopes of nearby points are higher than those of distant points and, as a result, they can be directly used to derive scene depth. Since the shape of a trajectory is explicitly known to be a straight line, conventional edge detection methods can be applied to find them.

Some points, however, may not be visible throughout the entire sequence, but are occluded by other object parts. The corresponding EPI lines are therefore also partly occluded, as shown on the left-hand side of Figure 8.11. Fortunately, there is a particular structure in these occlusions which can be considered for efficient occlusion handling. Object points can be occluded only by others which are closer to the camera and thus have a larger slope in the epipolar image. As a result, lines with larger slope always occlude those with a smaller one. By searching the epipolar image for lines with large slopes first, removing those lines from the image, and continuing the search with less tilted lines, occlusion handling is inherently performed. The entire process of epipolar image analysis is thus as follows:

- Calibrate the camera
- Process all epipolar images
- In each epipolar image, search for lines, starting with lines having large slopes
- Remove those lines and continue search for lines with smaller slope
- Compute depth of object point from slope of line

Since the entire length of the line is exploited to determine depth of the corresponding object point, accurate values can be determined, even in the presence of multiple occlusions. Figure 8.12 shows some results obtained with epipolar image analysis; 200 frames of a synthetic image sequence of CIF resolution are collated to an image cube. Lines are searched in the corresponding epipolar images and depth maps computed from the slopes. The right-hand side of Figure 8.12 shows one of these depth maps. No filtering, interpolation, and smoothness constraints are applied. Still, scene geometry is recovered with reasonable quality.

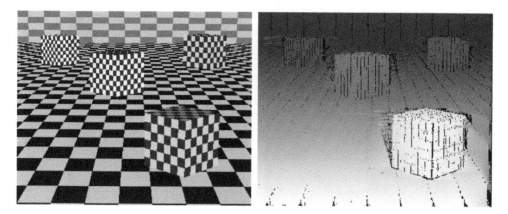

Figure 8.12   *Left*: synthetic sequence *Boxes*. *Right*: estimated scene depth

8.4.2   Image Cube Trajectory Analysis

Epipolar image analysis is a robust method for 3D reconstruction that jointly detects point correspondences for all available views of an image sequence. Occlusions as well as homogeneous regions can be handled efficiently. The big disadvantage of the algorithm is its restriction to linear equidistant camera movements, since the scene is viewed from one principle direction only. Other camera movements, for example circular camera configurations important for turntable or concentric mosaic acquisitions, cannot be handled.

One idea to overcome this problem is a piecewise linear EPI analysis where small segments of the object point trajectory are approximated by straight lines. This approach can also be applied to other camera movements (Li *et al.* 2001), but significantly reduces the amount of reference images and thus robustness of the 3D reconstruction.

This restriction to horizontal camera movements is overcome in *image cube trajectory (ICT) analysis* (Feldmann *et al.* 2003a) that can deal with circular camera configurations as well. For an inward-pointing, circularly moving camera, e.g., the trajectories in the image planes are no longer lines, but sinusoidal curves, as illustrated in Figure 8.13. Instead of searching for straight lines in the epipolar images, ICT analysis first discretizes the object space, similar to space-carving. Either regular voxel volumes can be built or more efficient irregular structures (Feldmann *et al.* 2004) can be set up. Given the camera motion, the trajectory of a particular 3D point through the image cube is computed. In a second step, colour constancy along the entire trajectory is evaluated. Trajectory parameters are varied and for the best matching ICTs the corresponding 3D positions are determined. Occlusion handling is hereby performed, similarly to the horizontal case, by using a special search order.

Although the approach can be applied to arbitrary known camera configurations, we will restrict ourselves to circular camera motion in the following. Three cases are distinguished as illustrated in Figure 8.14: an inward-facing camera for turntable measurements, and an eccentrically rotating camera facing in tangential or normal directions. The latter two cases are important for concentric mosaic analysis (Feldmann *et al.* 2003b).

For such circular motion, an arbitrary 3D point may be described in terms of its radius R to the centre of rotation, its rotation angle ϕ, and its height y. Assuming perspective

Figure 8.13 *Moche* sequence, circularly moving camera. *Left*: first camera frame. *Right*: image cube representation with sinusoidal shaped trajectories

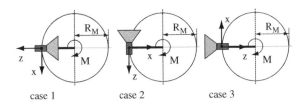

case 1   case 2   case 3

Figure 8.14 Camera configurations for three different cases. *Left*: turntable setup (inwards pointing camera). *Middle*: concentric mosaic (tangential direction). *Right*: concentric mosaic (normal direction)

projection, object point trajectories through the image cube can be computed analytically (see Feldmann *et al.* 2003b for more details). In contrast to the horizontal case described in Section 8.4.1, the trajectories are no longer restricted to lie in a 2D plane, but change in all three dimensions. The projection of the trajectory into the $X - \phi$ plane is illustrated for all three cases in Figure 8.15. Note that points outside and inside the camera radius R_m have to be treated differently.

Given the analytic trajectory description with the free parameters R, ϕ, and y, ICTs are computed for each 3D object point and a global search within the image cube is performed by evaluating colour constancy. The parameters of the best matching ICTs determine 3D positions of valid object points. Similar to the horizontal case, occlusions may occur which hide parts of the trajectories. To decide the correct processing order, two cases have to be distinguished. First, trajectory parts corresponding to points lying between the centre of rotation and the camera must be evaluated. For these parts, the slope of the curve is always positive and trajectories must be processed from large to small radii R. Trajectory parts being further away than the centre of rotation (with negative slope) are processed next with increasing radius. Each time a matching trajectory is found, it is removed from the image cube. In this way, it is assured that even partly occluded points are robustly found. For a detailed discussion of correct occlusion handling please refer to Feldmann *et al.* (2003b).

Figure 8.16 shows a result obtained for a turntable sequence with an inward-facing camera; 360 frames in CIF resolution are recorded that form the image cube. A 3D voxel model is created by searching through the image data for best matching trajectories. The right-hand

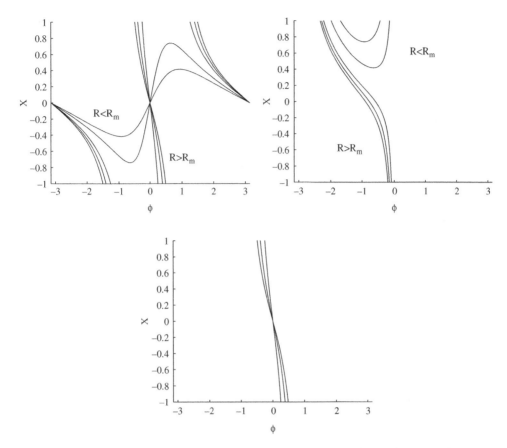

Figure 8.15 *X*-coordinate for points with varying radius. *Upper Left*: trajectory for concentric mosaics with inwards facing camera. *Upper Right*: tangentially facing camera. *Middle*: outward-facing camera

Figure 8.16 ICT reconstruction of the *Tree* sequence. *Left*: original frame of a tree on a turntable. *Middle*: reconstructed geometry displayed as depth map. *Right*: synthesized image from the voxel volume

side of Figure 8.16 gives an example of a reconstructed view rendered from the voxel dataset. The geometry of the object is illustrated in the middle image by means of a depth map. Although a very simple matching strategy is used in this particular experiment, finer details can also be recovered.

8.5 CONCLUSIONS

Volumetric reconstruction methods have recently gained increasing interest since modern computers allow the storage of large volumes and reconstruction of objects or scenes even in real-time. Compared with depth-based methods, a voxel representation often has the advantage that no point correspondences and registration of erroneous data are required. Due to the lack of smoothness constraints, very detailed structures can be recovered provided that accurate calibration information is available. One of the simplest methods is shape-from-silhouette that is computationally efficient and robust against lighting variations. Viewing cones computed from the contours are intersected to approximate the visual hull of the object that does not include any concavities. Higher shape accuracy can be obtained by using surface colour information instead of the binary silhouette mask. Space-carving or voxel colouring methods try to construct a coloured voxel volume whose back-projections into the image planes result in photo-consistent synthetic views. The methods mainly differ in the way the volume is traversed, how colour similarity is computed, and how visibility is determined. Occlusion handling is also elegantly solved by an occlusion-compatible ordering scheme in epipolar image analysis where trajectories of object points in an image cube are analysed. Again, all views of the scene are jointly exploited without establishing point correspondences. The classic linear camera motion assumption can also be generalized to other camera movements, enabling the usage of these methods to a wide range of applications.

REFERENCES

Baker H 1977 Three-dimensional modelling. *Proceedings of the 5th International Joint Conference on Artificial Intelligence*, Cambridge, MA, USA, pp. 649–655.

Bolles RC, Baker HH and Marimont DH 1987 Epipolar image analysis: An approach to determine structure from motion. *Intenational Journal of Computer Vision* **1**(1), 7–55.

Chein CH and Aggarwal JK 1986 Volume/surface octrees for the representation of three-dimensional objects. *Computer Vision, Graphics and Image Processing* **36**(1), 100–113.

Criminisi A, Kang SB, Swaminathan R, Szeliski R and Anandan P 2002 Extracting layers and analyzing their specular poperties using epipolar-plane-image analysis. *Technical Report, MSR-TR-2002-19*, Microsoft Research.

Culbertson WB, Malzbender T and Slabaugh G 1999 Generalized voxel colouring. *Proceedings of the International Conference on Computer Vision (ICCV)*, Corfu, Greece, pp. 67–74.

Eisert P 2004 3-D geometry enhancement by contour optimization in turntable sequences. *Proceedings of the International Conference on Image Processing (ICIP)*, Singapore, pp. 3021–3024.

Eisert P, Steinbach E and Girod B 1999 Multi-hypothesis, volumetric reconstruction of 3-D objects from multiple calibrated camera views. *Proceedings of the International Conference on Acoustics, Speech, and Signal Processing (ICASSP)*, Phoenix, USA, pp. 3509–3512.

Feldmann I, Eisert P and Kauff P 2003a Extension of epipolar image analysis to circular camera movements. *Proceedings of the International Conference on Image Processing (ICIP)*, Barcelona, Spain, pp. 697–700.

Feldmann I, Kauff P and Eisert P 2003b Image cube trajectory analysis for 3D reconstruction of concentric mosaics. *Proceedings of the International Workshop on Vision, Modeling, and Visualization*, Munich, Germany, pp. 569–576.

Feldmann I, Kauff P and Eisert P 2004 Optimized space sampling for circular image cube trajectory analysis. *Proceedings of the International Conference on Image Processing (ICIP)*, Singapore, pp. 1947–1950.

Fitzgibbon AW, Cross G and Zisserman A 1998 Automatic 3D model construction for turntable sequences. *Proceedings of the ECCV 98 Workshop on 3D Structure from Multiple Image in Large-Scale Environments*, Freiburg, Germany, pp. 154–170.

Goldlücke B and Magnor M 2003 Real-time, free-viewpoint video rendering from volumetric geometry. *Proceedings of the Visual Computation and Image Processing (VCIP)*, Lugano, Switzerland, **5150**(2), 1152–1158.

Grattarola AA 1992 Volumetic reconstruction from object silhouettes: A regularization procedure. *Signal Processing* **27**, 27–35.

Grau O 2003 Studio production system for dynamic 3D content. *Proceedings of Visual Computation and Image Processing (VCIP)*, Lugano, Switzerland, pp. 80–89.

Greenleaf JF, Tu TS and Wood EH 1970 Computer generated 3-D oscilloscopic images and associated techniques for display and study of the spatial distribution of pulmonary blood flow. *IEEE Transactions on Nuclear Science* **17**(3), 353–359.

Gross M 2003 blue-c: A spatially immersive display and 3D video portal for telepresence. *Proceedings of Computer Graphics (SIGGRAPH)*, San Diego, USA, pp. 819–827.

Horn BKP 1986 *Robot Vision*. MIT Press, Cambridge.

Kutulakos KN and Seitz SM 2000 A theory of shape by space carving. *International Journal of Computer Vision* **38**(3), 199–218.

Laurentini A 1994 The visual hull concept for silhouette-based image understanding. *IEEE Transactions on Pattern Analysis and Machine Intelligence* **16**(2), 150–162.

Li Y, Tang CK and Shum HY 2001 Efficient dense depth estimation from dense multiperspective panoramas. *Proceedings of the International Conference on Computer Vision (ICCV)*, Vancouver, Canada, pp. 119–126.

Lorensen WE and Cline HE 1987 Marching cubes: A high resolution 3D surface construction algorithm. *Proceedings of Computer Graphics (SIGGRAPH)*, vol. 21, pp. 163–169.

Martin WN and Aggarwal JK 1983 Volumetric description of objects from multiple views. *IEEE Transactions on Pattern Analysis and Machine Intelligence*, **15**(2), 150–158.

Matusik W, Buehler C and McMillan L 2001 Polyhedral visual hulls for real-time rendering. *Proceedings of the 12th Eurographics Workshop on Rendering*, pp. 115–126.

Niem W 1998 *Automatische Rekonstruktion starrer dreidimensionaler Objekte aus Kamerabildern.* PhD thesis University of Hanover Germany.

Niem W and Wingbermühle J 1997 Automatic reconstruction of 3D objects using a mobile monoscopic camera. *Proceedings of the International Conference on Recent Advances in 3D Imaging and Modelling*, Ottawa, Canada, pp. 173–180.

Pauly M, Keiser R, Kobbelt L and Gross M 2003 Shape modeling with point-sampled geometry. *Proceedings of Computer Graphics (SIGGRAPH)*, San Diego, USA, vol. 22, pp. 641–650.

Potmesil M 1987 Generating octree models of 3D objects models from their silhouettes in a sequence of images. *Computer Vision, Graphics and Image Processing* **40**(1), 1–29.

Prock AC and Dyer CR 1998 Towards real-time voxel colouring. *Proceedings of the Image Understanding Workshop*, pp. 315–321.

Rusinkiewicz S and Levoy M 2000 QSplat: A multiresolution point rendering system for large meshes. *Proceedings of Computer Graphics (SIGGRAPH)*, Los Angeles, USA, pp. 343–352.

Seitz SM and Dyer CR 1997 Photorealistic scene reconstruction by voxel colouring. *Proceedings of Computer Vision and Pattern Recognition*, Puerto Rico, pp. 1067–1073.

Shum HY and He LW 1999 Rendering with concentric mosaics. *Proceedings of Computer Graphics (SIGGRAPH)*, Los Angeles, USA, pp. 299–306.

Smolic A, Müller K, Merkle P, Rein T, Eisert P and Wiegand T 2004 Free viewpoint video extraction, representation, coding, and rendering. *Proceedings of the International Conference on Image Processing (ICIP)*, Singapore, pp. 3287–3290.

Srivastava SK and Ahuja N 1990 Octree generation from object silhouettes in perspective views. *Computer Vision, Graphics and Image Processing*, January, 68–84.

Steinbach E, Eisert P, Betz A and Girod B 2000 3-D reconstruction of real world objects using extended voxels. *Proceedings of the International Conference on Image Processing (ICIP)*, vol. I, Vancouver, Canada, pp. 569–572.

Szeliski R 1993 Rapid octree construction from image sequences. *Computer Vision, Graphics and Image Processing*, pp. 23–32.

Tarini M, Callieri M, Montani C, Rocchini C, Olsson K and Persson T 2002 Marching intersections: An efficient approach to shape-from-silhouette. *Proceedings of Vision, Modeling, and Visualization VMV'02*, Erlangen, Germany, pp. 283–290.

Veenstra J and Ahuja N 1986 Efficient octree generation from silhouettes. *Proceedings of Computer Vision and Pattern Recognition*, Miami Beach, USA, pp. 537–542.

Westover L 1990 Footprint evaluation for volume rendering. *Proceedings of Computer Graphics (SIGGRAPH)*, pp. 367–376.

Wu X and Matsuyama T 2003 Real-time active 3D shape reconstruction for 3D video. *Proceedings of the 3rd International Symposium on Image and Signal Processing and Analysis*, Rome, Italy, pp. 186–191.

9

View Synthesis and Rendering Methods

Reinhard Koch and Jan-Friso Evers-Senne

Christian-Albrechts-University, Kiel, Germany

Rendering and view synthesis are terms first coined in the computer graphics field. Novel views of a predefined synthetic 3D scene are visualized using render algorithms such as ray tracing or radiosity, where all elements of the scene, such as surface geometry and reflectance, light sources, and viewing camera are predefined and given. The colour values for each pixel of the synthesised view are formed by collecting the reflectance rays from the scene that are cast towards the camera onto the specific pixel. This works well in synthetic scenes where all elements are well defined.

In the context of video processing, this approach is reversed. All information that is available consists of a set of real camera images, taken from different view points at different times of an unknown dynamic real world scene. Thus, view synthesis in the field of videocommunication has the task to generate new views from existing images, called *image based rendering* (IBR).

One way would be to invert the rendering process and to reconstruct a full 3D scene model from the real images, using methods from computer vision. Fortunately, sometimes it is not necessary to fully recover the 3D scene structure of the scene, as we want to recover only the visual appearance from a novel view point and not the complete 3D scene geometry. Partial 3D information may be sufficient for this task.

In this chapter we therefore discuss methods to generate novel views from a given set of input images, without the need of full 3D reconstruction. We will first introduce the theoretical concept of the plenoptic function that captures fully the visual appearance of a scene. We will then derive a taxonomy and review methods from the literature that exploit the plenoptic function for view synthesis. The main concepts behind these methods will be explained.

3D Videocommunication — Algorithms, concepts and real-time systems in human centred communication
Edited by O. Schreer, P. Kauff and T. Sikora © 2005 John Wiley & Sons, Ltd

9.1 THE PLENOPTIC FUNCTION

The *plenoptic function* (*PF*) of a 3D scene, introduced by Adelson and Bergen (1991), is derived from the Latin *plenus* = full and *optic* which relates to vision. It describes the intensity of all irradiance observed at every point in the 3D scene, coming from every direction; for an arbitrary dynamic scene the plenoptic function is of dimension seven.

$$\text{Plen}_{\text{full}} : \mathbb{R}^3 \times [0, 2\pi)^2 \times \mathbb{R} \times \mathbb{R} \to \mathbb{R} \ , \ \text{Plen}_{\text{full}}(x, y, z, \phi, \theta, \lambda, t) = I \quad (9.1)$$

where I is the light intensity of the incoming light rays at any spatial 3D-point $(x, y, z)^{\text{T}}$ from any direction given by spherical coordinates (ϕ, θ) for any wavelength (colours) λ at any time t. If the *PF* is known to its full extent, then we can reproduce the visual scene appearance precisely from any view point at any time. Unfortunately, it is technically not feasible to record an arbitrary *PF* of full dimensionality, as we would need to simultaneously place light probes to fully cover the space over all times. The problem simplifies if we assume a static scene, which removes the time variable t. In a static scene we can move a single light probe over time to different spatial positions (x_t, y_t, z_t) and record the static plenoptic field sequentially.

$$\text{Plen}_{\text{static}} : \mathbb{R}^3 \times [0, 2\pi)^2 \times \mathbb{R} \to \mathbb{R} \ , \ \text{Plen}_{\text{static}}(x_t, y_t, z_t, \phi, \theta, \lambda) = I \quad (9.2)$$

9.1.1 Sampling the Plenoptic Function

To record an arbitrary scene we need to sample the high dimensional *PF*. The sampling problem can be simplified if we separate the four spatiotemporal dimensions (x, y, z, t) from the dimensions (ϕ, θ, λ) for viewing direction and colour sensing by using spherical colour image sensors. As *plenoptic sample (PS)* we define the three-dimensional subspace that forms a spherical image $I_{\text{ps}}(\phi, \theta, \lambda)$ at a particular spatiotemporal position (x_t, y_t, z_t).

$$I_{\text{ps}} : [0, 2\pi)^2 \times \mathbb{R} \to \mathbb{R} \ , \ I_{\text{ps}}(\phi, \theta, \lambda)|_{x,y,z,t} = I \quad (9.3)$$

The set of all plenoptic samples $I_{\text{ps}}(\phi, \theta, \lambda)$ for all spatial dimensions (x, y, z) and for all times t form the full *PF* (Equation 9.1). In a static scene, again the time t can be eliminated and we can define the spatial *PF* as collection of all *PS* in space.

$$\text{Plen}_{\text{spatial}} : \mathbb{R}^3 \to [0, 2\pi)^2 \times \mathbb{R} \ , \ \text{Plen}_{\text{spatial}}(x_t, y_t, z_t) = I_{\text{ps}}(\phi, \theta, \lambda) \quad (9.4)$$

A particular plenoptic sample is obtained by placing a spherical imaging sensor in space and recording the light intensity for each incoming ray. Direction angles ϕ and θ are discretized with pixel positions. Colour perception is obtained by recording three separate images for red, green and blue and interpolating the three colours according to the human tristimulus perception. This is equivalent to using λ in Equation (9.2), but discretising it with three samples.

A conventional perspective camera can be modelled as a tangent plane to the sphere with limited field of view, as shown in Figure 9.1. Thus, the sampling of the *PF* reduces to recording (spherical) colour images in each spatial position. If the scene is dynamic, then an

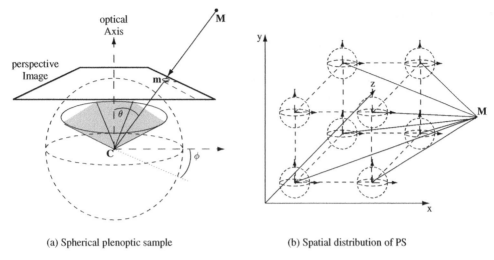

(a) Spherical plenoptic sample   (b) Spatial distribution of PS

Figure 9.1 (a) An ideal spherical sensor at position $\mathbf{C} = (x, y, z)^{\mathsf{T}}$ sampling a ray from $\mathbf{M}$ with (ϕ, θ). The image plane of a real sensor is tangent to the sphere giving an image point $\mathbf{m}$. (b) Eight plenoptic samples organized in a regular grid observing a point $\mathbf{M}$

image sequence can be taken to capture the time variation of a plenoptic sample. Sampling of Equation (9.4) with real sensors introduces discretization on two levels:

1. Angular sampling (ϕ, θ) of a single plenoptic sample due to the finite pixel resolution of the imaging sensor,

2. Spatial sampling (x_t, y_t, z_t) due to the finite sampling density between plenoptic samples.

It is therefore necessary to obey the sampling theorem to avoid aliasing. Angular sampling is usually not a problem, if high-resolution imaging devices are used. Resampling can be performed with simple interpolation filters on the 2D image data.

Spatial sampling may pose a problem if a large viewing volume needs to be sampled, as it is not possible to densely sample full 3D space. Therefore, the IBR systems have to distinguish between dense and sparse sampling of the *PF* in the following. Dense sampling eliminates parallax effects, but may cause unnecessary oversampling. Sparse sampling will violate the sampling theorem, hence additional information such as scene depth will be necessary. Chai *et al.* (2000) evaluated the effects of sampling density and depth influences on the rendering for the light field problem in detail and give bounds for the sampling density as function of the depth variations in the scene.

9.1.2 Recording of the Plenoptic Samples

The samples are recorded by capturing images at the specified spatial positions. Recently, a wealth of spherical and hemispherical imaging sensors have become available, such as catadioptric (mirror-optic) devices or wide-angle fish-eye lenses that directly capture $I_{(\phi, \theta, \lambda)}$ for a hemisphere. Also, the theory on the image formation to such sensing devices is now well understood, see (Baker and Nayar 1998; Bakstein and Pajdla 2003a;

Geyer and Daniilidis 2003). These hemispherical sensors allow the direct recording of the *PS* with low resolution. If high-resolution samples are needed, the spherical sample can be reconstructed by mosaicing of multiple rotated images into a spherical panorama. Many systems were developed for direct image mosaicing, called also rotational mosaics (Shum and Szeliski 1997), or with the use of motorised camera heads, as used for Quicktime VR systems (Chen 1995).Very-high-resolution cylindrical panoramas can be recorded with rotating high-resolution line-sensors that generate resolutions of up to $30\,000 \times 100\,000$ pixels for a single plenoptic sample (Klette *et al.* 2003). The spherical image can be mapped onto a perspective view for display with a conventional screen.

Plenoptic samples taken from the scene are denoted as *real views* in the following, to distinguish them from a *virtual view*, which is synthesized by reconstructing the *PF*. The word *camera* is used in this context to refer to the internal and external parameters (the projection) which have been used for a view. Thus, a *real camera* is associated with a specific *real view*, and a *virtual camera* is associated with every *virtual view*.

9.2 CATEGORIZATION OF IMAGE-BASED VIEW SYNTHESIS METHODS

Image-based view synthesis is essentially a resampling of the light rays of the plenoptic function. Synthesizing a virtual view is equivalent to generating the set of light rays that constitute the pixel of the new image. In the last decade several different approaches have been developed. A standard classification scheme has not yet evolved and several survey papers all use different categorisations.

Kang (1997) presented one of the first approaches for structuring IBR methods. He proposed four categories: non-physical-based image mapping, mosaicing, interpolation from dense samples and geometrically-valid pixel re-projection.

McMillan (1999) distinguished IBR systems based on their intermediate data representation: approximative scene geometry, images in databases to represent different environment locations and images as reference scene models from which to interpolate new views.

Shum and Kang (2000) classified IBR systems into three categories, depending on how much geometric information they use: systems not using any geometric information, systems using implicit geometry and systems using explicit geometry. The term *implicit geometry* is attributed to all systems that extract the geometrical information directly from the given images, while the term *explicit geometry* is used to describe systems that utilize additional geometric sources or externally given 3D surface data. IBR systems utilize implicit geometry only because it is derived from the image data, but they differ with respect to its use for depth compensation.

9.2.1 Parallax Effects in View Rendering

Geometric information is needed to compensate the parallax effects that occur in non-planar scenes when rendering from varying view points. As *parallax* or *disparity* we define the depth-dependent projection of the 3D scene point into the image when the view point changes. This situation is explained in Figure 9.2(a).

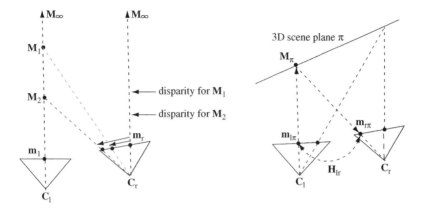

(a) Parallax effects in the image   (b) Homography mapping for planar scenes

Figure 9.2 (a) Projection of an infinitely distant point $\mathbf{M}_\infty$ has disparity zero. Finite points generate image parallax with non-zero disparity. (b) Correspondences of planar scene points can be computed with a planar homography without any disparity error

Image parallax or disparity describes the effects of scene depth in the image. A 3D point $\mathbf{M}$ will project onto a 2D image point $\mathbf{m}_j$ along the viewing ray between the 3D point and the camera centre. For a perspective camera j, the simple pinhole projection model can be applied:

$$\mathbf{m}_j = K_j R_j^\mathrm{T}[I| - \mathbf{C_j}]\mathbf{M} \tag{9.5}$$

with K_j defining the pixel calibration matrix, R_j the camera orientation and $\mathbf{C_j}$ the camera centre. $\mathbf{M} = (x, y, z, 1)$ is a homogenous scene point.

If a second, displaced camera views the 3D scene point, then the projected image point in the second camera will lie on the epipolar line that is formed by the projection of the viewing ray into the second image (for a discussion on epipolar geometry, see Chapter 6). The epipolar line itself is determined only by the relative pose between both cameras, while the position of the point on the epipolar line is determined by the scene depth relative to the first camera. The depth-dependent shift of the projected image point in the displaced camera is called the disparity, see Figure 9.2(a).

The disparity defines if direct visual interpolation from the plenoptic samples generates visual artifacts. Disparity is scene dependent and needs compensation with depth-dependent warping. In some instances, there may be no disparity between the real and synthetic views. In that case perfect reconstruction is possible without 3D scene knowledge. This is the case for infinitely distant or planar scenes where all geometry can be compensated with global homographies. For an infinitely distant point $\mathbf{M}_\infty = (x, y, z, 0)$, the projection is independent of the camera displacement, as the homogeneous point coordinate is zero. Hence the point projection Equation (9.5) simplifies to

$$\mathbf{m}_j = K_j R_j^T[I| - \mathbf{C_j}]\mathbf{M}_\infty = K_j R_j^T\mathbf{M}_\infty \tag{9.6}$$

Equation (9.6) applies to both cameras l and r in Figure 9.2(a). As the 3D point $\mathbf{M}_\infty$ is identical for both projections, it can be eliminated and a correspondence transfer between the image points $\mathbf{m}_l$ and $\mathbf{m}_r$ is found with the homography H_{lr}:

$$\mathbf{m}_r = K_r R_r^T \mathbf{M}_\infty = K_r R_r^T R_l K_l^{-1} \mathbf{m}_l = H_{lr} \mathbf{m}_l \tag{9.7}$$

The homograpy H_{lr} defines a planar projective mapping between both image planes. It maps each image point $\mathbf{m}_l$ onto the corresponding image point $\mathbf{m}_r$ via the relative rotation between the cameras. The correspondence transfer can also be seen as a mapping over the infinitely distant plane Π_∞, where the resulting disparity is zero. A similar relationship holds also for general 3D points in the following two cases:

1. The camera centres of both cameras coincide and both cameras define a single plenoptic sample, with rotated optical axes. This case applies to single-perspective panoramic imaging (see Section 9.3) .

2. All scene points $\mathbf{M}_\pi$ lie on a real 3D plane Π (see Figure 9.2b). In that case, a homographic mapping between any 3D point on the plane $\mathbf{M}_\pi$ and the corresponding image points $(\mathbf{m}_{l\pi}, \mathbf{m}_{r\pi})$ exist:

$$\mathbf{m}_{r\pi} = H_{\pi r} \mathbf{M}_\pi = H_{\pi r} H_{l\pi} \mathbf{m}_{l\pi} = H_{lr} \mathbf{m}_{l\pi} \tag{9.8}$$

In all other cases, there is disparity in the images and disparity compensation is needed for view synthesis. Disparity compensation can be achieved from implicit geometrical representations such as image flow and depth maps, or from explicit 3D shape representations.

9.2.2 Taxonomy of IBR Systems

We follow largely the categorization of Shum and Kang in this discussion, however, geometry alone does not suffice for categorization. An important additional factor is the spatial placement of the *PS* and the poses of the virtual views.

1. *Sampling density.* If samples are distributed densely over space, such that for each synthesized viewing ray there is a real plenoptic sample close by, than resampling is simplified to ray selection and colour interpolation between the nearby rays. The depth parallax will not distort the novel view and approximately geometry-free reconstructions are possible. Prominent examples of such reconstructions are panoramic viewing and light field approaches. This simplification does not hold for sparse sampling, where novel views are generated with possibly large parallax. For virtual views that are far from real views, parallax-dependent compensation is necessary and geometric warping must be employed.

2. *Distribution of sample positions.* The spatial arrangement of samples allow us to design special configurations where parallax-free rendering is possible in a restricted spatial range. It is clear that direct rendering can be used if novel views coincide with real sample positions. This is the case for spherical panoramas where free look-around capabilities exist, but no viewpoint change is possible. The samples can be distributed arbitrarily. Recently, a wealth of systems have been designed that exploit special dense and regular 1D or 2D arrangements of the samples. A typical 1D arrangement is the concentric mosaic that places dense sampling of real view points on a circular path by rotating a camera on

a fixed rod. A novel view can be generated if its position is in the plane of the camera movement and the viewing direction is restricted to lie inside the circle plane. The light field approach is a 2D arrangement where samples are placed in a 2D planar or spherical patch around the scene. Novel views can be generated from viewpoints inside a restricted bounded volume.

3. *Possible poses of virtual views.* The goal of IBR is rendering from freely chosen novel viewpoints. However, some systems impose restrictions on the possibles poses. In a panoramic viewer, full rotational interaction is possible, but one may only hop between discrete viewpoints. For a light field, free viewpoint selection is possible within a restricted viewing volume and viewing direction, and for 3D reconstructions, the virtual camera may move continuously with a full six degrees of freedom.

The taxonomy used here groups the different IBR approaches with respect to the categories mentioned. We have identified three dimensions of influence that are drawn in the category overview Figure 9.3.

1. *Geometry axis.* The use of geometry needed to compensate for scene parallax, ranging from systems without parallax compensation, over coarse geometrical approximations, local image correspondences to full 3D information. This axis is mostly correlated with

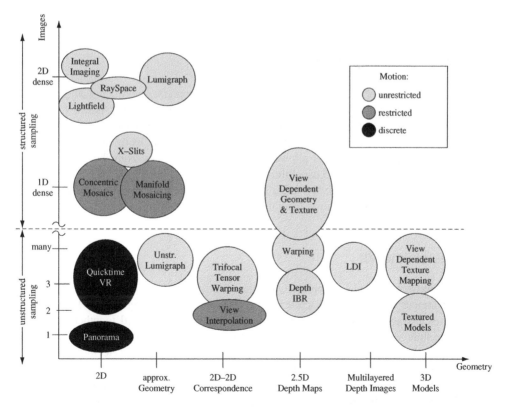

Figure 9.3 Categorization of different IBR methods

the complexity of the system, because the precise extraction of geometrical information from images is difficult.

2. *Image axis.* The spatial distribution of the *PS*, ranging from a few to many unstructured samples in the lower half, and a structured dense sampling in the top half of the graph. This axis is correlated with the memory demand of the system and the complexity of image acquisition, because very many images may be needed in a possibly very structured way.

3. *Virtual viewpoint selection (colour coded).* The ultimate goal of each IBR system is full freedom of the virtual camera pose, but some systems such as panoramas restrict the camera motion to discrete positions, others allow only restricted motion. We have coarsely categorized the methods into three motion categories in Figure 9.3 predetermined discrete positions (black), constrained motion along predetermined path (dark grey) and unrestricted motion (light grey).

9.3   RENDERING WITHOUT GEOMETRY

IBR Methods within this category use no geometry information at all. Since no parallax compensation is possible, the sampling must be either very dense, or the possible motion is restricted to lie near to the sample positions. Basically, these methods trade sampling density against geometric complexity.

9.3.1   The Aspen Movie-Map

The earliest system to obtain restricted interactive look-around capabilities was the Aspen Movie-Map by Lippman (1980). A car was driven along the streets of Aspen, Colorado, recording simultaneously four camera streams looking at right angles to cover a cylindrical view of the scene for every 3 m (see Figure 9.4a). The streams were captured on video disk for interactive playback. This allowed to render novel views from the given camera path with interactive look-around capabilities. As the video stream could be edited at street intersections, one could interactively select the route by switching between segments of the video stream. However, no deviation from the given camera path and positions was possible.

9.3.2   Quicktime VR

A similar approach is used in Apple's Quicktime VR (QVR) system (Chen 1995), where panoramic images are taken at discrete viewpoint positions (see Figure 9.4b). A camera is rotated on a tripod at the fixed position and all images are stitched together to form a plenoptic sample with a cylindrical or spherical panoramic view of the scene (Shum and Szeliski 1997). Each panorama can be stored as a plenoptic sample. By selecting the camera positions during capture properly, the user can interactively explore the scene by switching between different panoramas. The virtual viewpoints are restricted to lie exactly on the sampled real views, but look-around capability is provided by the cylindrical or spherical representation.

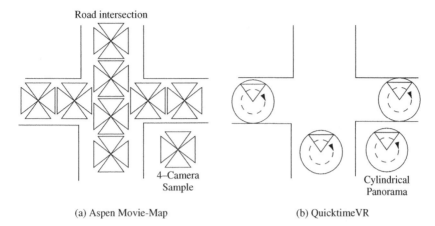

(a) Aspen Movie-Map   (b) QuicktimeVR

Figure 9.4 (a) The Aspen Movie-Map is based on dense sampling with four cameras. For each sample point, four orthogonal views were stored and could be selected interactively. (b) QuicktimeVR uses less dense sampling, but obtains high-resolution cylindrical or spherical panoramas

The Aspen Movie-Map system and QVR differ with respect to the spatial sampling. Both systems collect plenoptic samples of the scene, but the samples in the Aspen system are arranged in a systematic 2D linear grid, while QVR samples are located arbitrarily.

9.3.3 Central Perspective Panoramas

The panoramas discussed here use the single central perspective assumption. Therefore, there is no parallax between the images which allows high fidelity in the reconstruction. More recent implementations of such a system can be found in Antone and Teller (2002) and Teller *et al.* (2003), where additional spatial information and view interpolation is described. Kawasaki *et al.* (2001) use sequences of panoramic images (as in Takahashi *et al.* 2000) to render cities with this approach.

Despite their simplicity, panoramas are a very attractive means of IBR and versatile in use. Central perspective panoramas are easy to capture and mosaic stitching software nowadays comes with many digital cameras. They deliver perfect image quality for static scenes, and novel multi-camera hardware developments even allow panoramic video streams with high fidelity (Kang *et al.* 2003). Panoramas can also be used to capture environment maps of the incident light with high dynamic range images for integration of virtual objects in real scenes (Debevec 1998).

For videocommunication, panoramic images can serve to create and visualize static background, but due to the missing parallax they can not be used to generate novel views of nearby objects.

9.3.4 Manifold Mosaicing

More recently, a wealth of approaches have been introduced to overcome the limitation of fixed viewpoints, and alternative sampling and reconstruction approaches have been investigated. They are all based on the assumption that novel views may be generated from a dense sampling of the scene. As a full 3D sampling of the *PF* is not feasible, subspaces are sampled densely to allow bounded continuous view reconstruction.

A panorama has a fixed viewpoint and is parameterized by the two direction angles of the sphere. Peleg and Herman (1997) developed a new type of non-central panoramas called *manifold mosaics*. The camera is not fixed, but moves on some predefined trajectory while recording dense sequences. The camera motion is typically one-dimensional, in a linear or circular track. Unlike conventional central perspective mosaic stitching the resulting manifold panorama is composed of small image stripes from all the different views into one multi-perspective image manifold. Figure 9.5(a) shows the capture and strip selection for a manifold mosaic. The work was inspired by the 1D push-broom line scanner cameras that are used in aerial imaging of a flat terrain. All data can be stored together in a spatiotemporal image volume (Figure 9.5b). The manifold panorama can therefore be described by an MCOP image (multiple centre of projection images Rademacher and Bishop (1998)) where each image stripe (image lines perpendicular to the image motion) has another projection centre and is taken from another image. There is a strong relation to the EPI (epipolar plane image analysis by Bolles and Baker (1986)) that slices the spacetime image volume to analyse 3D scene parallax.

Typically, the camera motion will be linear in horizontal direction only, and from each image a central 1D slit column will be used to paste into the panorama. The resulting

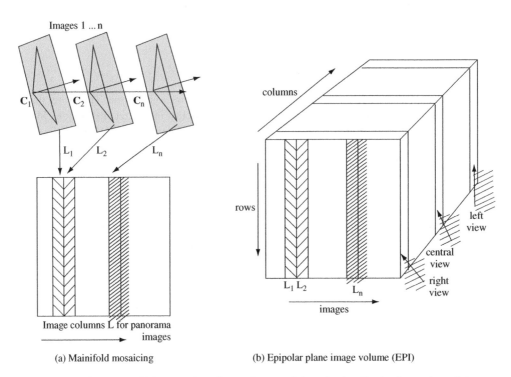

(a) Mainifold mosaicing   (b) Epipolar plane image volume (EPI)

Figure 9.5 (a) Top: manifold recording along a linear path by selecting fixed columns in each image; bottom: depending on the selected column, a sheared left or right view for stereo viewing can be generated. (b) Images are arranged in the epipolar plane image volume (EPI). Slicing the EPI along a fixed column over all images will generate a manifold mosaic

image is central perspective in the vertical column direction, but parallel perspective in the horizontal direction of camera motion. Therefore, a parallax compensation is needed for the vertical perspective image direction. It is easy to generate stereoscopic panoramas from this configuration by selecting sheared columns that control the gaze direction. See also (Rousso *et al.* 1997; 1998) for detailed description.

9.3.5 Concentric Mosaics

The concentric mosaic as proposed by Shum and He (1999) is a special configuration of a manifold mosaic. A camera is mounted on a horizontal arm and rotated on a circular path around an axis of revolution looking outwards, see Figure 9.6(a). A dense image sequence is recorded along the path. Similar to the linear manifold mosaic, vertical image stripes are cut from the image sequence to form MCOP images that are parameterized by the rotation angle and the elevation angle of the column height. Depending on the gaze angle of the column, the panorama for each slit i can be viewed as a bundle of rays that all lie tangential to a concentric circle of radius r_i. By selecting the image column, different radii are used, each forming a concentric mosaic. The central column generates a circle with radius 0 (a conventional cylindrical panorama).

Essentially, a concentric mosaic records a dense circular disk of view positions of the *PF*. The resulting plenoptic function is parameterized in three dimensions by the rotation and elevation angles as well as the radius r that describes the circle selected by the column.

$$I_{(\lambda)} = \text{Plen}_{\text{Concentric}}(\phi, \theta, r) \tag{9.9}$$

Virtual views can be rendered from the concentric mosaics if the viewing position is constrained to lie within the bound of the concentric circles. To compose a new view, the horizontal viewing direction of each image column is computed and the tangent ray for each light ray is computed from interpolation between the recorded concentric mosaics

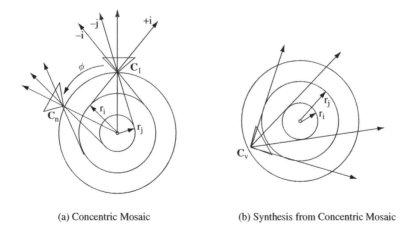

(a) Concentric Mosaic   (b) Synthesis from Concentric Mosaic

Figure 9.6 (a) A camera is rotated with angle ϕ on a circular path. Image columns $+i$ and $-i$ coincide with rays tangential to a circle with radius r_i. (b) Image columns for a virtual view are taken from the tangential rays of the concentric mosaic

(see Figure 9.6b). The interpolation is possible only for horizontal rays that lie within the concentric plane. The vertical elevation angle is not interpolated, but the stored image columns are stitched together directly. The resulting images are not perspective since parallax on the elevation angle is not compensated, but image deformations are minor as long as the elevation opening angle of the camera is small (Shum *et al.* 2002). Again, stereoscopic views can be created by rendering two displaced views from the same concentric mosaic. These stereo images can be used for direct stereoscopic viewing, or scene depth can be extracted with conventional binocular stereo analysis.

9.3.6 Cross-slit Panoramas

The cross-slit projection (X-slits) proposed by Weinshall *et al.* (2002) and Zomet *et al.* (2003) is a generalization of most of these techniques. A novel camera model is proposed that uses two displaced 3D lines (slits) to describe the projection process. In that general model, the possible projection rays are formed by all possible connections between both slits. If both slits intersect in a single point then the X-slit model degenerates to the central perspective model with a single focal point and a spherical ray representation. If the slits do not intersect, then some constraints on the projection hold. One interesting configuration is to use a horizontal and a vertical slit. Fixing a single point on the horizontal slit will generate a vertical 1D slice of the image (push-broom image). It can be used to render novel views from manifold panoramas where the horizontal slit defines the positions of the input images and the vertical slit defines the image slices. This is equivalent to cut slices in the spatiotemporal image volume of the input sequence. Bakstein and Pajdla (2003b) use this model to construct novel views with high fidelity. Concentric mosaics can also be described with this model. Combining a ray-space representation, omnidirectional plenoptic sampling and X-slit projection, Bakstein and Pajdla (2003c) generate new omnidirectional views and stereo mosaics from high-resolution concentric mosaics.

9.3.7 Light Field Rendering

Levoy and Hanrahan (1996) introduced the *light field* which allows free 3D motion of the virtual view in a bounded volume. The light field interpolates new views using a 4D representation of the plenoptic function. It is based on the assumption that light rays are emitted from the surface of an object that is enclosed by a rectangular bounding box. The system records all rays that leave one side of the bounding box by placing a very dense regular 2D grid of cameras looking into the bounding box. Two coplanar planes (u, v) and (s, t) are placed such that the optical centres of the real cameras are located in the (u, v) camera plane (the sample positions x, y) while the image planes are all rectified to the (s, t) pixel plane (the viewing directions ϕ, θ). This setup is shown in Figure 9.7(a). To capture the appearance of an object from all sides, six (u, v, s, t) configurations have to be placed around it.

Because of the specialized capturing geometry the plenoptic function can be re-parameterized from the 5D function (Equation 9.4) to a 4D representation (Figure 9.7a). Each light ray passing through the volume between the planes can be described by its intersection points with the planes. Thus the plenoptic function can be written as:

$$I_{(\lambda)} = \text{Plen}_{\text{Light field}}(u, v, s, t) \tag{9.10}$$

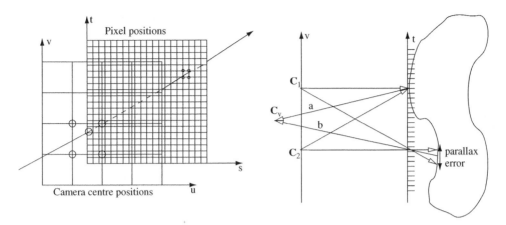

Figure 9.7 (a) 4D Light field parameterization with two planes. Each light ray is defined by the intersection with (u, v)-(s, t)-coordinates. (b) Ray interpolation for novel view synthesis. Ray **a** is interpolated correctly, while ray **b** has parallax artefacts due to displaced scene geometry

During recording, for all pixels of all real views, light rays are computed and parameterized as described. All these light rays are stored in a ray database called the Light Field.

A new view can be generated if it is placed outside of the bounding box looking inside or vice versa. For each pixel i of the new view, a viewing ray $\mathbf{r}_i$ is computed that passes through the two-plane parameterization and generates a particular sample (u_i, v_i, s_i, t_i). If such a sample exists in the database, then the appropriate colour value is assigned to the pixel i. If no matching ray can be found, the nearest ones are selected and blended. In general, a desired ray may not pass though the optical centre of one real camera, but it passes between four centres of real cameras. This gives four potential (u, v)-coordinates. When the desired ray passes through the (s, t) plane in general it will not hit one pixel, but will pass between four pixels giving four (s, t)-coordinates. For these four corresponding pairs of (u, v)-(s, t)-coordinates, rays can be looked up in the ray database and quadrilinear interpolation is performed.

The chosen parameterization assumes a flat scene lying in the (s, t) plane. The visual reconstruction can achieve high accuracy, if a given scene matches this criterion or if the sampling is very dense such that the distance between adjacent camera centres is small. Less dense sampling of the viewing space and non-flat structures result in visual artifacts when interpolating between views, as no parallax compensation is used. In Figure 9.7(b) the effect of scene parallax in the lightfield is shown. A virtual camera is placed at position $\mathbf{C}_v$ and novel rays **a** and **b** are interpolated from the nearby cameras at positions $\mathbf{C}_1$ and $\mathbf{C}_2$. For ray **a**, a valid interpolation is found since the scene surface coincides with the pixel plane. For ray **b**, however, parallax effects will occur as the wrong viewing rays are interpolated.

9.3.8 Lumigraph

Simultaneously to the light field, Gortler *et al.* (1996) developed their *lumigraph* system. Very similar to the light field, a ray database parameterized by two planes is built and view synthesis selects rays from it. To handle image acquisition from a hand-held camera, images from cameras which are not coplanar to the (s, t)-plane are rectified and mapped onto the light field plane. This rectifying interpolation uses a convex 3D shape approximation of the scene, the visual hull obtained from silhouette intersection. Thus we categorize the lumigraph with some approximate geometry for preprocessing. The rendering itself does not use any geometry for scene warping.

9.3.9 Ray Space

Another 4D re-parameterization of the plenoptic function is the *ray space*. First published by Fujii (1994) it uses a plane in space to define bundles of rays passing through this plane. For the (x, y)-plane at $z = 0$ each ray can be described by its intersection with the plane at (x, y) and two angles (θ, ϕ) giving the direction:

$$I_{(\lambda)} = \text{Plen}_{\text{RaySpace}}(x, y, \theta, \phi) \tag{9.11}$$

After capturing a scene with several cameras the re-parameterized data can be stored in a 4D structure. New views can then be synthesized by looking up intensities from this ray database. This method is a hybrid mixture between 4D light field and EPI. Fujii and Tanimoto (2002) exploit this approach in their Free-Viewpoint TV (FTV) system. Chapter 4 is dedicated to this method.

9.3.10 Related Techniques

Several extensions to the light field and lumigraph methods have been presented in the last years. Buehler *et al.* (2001) proposed *unstructured lumigraph* rendering which is a hybrid design between view dependent texture mapping (VDTM) and light field rendering. Unlike VDTM, they do not rely on a high-quality geometric model, but they need a geometrical approximation of the scene.

Tong *et al.* (2001) enhanced the lumigraph by layers of different resolution for effective level-of-detail control based on the optimal sampling for a given scene complexity. Via epipolar plane image analysis (Booles and Baker 1986) surface planes are identified in the scene and a geometry proxy is generated for VDTM texturing.

Kurashima *et al.* (2002) set up a videoconferencing system using a geometry proxy and view-dependent texture mapping to generate perspectively corrected views. From up to four cameras observing the user, a plane+parallax proxy is calculated.

Isaksen *et al.* (2000) propose a framework for dynamic re-parameterization of 4D light fields. It allows arbitrary scenes and camera setups and opens up several other possible effects such as aperture or depth of field.

Takahashi *et al.* (2000) use image sequences from hemispherical cameras taken from a driving car and analyse where the virtual camera can be placed. New views are then synthesized by collecting slits (rays) from different images.

Naemura *et al.* (2001) describe an approach to synthesize arbitrary new views from integral photography images for autostereoscopic displays. Integral photography (IP) is a technique for dense spatial sampling of the plenoptic function by using micro-lens arrays in front of one single standard CCD sensors. Each micro-lens has its own centre of projection which makes this setup similar to a grid of small classic cameras. Due to the limited resolution of the sensor which is shared by all lenses, the angular resolution is small compared with a standard camera. Autostereoscopic displays use the inverse method by displaying multiple-perspective images on screens with lens or prism arrays, see Chapter 13 for more details.

An efficient light field representation for known or estimated geometry has been proposed by Chen *et al.* (2002). By factorization of 4D light field data into surface maps and view maps, compression and rendering is performed using programmable graphics hardware. The restriction of a flat scene is circumvented.

9.4 RENDERING WITH GEOMETRY COMPENSATION

So far we have described methods that do not handle scene parallax explicitly because it is assumed that a dense sampling or planar scene will avoid parallax effects. The trade-off is that one is restricted in the interactive camera pose selection for the virtual view. If full 3D interaction is desired, then parallax effects must be compensated during rendering. We can distinguish between methods that *interpolate* between given views along a predefined path, and the *extrapolation* of virtual views by freely selecting the pose.

9.4.1 Disparity-based Interpolation

Disparity based interpolation methods, or 2D morphing, interpolate novel perspective views from a given pair of real views and a given dense correspondence map between the views. The relative pose of both real views is known, hence we can compute the epipolar geometry between the views which encodes the direction of the correspondence vector for each image point. The disparity encodes the length of the epipolar displacement for each image correspondence. Chen and Williams (1993) introduced view interpolation by linear interpolation of the length of the 2D correspondence vectors, the disparity. With this method, the virtual view is placed on the line between the real camera centres. Correct perspective views will be generated only if the real cameras are in rectified standard stereo geometry. For a detailed account of stereo rectification, see Chapters 6 and 7. Seitz and Dyer (1996) extended this method to *view morphing*, where the real views are pre-rectified to standard stereo geometry before disparity interpolation. This allows for perspectively correct linear interpolation between tilted cameras P_l and P_r (Figure 9.8a).

In the rectification step, each real camera image is rectified by homography mapping onto a rectified image plane that is parallel to the baseline $\mathbf{C}_r - \mathbf{C}_l$ between the cameras, and epipolar lines correspond to horizontal pixel coordinates. The disparity between corresponding image pixel $\mathbf{m}_l$ and $\mathbf{m}_r$ is simplified to the displacement of x-coordinates $d = x_r - x_l$. Linear interpolation of the rectified image onto a virtual camera centre $\mathbf{C}_v$ is performed on the x-coordinates only:

$$x_v = x_l + \frac{|\mathbf{C}_v - \mathbf{C}_l|}{|\mathbf{C}_r - \mathbf{C}_l|} d \qquad (9.12)$$

Proper interpolation is possible if the disparity of each pixel is known. Finding correspondences from the image data is the most critical task, and disparity estimation for image pairs is described in Chapter 7. Problems occur at object boundaries where background is

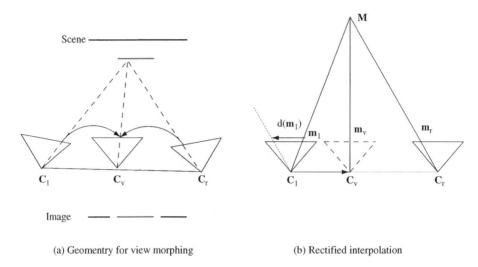

(a) Geomentry for view morphing   (b) Rectified interpolation

Figure 9.8 (a) View morphing interpolates along a linear camera path for arbitrary camera orientation. The (dark) background object is partially occluded in the real images due to the (light) foreground object, hence some image regions may not be interpolated. (b) Interpolation geometry for rectified images is simplified to horizontal interpolation

uncovered or occluded in one of the images. In those regions no correspondence can be established and holes may occur in the interpolated image.

Cooke *et al.* (2002) use view morphing in 3D video conferencing. From segmented disparity maps and images, special representations with low redundancy are assembled for transmission.

9.4.2 Image Transfer Methods

The fundamental geometric relations between two and three images can be exploited for view synthesis. Disparity interpolation actually is such a method where the 2D correspondence is separated into epipolar line (direction) and disparity (length). In uncalibrated systems, the fundamental matrix (see Chapter 6) can also be exploited to define the epipolar geometry.

Image transfer can also be used to extrapolate view, thus allowing more freedom in virtual view point selection. Laveau and Faugeras (1997) use epipolar transfer to extrapolate novel views from a set of given reference views. For the virtual view, epipolar geometry is computed with two reference views and correspondences are searched for in those views. A combination of disparity estimation and image warping has been proposed by Schaufler and Priglinger (1999). Restricting the disparity search along epipolar lines reduces the complexity to $O(n^2)$ instead of $O(n^3)$.

Avidan and Shashua (1997) introduced *trifocal tensor warping* to compute correspondences with the trifocal tensor $\mathcal{T}_k^{ij}$, (see Chapter 6). $\mathcal{T}_k^{ij}$ defines the fundamental geometry for three views which describes the image transfer between these views. If the tensor between three views and the correspondence between two views is known for calibrated cameras, then one can directly specify the correspondence in the third image and transfer the image intensity to the third image. The third view can be used to define the virtual camera and trifocal image transfer can be applied to synthesize the novel view.

9.4.3 Depth-based Extrapolation

So far, image correspondences have been used to compensate scene parallax. As discussed in Section 9.4.1 (Figure 9.8a), occluding boundaries will block parts of the view and leave unmodelled regions that cannot be interpolated from the image data.

If a dense depth map is given, then novel views can be synthesised using depth-compensated warping. A depth map contains the scene depth for every image point of a given view. Sometimes it is called 2.5D view, as it contains full 3D information, but from a single viewpoint only.

Depth could come from additional sensors such as a range scanner, or from stereoscopic depth estimation from multiple views. Dense correspondence estimation from a single image pair will always leave some background regions undefined, but the fusion of multiple views into a unique depth map allows dense depth computation (Koch *et al.* 1998).

Knowing the disparity of a pixel and the relative pose between the real cameras, the depth can be computed. Having more then two views, multi-view stereo approaches can be used to handle occlusions, and holes in the disparity maps can be filled from other views. The typical result of a such a stereo algorithm is a dense depth map: an image which codes the distance of each pixel of the original image. Having a per-pixel depth, the movement of the virtual camera is not restricted and allows one to extrapolate novel views from one single image and the associated depth map.

The simplest method is forward-warping of the pixel from the real view into the virtual view. A scene point $\mathbf{M} = (x, y, z)^\mathrm{T}$ in Euclidean notion is projected to an image point $\mathbf{m}$ with the given pose $(R, \mathbf{C})$

$$z \cdot \mathbf{m} = KR^\mathrm{T}(\mathbf{M} - \mathbf{C}) \qquad (9.13)$$

z is the pixel depth value which is lost during projection. Having the projective depth z for the pixel from depth estimation, the 3D scene point $\mathbf{M}$ can be reconstructed:

$$\mathbf{M} = z(KR^\mathrm{T})^{-1}\mathbf{m} + \mathbf{C} \qquad (9.14)$$

The desired point $\mathbf{m}_\mathrm{v}$ in the virtual camera P_v is then determined by projecting $\mathbf{M}$ into the virtual camera (see Figure 9.9a):

$$\mathbf{m}_\mathrm{v} = P_\mathrm{v}\mathbf{M} = z(K_\mathrm{v}R_\mathrm{v}^\mathrm{T})(KR^\mathrm{T})^{-1}\mathbf{m} + (\mathbf{C} - \mathbf{C}_\mathrm{v}) \qquad (9.15)$$

Equation (9.15) is another form of epipolar pixel transfer from the real to the virtual view. In fact, it directly describes the epipolar line with z as a free parameter.

One implementation of the described warping is *depth-image-based rendering* as proposed by Fehn (2004). It generates stereo image pairs from one monoscopic image and a depth map. The method is described in detail in Chapter 2.

Depth extrapolation does not solve the occlusion problem, because parts that are occluded in the real view may become visible for a new view. Without any further information this results in holes or artifacts, even if the depth map is completely filled. Having depth maps for adjacent views, forward mapping can be applied for these too, which can solve the occlusion problem. But forward mapping from multiple images is redundant for non-occluded regions and also can produce holes if the resolution of the source and the destination image are different.

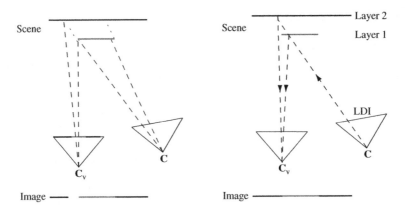

Figure 9.9 (a) Using depth maps allows extrapolation, occluded regions have to be filled from other cameras. (b) Fusing depth from multiple cameras into a multilayer representations allows extrapolation from the LDI without gaps

9.4.4 Layered Depth Images

To overcome the problem of occlusions, *layered depth images* (LDI) were introduced by Shade *et al.* (1998). The idea is to generate a multi-valued depth map, combining depth information from several real views into a single LDI view by fusing depth information and adding additional layers for each pixel corresponding to the different layers in the scene. In this way, both foreground and background object are stacked in the same representation and occlusion artifacts can be avoided.

Similar multi-valued representations have also been proposed by Chang and Zakhor (1999). In an acquisition phase, dense depth maps are computed and planes are fitted in low-contrast regions. These planes are identified and tracked through the image sequence. After transformation of all depth maps into one reference frame, a multilayered representation with both colour and depth information is constructed. View generation from multilayered images can be performed by forward-mapping each pixel of each layer via its depth information. Care has to be taken to ensure that nearby objects are not overdrawn by background objects. The layers have to be traversed in back-to-front order and depth-tests are needed when updating pixels in the destination image (see Figure 9.9b).

LDI representations can be computed from multiple real views simultaneously if a multi-camera plane-sweep algorithm is used (Collins 1996). A plane is positioned in space, all real images are projected onto the plane, and the image consistency of the projection for all images is tested. If all projections give a consistent photometric measure (the photo-consistency), then the colour and depth of the projection is saved in the LDI.

Yang *et al.* (2002) use a plane-sweep directly for rendering from given real views based on photo-consistency. Instead of computing a LDI, they directly render the colour and depth values of the plane sweep into the new view. For each pixel, photo-consistency decides if it is accepted or rejected. Then the plane is moved along the optical axis of the virtual camera. This process is repeated until the virtual view is completely filled.

A slightly different method of *texture slicing*, also known as elevation maps, is proposed by Vogelgsang and Greiner (2003). The per-pixel depth is stored in the alpha-channel of the textures and a plane-sweep algorithm uses an alpha-test to decide which pixel to render. The planes are rendered in depth intervals which are calculated when taking projection errors into account. If a pixel on a plane corresponds to a depth map pixel, the colour is taken from the associated pixel in the real view.

9.5 RENDERING FROM APPROXIMATE GEOMETRY

Multilayered depth images are a first step towards globally consistent models of the scene. Many other methods have been proposed to generate intermediate representations from the images and depth maps. This can be called 'modelling', because a static 3D model is created. If the resolution of the model is quite rough it is sometimes referred to as a 'geometry proxy', a geometrical approximation.

9.5.1 Planar Scene Approximation

LDI representations use plane hypotheses that are perpendicular to the optical axes. Often, the scene is rather restricted in depth. If the region of interest is roughly of continuous depth to the camera, then a planar approximation may help to compensate parallax effects. One example is the lumigraph approach where the scene geometry is approximated by a fitting plane that coincides with the (s, t) image plane. The scene geometry can now be thought of as decomposed into some fitting planes and additional local parallax components. Virtual views are interpolated by rectifying the real views into the virtual view by way of planar homography mapping. The deviations of the real scene depth from the fitting planes will cause image distortions due to the local parallax effects. If the fitting planes describes the dominant scene geometry well, then global homography mapping is a suitable rendering method. Buehler *et al.* (2001) and Bolles and Baker (1986) use this method.

If a consistent 3D surface is available, traditional polygon rendering techniques can be applied. In conjunction with dense input imagery, the polygon model can be used as approximate global geometry, and the real views that are nearest to the virtual view are mapped onto the surface to create *view-dependent texture mapping* (VDTM, Debevec *et al.* 1998). VDTM bridges the gap between traditionally textured surface models and IBR with approximate geometry.

9.5.2 View-dependent Geometry and Texture

Computing consistent 3D models from real views is not always possible. Therefore, methods have been proposed to generate intermediate representations or models that adapt the geometry for each new virtual view. These local models can avoid some of the problems of depth map representations. Depth maps cannot be used for backward mapping, and forward mapping of the discrete depth values may lead to holes. To circumvent this problem, Heigl *et al.* (1999) presented a modelling approach which constructs a local mesh that approximates the surface of the scene for each view. Using this mesh, the colour of each pixel is determined by backward mapping.

Evers-Senne and Koch (2003) enhanced this technique of view-dependent geometry as warping surface. Starting with real cameras and associated depth maps, in a first preprocessing

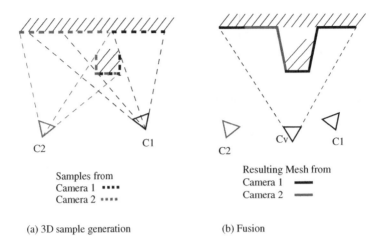

(a) 3D sample generation   (b) Fusion

Figure 9.10 View-dependent geometry: (a) 3D sample points are generated by backprojection; (b) fusing samples from multiple cameras results in a local mesh approximating the surface

step, for each real camera a set of 3D samples is generated by back-projecting and sampling the depth maps, as shown in Figure 9.10(a). During rendering, the samples of some selected real cameras are projected into the virtual camera. In the image plane of the virtual camera, a regular grid of 2D samples is used to fuse 3D samples and handle occlusions correctly. 2D sample points not hit by any 3D sample are removed and the remaining 2D sample points are meshed by Delauny triangulation. The resulting 2D mesh is then transfered into a 3D mesh using the associations between 3D samples and 2D samples. The result as shown in Figure 9.10(b) is an approximation of the scene only valid for this particular virtual camera. This approximation is used as warping surface for projective texture mapping from the real to the virtual view. The rendering approach is embedded into a complete system for plenoptic modeling from hand-held cameras (Evers-Senne *et al.* 2004) which has several key features:

- It can be used for view synthesis from dense or sparse plenoptic sampling obtained from hand-held uncalibrated camera images.
- The quality of the rendered images and the time required for rendering can be controlled by the number of 3D sample points per camera and by the number of real cameras to interpolate from.
- Inserting virtual objects is easy, because the geometry approximation in conjunction with the depth-buffer ensures correct occlusion handling.

9.6 RECENT TRENDS IN DYNAMIC IBR

The rendering from static scenes is now well understood. Challenges lie in the domain of dynamic scene modelling for 3D television and free-viewpoint video.

Generation of virtual views for dynamic scenes is not totally different from static scenes. Each of the static methods discussed could be used for dynamic scenes, assuming it can be implemented to run fast enough and to handle the amounts of data. One trend in recent years has been to transfer significant computational load to the graphics GPU hardware.

For visualization of polygonal models this is obvious, but today's programmable GPUs allow much more complex algorithms to be executed. For nearly every method presented, a real-time version using GPU support exists. Real-time interaction in this case means, that the user can move the virtual camera interactively and the virtual views are generated with 10–50 frames per second. Most often, precomputation requires much more time, for example dense depth estimation, camera pose estimation, calculation of optimized intermediate data structures. The step towards dynamic scenes requires to reduce and speed up these steps so that video sequences from several cameras can be processed and the user can control the virtual camera at interactive rates.

Some recent real-time systems use volumetric modelling. Volumetric models are similar to medical computer tomography datasets. Typically, the 3D space is partitioned into volume elements called voxels. For each voxel it is determined if it belongs to an object or not. This representation can also be constructed from depth maps, but the more popular approach is called 'shape-from-silhouette'. A visual hull is computed for objects by intersecting the silhouette cones from different real views. Li *et al.* (2003) use shape-from-silhouette to construct the visual hull by back-projecting images and using alpha and stencil calculation. Chapter 8 is devoted to volumetric reconstruction. To generate novel views from volumetric models different methods are available.

One common approach for IBR view synthesis of dynamic scenes is shown in Figure 9.11. The pipeline processing starts by capturing the scene using multiple calibrated fixed cameras. After object segmentation, shape-from-silhouette algorithms are used to create a volumetric model on the fly. After conversion to a surface model using polygon meshing, the standard polygonal rendering with view dependent texture mapping is used for visualization.

Saito *et al.* (1999) use 49 calibrated cameras, and compute a volumetric model. This is transferred into a polygonal surface model and during rendering it is used to generate correspondences in selected real views. From these correspondences per-pixel interpolation is performed after the determination of the disparity vectors.

Yamazaki *et al.* (2002) introduced *billboards*. A billboard or *micro-facet* is a small polygon, always facing the virtual camera. They approximate surface models, resample them to

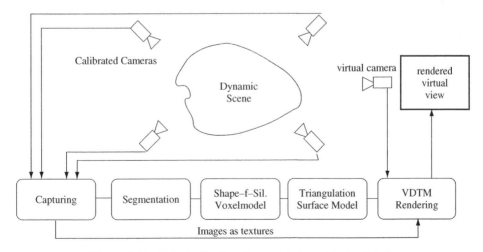

Figure 9.11 IBR pipeline for view generation with visual hulls

binary voxel-models and create a multi-resolution octree. Depth maps are used for per-pixel visibility culling to prevent texturing facets with inappropriate pixel.

Goldluecke and Magnor (2003) calculate volumetric models from different views with a shape-from-silhouette approach. This model is then rendered using billboards textured from original views. It is also possible to convert volumetric models into surface models, one approach is discussed in Chapter 3.

Recently, direct depth-based view interplation for dynamic scenes was proposed by Zitnick *et al.* (2004). They capture dynamic scenes from a set of fixed video cameras and compute depth-compensated view interpolation from multiple views interactively. The results look promising and show that indeed the challenge of interactive free-viewpoint video can be mastered in the near future.

REFERENCES

Adelson E and Bergen E 1991 The plenoptic function and the elements of early vision. *Computation models of visual processing*. MIT Press, pp. 385–394.

Antone M and Teller S 2002 Extrinsic calibration of omni-directional image networks. *International Journal of Computer Vision* **49**(2–3), 143–174.

Avidan S and Shashua A 1997 Novel view synthesis in tensor space, p. 1034.

Baker S and Nayar SK 1998 A theory of catadioptric image formation. *Proceedings of the IEEE Inter-national Conference on Computer Vision*, Bombay, pp. 33–42.

Bakstein H and Pajdla T 2003a Non-central cameras for 3d reconstrution. *Proceedings of Workshop 2003*, Czech Technical University in Prague. CTU Publishing House, Faculty of Architecture of CTU, Prague, Czech Republic, pp. 240–241.

Bakstein H and Pajdla T 2003b Ray space volume of omnidirectional 180×360 deg. images. In *Computer Vision — CVWW'03 : Proceedings of the 8th Computer Vision Winter Work-shop* (ed. Drbohlav O), Czech Pattern Recognition Society, Prague, Czech Republic, pp. 39–44.

Bakstein H and Pajdla T 2003c Rendering novel views for a set of omnidirectional mosaci images. *Proceeding of Omnivis 2003: Workshop on Omnidirectional Vision and Camera Networks*, IEEE computer Society Press, Los Alamitos, USA, p. 6.

Bolles RC and Baker HH 1986 Epipolar-plane image analysis: A technique for analyzing motion sequences. *Technical Report 377, AI* Center, SRI International, 333 Ravenswood Ave., Menlo Park, CA 94025.

Buehler C, Bosse M, McMillan L, Gortler SJ and Cohn Mf 2001 Unstructured lumigraph rendering. In *SIGGRAPH 2001, Computer Graphics Proceedings* (ed. Fiume E), ACM Press/ACM SIGGRAPH, pp. 425–432.

Chai JX, Chan SC, Shum HY and Tong X 2000 Plenoptic sampling. *Proceedings of the 27th annual conference on Computer graphics and interactive techniques*, ACM Press/Addison–Wesley, pp. 307–318.

Chang NL and Zakhor A 1999 A Multivalued Representation For View Synthesis. *Proceedings of the International Conference on Image Processing (ICIP)* Kobe, Japan, pp. 505–509.

Chen S and Williams L 1993 view interpolation for image synthesis *Siggraph 1993, Computer Graphics Proceedings*, pp. 279–288.

Chen SE 1995 QuickTime VR — an image-based approach to virtual environment navigation. *Computer Graphics* **29**(Annual Conference Series), 29–38.

Chen WC, Bouguet JY, Chu MH and Grzeszczuk R 2002 Light field mapping: Efficient representation and hardware rendering of surface light fields. In *SIGGRAPH 2002 Conference Proceedings* (ed. Hughes J), Annual Conference Series. ACM Press/ACM SIGGRAPH, pp. 447–456.

Collins. R 1996 a space-sweep approach to true multi-image matching. *Proceedings of Computer Vision and Pattern Recognition Conference*, pp. 358–363.

Cooke E, Kauff P and Schreer O 2002 Imaged-based rendering for tele-conference systems. *Proceedings of WSCG 2001, 9th Int. Conference on Computer Graphics, Visualization and Computer Vision*, Plzen, Czech Republic, p. 119.

Debevec P 1998 Rendering Synthetic objects into real scenes: Bridging traditional and image-based graphics with global illumination and high dynamic range photography. *Computer Graphics* **32** (Annual Conference Series), 189–198

Debevec P, Yu Y and Boshokov G 1998 Efficient view-dependent image-based rendering with projective texture-mapping. *Technical Report CSD-98–1003*, University of California, Berkelely.

Evers-Senne JF and Koch 2003 Image based interactive rendering with view dependent geometry. *Eurographics 2003* Computer Graphics Forum Eurographics Association, pp. 573–582.

Evers-Senne JF, Woetzel J and Koch R 2004 Modelling and rendering of complex scenes with a multi-camera rig, pp. 11–19.

Fehn C 2004 Depth-image-based rendering (dibr), compression and transmission for as new approach on 3D-TV. *Proceedings Stereoscopic Displays and Applications*, San Jose, CA, USA.

Fujii T 1994 *A Basic Study in the Integrated 3-D visual Communication*. PhD Thesis University of Tokyo.

Fujii T and Tanimoto M 2002 Free Viewpoint TV system based on ray-space representation. *Three-Dimensional TV, Video, and Display* **4864**, 175–189.

Geyer C and Daniilidis K 2003 Omnidirectional Video. *The Visual Computer*, pp. 405–416.

Goldluecke B and Magnor M 2003 Real-time, free-viewpoint video rendering from volumetric geometry. In *Visual Communications and Image Processing 2003* (ed. Ebrahimi T and Sikora T), *Proceedings of SPIE*, **5150**, pp. 1152–1158.

Gortler SJ, Grzeszczuk R, Szeliski R and Cohen MF 1996 The Lumigraph. *Proceedings SIGGRAPH '96* **30** (Annual Conference Series), 43–54.

Heigl B, Koch R and Pollefeys M 1999 Plenoptic modeling and rendering from image sequences taken by a hand-held camera *Proceedings of DAGM 1999*, pp. 94–101.

Isaksen A, McMillan L and Gortler SJ 2000 Dynamically reparameterized light fields. In *Siggraph 2000, Computer Graphics Proceedings* (ed. Akeley K), ACM Press /ACM SIGGRAPH /Addison Wesley Longman, pp. 297–306.

Kang SB 1997 A survey of image-based rendering techniques. *Technical Report*, DEC Cambridge Research.

Kang SB, Uyttendaele M, Winder S and Szeliski R 2003 High dyunamic range video. *ACM Trans. Graph.* **22**(3), 319–325.

Kawasaki H, Ikeuchi K and Sakauchi M 2001 Light field rendering for large-scale scenes. *Computer Vision and Pattern Recognition (CVPR)*, Hawaii, USA, p. 2.

Klette R, Gimel'farb G, Wei S, Huang F, Scheibe K, Scheele M, Börner A and Reulk R 2003 On design and applications of cylindrical panoramas. *Proceedings, Computer Analysis of Images and Patterns*, Groningen, The Netherlands, pp. 1–8.

Koch R, Pollefeys M and Gool LV 1998 Multi viewpoint stereo from uncalibrated video sequences *Proceedings ECCV'98* number 1406 in *LNCS*. Springer, Freiburg, pp. 55–71.

Kurashima C, Yang R and Lastra A 2002 Combining approximate geometry with view-dependent texture mapping. *XV Barzilian Symposium on Computer Graphics and Image Processing*, Fortaleza, CE, Brazil, pp. 112–120.

Laveau S and Faugeras O 1997 3D representation as a collection of images. *Proceeding of the IEEE Int. Conf. on Pattern Recogniton (CVPR'97)* IEEE Publishers, pp. 689–691.

Levoy M and Hanrahan P 1996 Light field rendering. *Proceedings SIGGRAPH '96* **30** (Annual Conference Series), 31–42.

Li M, Magnor M and Seidel Hp 2003 Improved hardware-accelerated visual hull rendering. *Proceedings, Vision, Modeling, and Visualization (VMV-2003)*, Munich, Germany, pp. 151–158.

Lippman A 1980 Movie-maps: An application of the optical videodisc to computer graphics *Proc. ACM SIGGRAPH*, pp. 32–42.

McMillan L 1999 Image-based rendering using image-warping — motivations and background. *computer Graphics (SIGGRAPH'99), course n. 39*, pp. 61–64.

Naemura T, Yoshida T, and Harashima H 2001 3D computer graphics based on intergal photography. *Optics Express* **8**, 255–262.

Peleg S and Herman J 1997 Panoramic mosaics by manifold projection. *CVPR97*, pp. 338–343.

Rademacher P and Bishop G 1998 Multiple-centre-of-projection images. *Computer Graphics (SIGGRAPH'98)*, pp. 199–206.

Rousso B, Peleg S and Finci I 1997 Mosaicing with genralized strips. *DARPA97*, pp. 225–260.

Rousso B, Peleg S, Finci I and Rav-Acha A 1998 Univesal mosaicing using pipe projection. *ICCV 98*, pp. 945–952.

Saito H, Baba S, Kimura M, Vedual S and Kanade T 1999 Appearance-based virtual view generation of temporally-varying events from multi-camera images in the 3d room. *Proceedings of 2nd International Conference on 3D Digital Imaging and Modeling*, pp. 516–525.

Schaufler G and Priglinger M 1999 Efficent displacement mapping by image warping, *Proceedings of the 10th Eurographics Workshop on Rendering*, pp. 175–186.

Seitz SM and Dyer CR 1996 View morphing. *SIGGRAPH 96*, pp. 21–30.

Shade J, Gortler S, He LW and Szeliski R 1998 Layered depth images. *Proceedings ACM SIGGRAPH*, ACM Press / ACM SIGGRAPH, pp. 231–242.

Shum HY and Szeliski R 1997 Panoramic image mosaics. *Technical Report*, Microsoft Research.

Shum HY, Wang L, Chai J and Tong X 2002 Rendering by manifold hopping. *International Journal of Computer Vision (IJCV)*, 185–201.

Shum HY and He LW 1999 Rendering with concentric mosaics. *Proceedings of the 26th Annual Conference on Computer Graphics and Interactive Techniques*, ACM Press/Addison–Wesley Publishing, pp. 299–306.

Shum HY and Kang SB 2000 A review of image-based rendering techniques. *Proceedings, Visual Communications and Image Processing*, pp. 2–13.

Takahashi T, Kawasaki H, Ikeuchi K and Sakauchi M 2000 Arbitrary view position and direction rendering for large-scale scenes. *Proceedings CVPR 2000*, pp. 296–303.

Teller S, Antone M, Bodnar Z, Bosse M, Coorg S, Jethwa M, and Master N 2003 Calibrated, registered images of an extended urban area. *International Journal of Computer Vision*, 93–107.

Tong X, Chai J and Shum HY 2002 Layered lumigraph with lod control. *Journal of Visualization and Computer Animation*, 249–261.

Vogelgsang C and Greiner G 2003 Interactive range map rendering with depth interval texture slicing. *Vision, Modelling and Visualization (VMV)*, Munich, Germany, pp. 477–484.

Weinshall D, Lee MS, Brodsky T, Trajkovic M and Feldman D 2002 New view generation with a bi-centric camera. *Proceedings of the 7th European Conference on Computer Vision*, Copenhagen, DK, pp. 614–618.

Yamazaki S, Sagawa R, Kawasaki, H and Ikeuchi K and Sakauchi M 2002 Microfacet billboarding. *Rendering Techniques 2002 (Eurographics Workshop Proceedings)*, pp. 169–179.

Yang R, Welch G and Bishop G 2002 Real-times consensus-based scene reconstruction using commodity graphics hardware. *Proceedings of Pacific Graphics*, Tsinghua University, Beijing, China, pp. 207–214.

Zitnick C, Kang S, Uyttendaele M, Winder S and Szeliski R 2004 High-quality video view interpolation using a layered representation. *Proceedings, ACM SIGGRAPH*, Los Angeles, CA, pp. 600–608.

Zomet A, Feldman D, Peleg S and Weinshall D 2003 Mosaicing new views: The crossed-slits projection. *IEEE Trans. on PAMI* pp. 741–754.

10
3D Audio Capture and Analysis

ÜBERARBEITET
14:03, 23.05.2005

Markus Schwab and Peter Noll

Technische Universität Berlin, Germany

10.1 INTRODUCTION

In human communication, speech is very important and communication without speech is hard to realize. Natural and comfortable communication in a video conference depends strongly on the quality of the acquisition and reproduction of speech. High-quality speech acquisition can be achieved by decreasing the distance between the microphone and the speaker, e.g., by using a lip microphone. This maximizes the power of the captured signal and results in a high signal-to-noise ratio. In the reproduction chain, earphones provide a perfect coupling. Additionally, microphone and loudspeaker are decoupled and feedback (acoustic echo) is minimized. On the other hand, this has the disadvantage that every videoconference participant has to wear a microphone device and earphones, resulting in uncomfortable communication. Another drawback is that it is very difficult to create a virtual 3D acoustic environment with a headset because the position and orientation of the user's head are unknown.

A high-quality videoconferencing system should allow users to move freely in the conference environment and should create a virtual acoustic environment which contains 3D information (e.g., position of a speaker) and enables multimedia functionalities such as speech-driven devices or playback of music videos. Such features can be realized with microphone arrays and loudspeaker arrays. Microphone arrays are capable of picking up sound in high quality and also extracting information about the position of the sound source. With a loudspeaker array sound sources can be virtually placed in the conference environment and create a 3D virtual acoustic environment.

Figure 10.1 shows an overview of a high-quality 3D audio conference system. Whereas the noise reduction, echo canceler, and coding/decoding modules are vital for a communication

3D Videocommunication — Algorithms, concepts and real-time systems in human centred communication
Edited by O. Schreer, P. Kauff and T. Sikora © 2005 John Wiley & Sons, Ltd

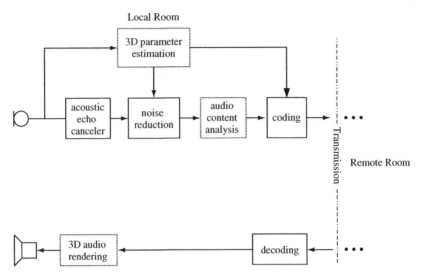

Figure 10.1   Audio system overview for 3D telecommunication

system, the estimation of 3D parameters and the audio rendering and playback are essential for 3D video communication in order to create a virtual 3D audio environment. Content analysis of the audio signal can enable special features such as voice-controlled virtual presentations or enhancement of an audio playback system which depends on the nature of the sound, e.g., speech or music. 3D audio rendering will be discussed in Chapter 15.

The scope of this section does not allow us to go into details of each module. The interested reader is recommended to consult the references. The intention of this chapter is to emphasize which tools are necessary for a 3D audio communication system and how they perform. First, we will survey the state-of-the-art of acoustic echo control. Then, issues on the sensor placement are discussed which are relevant for a 3D speaker localization. After this, a speaker localization and tracking system will be presented. The location of the sound sources are important for multi-channel speech enhancement as well as for virtual 3D audio playback. Finally, an overview of multi-channel and single-channel noise reduction systems will be presented, including some results.

10.2   ACOUSTIC ECHO CONTROL

Acoustic echo control is the first module in the signal processing chain. Combinations with noise reduction modules, e.g., beamformer, have shown that it is preferable to place the acoustic echo control modules directly in the sensor channels to obtain optimal performance (Herbordt and Kellermann 2002). The acoustic echo problem is visualized in Figure 10.2. Consider speaker A in the local room. Speaker A's speech is transmitted to the loudspeaker in the remote room and is picked up by a microphone in the remote room. Without any precautions, the signal of speaker A will be retransmitted to the local room. Therefore, conference participants will be annoyed by listening to their own speech delayed by the round-trip of the audio system. Such a system may also become instable if the amplification of a signal during a round trip is larger than one. This problem is resolved by placing acoustic echo control units in the signal path. The acoustic echo control units attenuate the remote speaker signal coming from the loudspeaker and picked up by the microphone before being transmitted to the remote speaker. This problem has been tackled by a large number of publications over the last two decades and many techniques have been proposed to solve it. Good overviews and bibliographies are given in Hänsler (1992), Gilloire *et al.* (2000), Breinig *et al.* (1999), Gilloire *et al.* (1996), and Hänsler and Schmidt (2004).

10.2.1 Single-channel Echo Control

In this section, we outline the state-of-the-art of acoustic echo control in single-channel systems. Then we will extend to multi-channel cases.

An early attempt to reduce acoustic echo was to decouple the loudspeaker from the microphone. This can be achieved by using a headset incorporating a microphone . This will inconvenience the user in that it is necessary to wear a headset with a mounted microphone. In order to enable hands-free speech communication an electronic solution is favoured.

In the past, two major electronic solutions have been proposed to reduce acoustic echo in the return path of a communication system. The oldest is a so-called loss control unit which attenuates the signals, depending on a voice activity detection (Figure 10.2). When the local speaker A is active an attenuation is put into the signal path from the remote to the local room and vice versa. In the case of double talk, i.e., local and remote speaker are simultaneously active, attenuation is applied to both signal paths. In essence such an echo control provides a half-duplex communication. But this disadvantage is balanced by very good echo attenuation which satisfies the requirements of the International Telecommunication Union (ITU) and European Telecommunication Standards Institute (ETSI). Loss control is usually applied in telephone systems where low delays and low computational complexity play an important role.

The second solution is to place adaptive filters in parallel with the acoustic system (Figure 10.3). Ideally, the echo signal $d(n)$ of the incoming signal $x(n)$ is canceled by subtracting an echo estimate $\hat{d}(n)$ from $d(n)$. The adaptive filter $\hat{h}(n)$ is an estimate of the transfer function between the digital input signal of the loudspeaker $x(n)$ and the digital output signal of

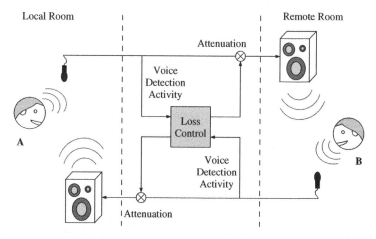

Figure 10.2 Acoustic echo scenario with a loss control unit, leading to a half-duplex system

$$ERLE(n)= \frac{E\{d^2(n)\}}{E\{(d(n)-\hat{d}(n))^2\}}$$

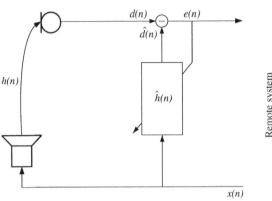

Figure 10.3 Acoustic echo canceller

the microphone $d(n)$. This implies that the overall transfer function includes the room transfer function $h(n)$ as well as the transfer functions of the loudspeaker with its D/A converter and the microphone, including its A/D converter. The adaptation is a system identification problem. If the estimate is accurate, the signal $e(n)$ does not contain any components of $x(n)$, thus, $e(n)$ and $x(n)$ are perfectly decoupled. This echo control solution is called adaptive echo cancellation. In real-world situations, the impulse responses can have up to several thousands of samples, depending on the reverberation time. In general, larger rooms exhibit longer impulse responses, but the wall reflection coefficients and the interior decoration of the room have a great impact on the resulting impulse responses. Due to realization aspects the estimated filter length N has to be limited. Filter length limitations and estimation errors result in an echo suppression that is not perfect. A measure of the efficiency of the acoustic echo cancellation is the so-called echo return loss enhancement (*ERLE*):

$$ERLE(n) = \frac{\mathbf{E}\{d^2(n)\}}{\mathbf{E}\{(d(n) - \hat{d}(n))^2\}} \tag{10.1}$$

The system identification, i.e. the estimation of $\hat{h}(n)$, is usually accomplished by adaptive algorithms because it is necessary to track changes. The three main adaptive algorithms are:

- normalized least-mean-square (NLMS) algorithm;
- affine projection (AP) algorithm;
- recursive least-squares (RLS) algorithm.

In the following, we will briefly describe some properties of the different algorithms.

Most acoustic echo canceling systems rely on the NLMS algorithm. This is a gradient-based algorithm which minimizes the mean square error $E\{e(n)^2\}$. The filter update is carried out by taking only the current signal vector into account. This leads to a slow convergence speed in the case of a correlated input signal, e.g., speech. But the simplicity of this algorithm makes it attractive for real-time applications in communication systems.

Table 10.1 Properties of adaptive algorithms NLMS, AP, and RLS

Algorithm	Complexity	Convergence
NLMS	$O_{\mathrm{NLMS}}(N) \propto 2N$	Slow
AP	$O_{\mathrm{AP}}(N) \propto 2LN$	Fast
RLS	$O_{\mathrm{RLS}}(N) \propto N^2$	Fast

The AP algorithm uses instead the last M signal vectors to update the estimation. This increase in computational load is compensated by a faster convergence speed. The computational cost is due to the inversion of a $M \times M$ matrix at each coefficient update. Attention should also be paid to the robustness constraints. Faster convergence implies a faster divergence in the case of errors, e.g., misdetection of double talk.

As the most powerful algorithm, the RLS algorithms minimizes the weighted sum of the squared error $\sum_{k=1}^{n} = \beta(k)e(n)^2$ with $0 < \beta < 1$. At each coefficient update an $N \times N$ matrix has to be inverted. If an exponential window $\beta(k) = \lambda^{n-k}$ with the forgetting factor $0 < \lambda < 1$ is used, the matrix inversion can be approximated by a recursive update. This leads to a complexity of $O(N^2)$. A major problem of the RLS algorithm is stability especially for the inversion of the matrix. To overcome this problem, matrix regularization is often used.

It should be noted that fast versions of the AP algorithm (Gay and Travathia 1995) and the RLS algorithm (Kalouptsidis and Theodoridis 1993) exist. The properties of these three adaptive algorithms are summarized in Table 10.1.

The convergence speed can be increased by decorrelating the input signals. This can be achieved by a decorrelation filter, e.g., a linear prediction error filter, applied to the loudspeaker signal $x(n)$ to estimate the room transfer function. More details can be found in Chapter 9 of Hänsler and Schmidt (2004).

10.2.2 Multi-channel Echo Control

Multi-channel echo control is based on several microphones and loudspeakers (Figure 10.4). In the case of L loudspeakers and M microphones the multi-channel acoustic echo cancellation algorithm has to identify L impulse responses for M microphones. This is shown in (Figure 10.5). In total LM room transfer functions have to be estimated.

The main problem in multi-channel acoustic echo cancelling is that the input signals coming from the loudspeaker are correlated, i.e., filtered versions of the same signal. These correlations degrade the performance of the estimation algorithms of the impulse responses and lead to an ill-conditioned problem where the optimal impulse responses are not uniquely defined. Benesty *et al.* (1998) have shown that it is advantageous to use the RLS algorithm instead of the NLMS algorithm.

The solution for this problem is a decorrelation of the input signals. Three common methods have been proposed in the past:

- insertion of noise;
- nonlinear transformation;
- time-variant filtering.

All these methods are applied to the loudspeaker input signal, thus introducing more or less audible distortions in the reproduced speech signal. In Figure 10.5 nonlinear functions are

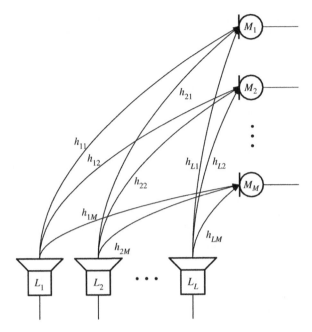

Figure 10.4   Signal paths in a multi-channel system

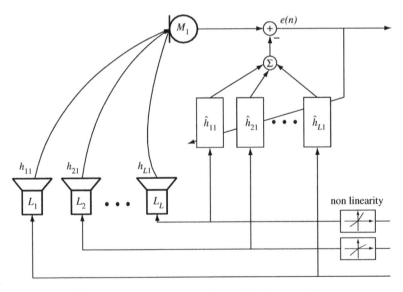

Figure 10.5   Multi-channel acoustic echo canceller

indicated in the loudspeaker signal paths as a nonlinear transformation method. A real-time implementation of a stereophonic acoustic echo cancellation system has been described in Fischer *et al.* (2001). This realization was based on a nonlinear transformation and the RLS algorithm.

10.3   SENSOR PLACEMENT

Sensor placement is the first issue in the design of a microphone array. Acoustic source localization algorithms generally exploit the coherence between the acquired microphone signals. Thus, we will group the microphones into K microphone pairs whose correlations will be used for the source localization. $K = \frac{1}{2}M(M-1)$ is the maximum number of possible microphone pairs if M is the number of microphones. The distance between microphones grouped to a pair should be small because the coherence of the microphone signals decreases with an increase in the distance between the sensor positions (Brandstein *et al.* 1996; Drews 1999). In the case of a time-difference of arrival (TDOA)-based maximum likelihood localization algorithm (described in the following section), the expected localization error Δp can be estimated with knowledge of the error variances of the TDOA estimates, the source location, and the microphone positions (Brandstein *et al.* 1996):

$$cov(\Delta p) = (\mathbf{H}^T \mathbf{V} \mathbf{H})^{-1} \tag{10.2}$$

where $\mathbf{V}$ is a diagonal (K, K) matrix. The diagonal elements are the reciprocal variances of the TDOA estimates $v_{ii} = 1/\sigma_i^2$. $\mathbf{H}$ is a $(K, 3)$ matrix which relates the error of the localization Δp to the errors in the estimated TDOAs $\Delta \tau$:

$$\Delta \tau = \mathbf{H} \Delta p \tag{10.3}$$

In (Figure 10.6) the predicted location errors are depicted for two different array geometries. In the left figure the array geometry is a compact eight-microphone array and in the right figure the array geometry is more sparse, i.e., a microphone pair is placed at each border of the sensed area. It is assumed that the variances for the TDOA estimation σ_i^2 are affected only by the signal-to-noise ratio (SNR) at the microphones. Since the spacing between the microphones of a pair is small the same SNR can be assumed for both microphones. If we additionally assume free-field propagation, the variances of the TDOA estimates are reciprocally proportional to the square of the distance r_i between the microphone pair and the sound source $\sigma_i^2 \propto 1/r_i^2$.

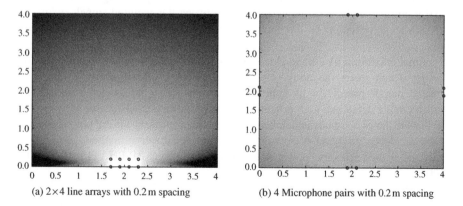

(a) 2×4 line arrays with 0.2 m spacing   (b) 4 Microphone pairs with 0.2 m spacing

Figure 10.6   Logarithmically scaled predicted localization error (darker dots have a greater predicted error): (a) one compact eight-microphone array with a spacing of 0.2 m between the microphones; (b) array with four microphone pairs, each with a spacing of 0.2 m between the microphones

The brightness in Figure 10.6 is related to the predicted error at a given position. A brighter dot indicates a smaller predicted error. Figure 10.6 (a) shows that a precise localization is possible only close to the microphone array whereas in Figure 10.6 (b) the predicted localization error is better balanced and a localization is feasible for all positions in the sensed area $[4 \times 4\,\mathrm{m}^2]$. This 2D representation can easily be extended to 3D. Thus, the result of Figure 10.6 illustrates that a 3D source localization with a microphone array is feasible only in the near field of the array. The conclusion from this example is that the microphone pairs have to be widely distributed in the sensed area to be able to perform a 3D source localization.

10.4 ACOUSTIC SOURCE LOCALIZATION

10.4.1 Introduction

Acoustic source localization and tracking is an important task in a 3D audio conference system. The information of the position of sound source can be exploited by a multi-channel speech enhancement algorithm which steers to the direction of the sound source and attenuates sources coming from other positions. The 3D information about the position of the sound source is also needed by the audio rendering algorithm because with this information the rendering system is capable of virtually placing the sound at the right position in the remote room.

The problem of acoustic source localization based on microphone arrays has been a mainstream research topic for over two decades. The solutions available in the literature can be coarsely classified into three broad categories:

- those based on maximizing the steered response power (SRP) of a beamformer;
- those based on high-resolution spectral estimation (HRSE) methods;
- those based on Time-Difference Of Arrival (TDOA) algorithms.

Steered beamforming is a well-known method for deriving information on the source locations directly from a filtered linear combination of the acquired signals. Methods based on HRSE imply the analysis of the correlations between the acquired signals, while TDOA methods extract information on the source location through the analysis of a set of delay estimates.

The optimal maximum likelihood (ML) SRP-based localization methods rely on a focused beamformer which steers the array to various locations in space and looks for peaks in the detected output power. In its simplest implementation, the steered response can be obtained through a delay-and-sum process performed on the signals acquired by two micro-phones. One of the two signals is delayed in order to compensate for the propagation delays due to the incidence direction of sounds. SRP based localization methods are very robust especially the steered response power with phase transform (SRP-PHAT) method (DiBiase *et al.* 2001).

Source localization methods of the second category are all based on the analysis of the spatial covariance matrix (SCM) of the array sensor signals. The SCM is usually unknown and needs to be estimated from the acquired data. Such solutions rely on high resolution spectral estimation techniques (Stoica and Moses 1997). Popular algorithms based on HRSE are the minimum variance beamformer (MVB) (Krim and Viberg 1996) and the multiple signal classification (MUSIC) algorithms (Schmidt 1986). These algorithms can be extended to wideband signals, e.g., speech, by transforming the signal into a narrow band signal. Each narrowband signal can be processed individually (incoherent method) or an universal focusing SCM (Wang and Kaveh 1985) can be generated to perform a coherent localization.

Methods based on TDOA algorithms have two steps. First, they estimate the TDOAs. The most popular method for TDOA estimation is the cross-correlation approach. A more robust estimation can be achieved with the generalized cross-correlation GCC (e.g., GCC-PHAT) (Knapp and Carter 1976). A more recent approach is the so-called adaptive eigenvalue decomposition (AED) algorithm (Benesty 2000). The AED algorithm performs better than the GCC-PHAT algorithm in highly reverberant environments. The second step is to estimate the position of the source with knowledge of the TDOAs. For each TDOA (note that there are $M - 1$ independent TDOAs, where M is the number of sensors) a nonlinear equation can be established:

$$\tau_{ij} = \frac{|\mathbf{M_i} - \mathbf{S}| - |\mathbf{M_j} - \mathbf{S}|}{c} \tag{10.4}$$

This set of equations can be solved by a minimization algorithm. Abel and Smith (1987) proposed a closed-form solution called spherical interpolation. From the $M - 1$ Equations (10.4) and a reference equation which relates the distance between the first microphone M_1 and the source position S a set of $M - 1$ linear equations can be formed and a closed-form solution can be expressed. Huang *et al* (2001) extended this approach by taking the reference equation into account 2001.

10.4.2 Real-time System and Results

In this section we describe a real-time acoustic source localization and tracking system for multiple sound sources. First, candidates for source locations are calculated with the previously mentioned SRP-PHAT method. These candidates are fed into a logical system which allocates the candidates to previously found sound sources. A Kalman filter for each sound source optimally estimates the trajectories of the sound sources.

As mentioned before, microphone pairs have to be selected whose correlations contribute to the localization. For each microphone pair, the normalized (whitened) cross-power spectrum is calculated and then transformed into the time domain to obtain the generalized cross correlation with phase transform (GCC-PHAT) for each selected microphone pair.

With the GCC-PHAT sequences from all microphone pairs we define a probability function for the presence of a sound source Q depending on a virtual speaker position S:

$$Q(S) = \sum_{\text{pairs}(i,j)} R^{i,j}_{\text{GCC–PHAT}} \delta(S, i, j), \tag{10.5}$$

where $R^{i,j}_{\text{GCC–PHAT}}$ is the generalized cross-correlation with phase transform between the ith and jth microphone signal. $\delta(S, i, j)$ denotes the TDOA between the ith and jth microphone

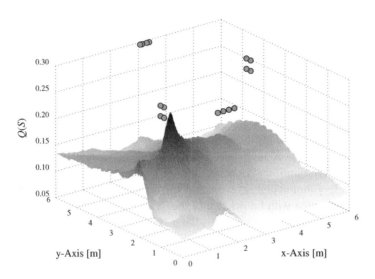

Figure 10.7   Example of the probability function for the presence of a single sound source $Q(S)$ obtained with the SRP-PHAT method, Darkness indicates a high probability of the presence of a sound source

signal related to a virtual sound source location at position S. If M_i and M_j are the microphone positions, δ can be calculated as follows:

$$\delta(S, i, j) = \frac{|S - M_i| - |S - M_j|}{c} \tag{10.6}$$

where c represents the velocity of sound in air.

Figure 10.7 shows an example of the probability function of the presence of a sound source over the sensed area. The black circles with a grey colour represent the microphones. The dark elevation in the figure indicates the high presence probability of a sound source at $x = 1.18$ m and $y = 2.10$ m.

A candidate speaker position is given at the maximum of the probability function $P(s)$. The SRP-PHAT algorithm is only able to estimate a single source position. In order to be capable of tracking multiple sound sources, a logical system is employed. The logical system has the task of associating the candidate source position to previously found sound sources. First, the peak in the probability function of the presence of a sound source $P(S)$ is evaluated to accept or reject the position as a candidate. The data association can be accomplished by the use of acceptance regions (Bar-Shalom and Fortman 1988). If a candidate source position cannot be associated to an existing sound source, the candidate position is used to initialize a new sound source. If there is a certain number of candidate positions within a time slot and an acceptance region, the new sound source is accepted. A sound source is lost if no new candidate is associated with this sound source within a predefined time.

As already mentioned, a Kalman filter is used to improve the quality of the estimated speaker position. The process noise is modelled by uncorrelated zero-mean noise. The covariance of the measurement noise is calculated as a function of source location, sensor positions, and the generalized cross-correlations of the microphone pairs (Brandstein *et al.* 1996).

Figure 10.8 shows an overhead view of the experimental setup. The room has a size of 10×6 m, its height is 4 m and the room reverberation is moderate. The sensed area is

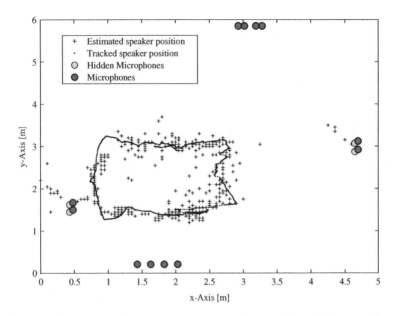

Figure 10.8 Overhead view of the scenario with a speaker tracking example

5×6 m. We have used 16 microphones subdivided into 4 subarrays with 4 microphones each. The microphones are marked with bold circles with grey faces. For two subarrays only two microphones can be seen in the overhead view since the four microphones in these subarrays are arranged in vertical rectangle. The four actually hidden microphones are indicated by circles with lighter gray faces. A result of a speaker tracking is shown. The crosses mark the estimated position as derived by the SRP-PHAT method and the dots (note that the sequence of dots appears almost as a solid line) represents the improved results after Kalman filtering.

10.5 SPEECH ENHANCEMENT

Audio signals recorded in the real world are often contaminated by environmental noise. This effect degrades the quality of the audio signals. Speech coders are often sensitive to noise and the quality of the reconstructed speech at the output of the speech decoder can decrease drastically. Also, an analysis of an audio signal will be more difficult if the performance of analysis tools decreases due to noise. As already shown in Figure 10.1, a noise reduction module is often needed and usually placed after the acoustic echo cancellation (Herbordt and Kellermann 2002).

In the following, we will distinguish between stationary and non-stationary noise and between directed and diffuse noise. For example, environmental background noise in a conference room is normally stationary and diffuse whereas a door opening is non-stationary and directed noise.

Two major strategies exist in reducing the noise. Multi-channel approaches, also called beamforming, combine several microphone signals in such a way that signals from a desired position are amplified, whereas signals coming from other positions are added diffusely or even destructively. Beamforming is a spatial filtering where the beamformer output has a spatial directivity and emphasizes the signals from the desired position. This method is especially suited for directional noise because the microphone signals can be combined in such a way that interference signals from a direction are deleted, so-called null-steering. Such beamforming techniques are able to separate signalsn be separated by aiming at each signal and setting the directivity at the other signal direction to zero.

Single-channel approaches cope with the problem of stationary noise. The noise power is usually estimated in speech pauses and an optimally weighting factor is calculated to estimate the speech from the noisy speech. It is assumes that the noise power does not change during speech periods. Single-channel post-filters are often applied after beamforming.

10.5.1 Multi-channel Speech Enhancement

Delay-and-Sum beamforming (DSB) is a simple multi-channel speech enhancement technique. The microphone signals are time-aligned and then added up. In the ideal case the time-aligned signals match perfectly and their amplitudes are added whereas the noise signals at each microphone are completely uncorrelated and only their power is added. Assuming M microphones, the signal-to-noise ratio (SNR) can be enhanced theoretically by a factor M. In real-world scenarios the SNR enhancement is much smaller than this theoretical value because of two effects. First, the noise signals at each microphone are partly correlated, depending on the distance between the microphones (Kuttruff 1990). By increasing the distance between the microphones, the correlation decreases. Secondly, since the correlation between the desired signals at each microphone decreases and the performance of the DSB decreases because the signals cannot be added coherently. Superdirective beamformers take the correlation properties of the noise into account to achieve a higher directivity. Figure 10.9 shows the directivity pattern for different frequencies in the DSB case (solid curves) and in the superdirective case (dashed curves). The microphone considered array is a uniform linear array with four microphones and a spacing 0.1 m between the microphones. At low frequencies the directivity for the DSB is unsatisfying. The superdirective array has a better performance at low frequencies. This increase in directivity is at the cost of a high susceptibility to errors in the microphone positions and in the directivity patterns of the microphones doerbecker. This problem arrises especially at low frequencies. Therefore, the superdirectivity has to be reduced by regularization of the cross correlation matrix of the noise field. At high frequencies the directivity patterns of the superdirective beamformer and the DSB are similar. In order to obtain an acceptable directivity, i.e. suppression of interference signals coming from other directions, many microphones are needed Silverman *et al.* 1996. This makes microphone arrays not very suitable to cope with the problem of diffuse noise.

In the case of directed noise or interference speakers adaptive beamforming promises a good solution. The most popular approach for adaptive beamforming is the *generalized sidelobe-canceller* method (GSC) (Griffithes and Jim 1982). Figure 10.10 shows its structure. The upper path consists of a fixed beamformer, e.g., DSB, which generates a first approximation of the desired signal. In the lower path, noise reference signals are generated by a blocking matrix which eliminates the target signal. An adaptive input canceller removes all signal components correlated with the noise reference signals in the fixed beamformer output signal.

Recent implementations of the GSC take advantage of adaptive filters for the blocking matrix and the multiple input canceller. Robustness constraints contribute to a more stable system and small errors in the sound source localization are tolerated. Herbordt and

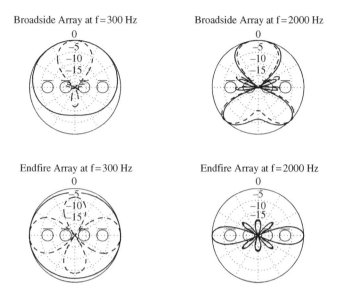

Figure 10.9 Directivity patterns in dB for a Delay-and-Sum beamformer (solid) and a superdirective beamformer (dashed). The results are given for an uniform linear array (ULA) with four microphones and a spacing of 0.1 m

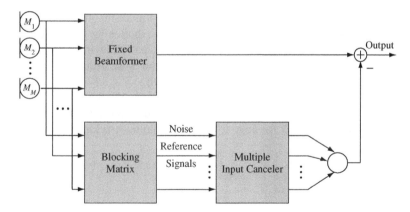

Figure 10.10 Structure of the generalized sidelobe canceller

Kellermann (2001) report an interference rejection of more than 25 dB in a room with low reverberation (reverberation time $T_{60} = 50$ ms).

10.5.2 Single-channel Noise Reduction

Methods of single-channel noise reduction are often applied as post-filters to the output of a multi-channel speech enhancement system. Single channel algorithms remove stationary diffuse noise. This is complementary to multi-channel algorithms which are less effective in the presence of diffuse noise.

Many noise reduction algorithms use some form of a statistical signal model and apply some form of short-time spectral analysis/synthesis. The noisy signal is then decomposed into spectral components by means of a spectral transform, a filter bank, or wavelet transform. Figure 10.11 depicts a typical implementation of a single channel noise reduction system based on a spectral transform. This system has two major components: (i) the noise estimation, and (ii) the signal estimation module.

Noise estimation. The first algorithms for the noise estimation module were based on voice activity detectors (VAD) (Boll 1979; McAulay and Malpass 1980). The noise estimation is updated only when speech is absent. The update is usually done by recursive smoothing over the past frames detected as noise. While a speech signal is present, the noise estimation is frozen. This approach has two major disadvantages. First, the noise estimation cannot be achieved while speech is active, secondly, the noise estimation relies fundamentally on the decisions of a VAD. If the VAD fails, the noise estimation takes the speech signal also into account while updating. This causes speech signal cancellation and the quality of the speech decreases drastically. A better solution was proposed by Martin (1994). He suggested tracking the minima of the spectral power in sub-bands over the last D signal vectors. This circumvents the necessity for voice activity detection. If the search window for the minima (last D frames) is large enough, the speech components will not contribute to the minima. The fact that the minima are searched in sub-bands allows updating of the noise estimate, even if speech is present, but not in all frequencies. Such a method produces a biased estimate of the noise power. Therefore, a bias compensation is needed as proposed in Martin (1994; 2001). Furthermore, in Martin (2001) an optimal smoothing for the estimation of the noisy power spectrum was introduced. It enables shortening of the search window length for the minima and therefore the noise estimation can follow more rapidly increasing noise levels. These methods still have the problem that the noise estimation lags behind the true noise and the variance of the noise estimation is high.

A combination of the minima tracking algorithm from Martin with a simple recursive averaging for the noise estimation called the minimum controlled recursive averaging (MCRA) (algorithm Cohen and Berdugo 2002). The minima tracking is used to define a

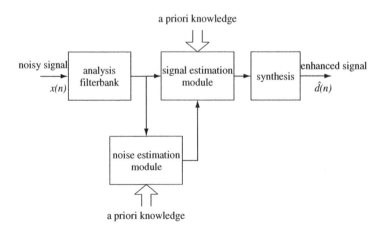

Figure 10.11 Schema of a single-channel noise reduction system

voice activity detector in each sub-band which controls the smoothing parameter for the noise estimation. An improved version of the MCRA algorithm is presented in Cohen (2003). Some noise estimation techniques take advantage of the special structure of speech, in particular the harmonic structure, to improve the noise estimation, especially for non-stationary noise. An example is the harmonic tunnelling technique (Ealey *et al.*, 1999 Kim *et al.* 2003).

Figure 10.12 shows an example of a noise tracking in a sub-band (centre frequency is at 687.5 Hz) based on the MCRA algorithm. The thin line represents the smoothed power of the noisy speech and the thick line is the estimated noise power. While speech is active in this sub-band the noise estimation is frozen. Between the speech activity periods the algorithm tries to follow the power and updates the noise power estimation.

When the single-channel algorithm is applied as a post-filter of a multi-channel algorithm, the noise estimation module can take advantage of the multi-channel system. Cohen *et al.* (2003) used the noise reference signals (lower path in Figure 10.10) in the GSC to detect non-stationary noise and a suppression of this noise becomes possible.

Signal estimation.   The most popular signal de-noising filter is the Wiener filter. It is a linear filter optimizing the minimum mean-square error (MMSE) between the enhanced signal and the clean signal. Unfortunately, Wiener filtering introduces 'musical noise' phenomena which are very annoying for the listener.

Spectral subtraction is a widely used algorithm in acoustic noise reduction, mainly because of its simple implementation. It was introduced in the late 1970s by Boll (1979) and is still used in more recent publications (Martin 1994). A milestone in speech enhancement are the papers by Ehraim and Malah (1984; 1985). In these, the authors used a decision-directed method for estimating the a priori signal-to-noise ratio (McAulay and Malpass 1980) and applied their newly developed speech estimation rules. The method proved to be very efficient in reducing the musical noise phenomena (Cappe 1993). The next advance in speech enhancement was achieved by Malah *et al.* (1999) and, Cohen (2001) they modified

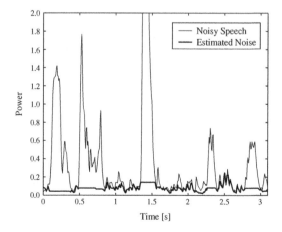

Figure 10.12   Noise estimation example based on the MCRA method. The sub-band power spectrum for the frequency bin $k = 20$ equivalent to a frequency of 687.5 Hz. The thin line is the smoothed power of the noisy speech and the thick line is the estimated power spectrum of the noise

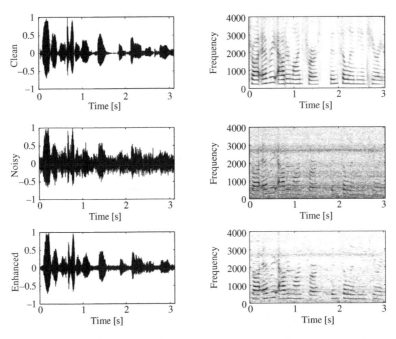

Figure 10.13   Single-channel noise reduction example. Noise estimation method is minima controlled recursive averaging (MCRA) and the speech estimation is according to optimally modified logarithmic spectral amplitude (OM-LSA) estimation

the original gain function from Ephraim and Malah (1985) with a presence probability of speech. In recent works several probability distributions for speech and the noise have been investigated, examples are, gamma, supergaussian, and Laplacian distributions. The reported results in Martin (2002) are encouraging as are combined and cascaded noise reduction schemes (Kim *et al.* 2003).

Figure 10.13 illustrates an example of the performance of single-channel speech enhancement system. The example is a female speech signal sampled with 8000 Hz. Cockpit noise is added to the clean speech signal in order to simulate non-stationary noise conditions. The resulting signal-to-noise ratio is 2 dB. Speech enhancement system is composed of a MCRA-based noise estimation module and a OM-LSA-based speech estimation module. The segmental SNR enhancement, (Noll 1974), is more than 11 dB.

10.6   CONCLUSIONS

Many audio signal processing modules are necessary to ensure natural and comfortable 3D audio communication. The modules acoustic echo control, noise reduction, and 3D parameter estimation have been discussed in this chapter. Powerful algorithms have been described which enable high-quality speech communication from the acquisition point of view. Audio analysis is beyond the scope of this chapter, but it is a mainstream of current research. Especially in multimedia applications, an audio content analysis can be very useful. A good overview of audio content analysis can be found in Kim *et al.* (2004).

To complete the audio system, the captured audio signals have to be transmitted and reproduced in the remote communication location. The reproduction of the audio signal should preserve the spatial information of the audio signal in order to increase the feeling of immersion of the videoconference participants. In Chapter 15, multi-channel and multi-object based audio reproduction algorithms and techniques will be presented. The discussion of the problem of spatial coherence of audio and video reproduction will further give indications how to improve the quality of 3D audiovisual communication.

REFERENCES

Abel J and Smith J 1987 The spherical interpolation method for closed-form passive source localization using range difference measurements. *IEEE International Conference on Acoustics, Speech, and Signal Processing,* vol. 12, pp. 471–474.

Bar-Shalom Y and Fortman T 1988 *Tracking and Data Association.* Academic Press.

Benesty J 2000 Adaptive eigenvalue decomposition algorithm for passive acoustic source localization. *Journal of the Acoustical Society of America* **107**(1), 384–391.

Benesty J, Morgan D and Sondhi M 1998 A better understanding and an improved solution to the specific problems of stereophonic acoustic echo cancellation. *IEEE Transactions on Speech and Audio Processing* **6**(2), 156–165.

Boll S 1979 Suppression of acoustic noise in speech using spectral subtraction. **27**(2), 113–120.

Brandstein M, Adcock JE and Silverman HF 1996 Microphone array localization error estimation with application to optimal sensor placement *Journal of the Acoustical Society America* **29**(2), 3807–3816.

Breining C, Dreiseitel P, Hänsler E, Mader A, Nitsch B, Puder H, Schertler T, Schmidt G and Tilp J 1999 Acoustic echo control. an application of very-high-order adaptive filters. *IEEE signal Processing Magazine* **16**(4), 42–69.

Cappe O 1993 Elimination of the musical noise phenomenon with the Ephraim and Malah noise suppressor. *Euroseech,* 1093–1096.

Cohen I 2001 On speech enhancement under signal presence uncertainty *International Conference on Acoustic and Speech Signal Processing.* pp. 167–170.

Cohen I 2003 Noise spectrum estimation in adverse environments: Improved minima controlled recursive averaging. *IEEE Transactions on Speech and Audio Processing* **11**(5), 466–475.

Choen I and Berdugo B 2002 Noise estimation by minima controlled recursive averaging for robust speech enhancement. *IEEE Signal Processing Letters* **9**(1), 12–15.

Cohen I, Gannot S and Berdugo B 2003 An integrated real-time beamforming and postfiltering system for non-stationary noise environments. *EURASIP Journal on Applied Signal Processing* (11), 1064–1073.

DiBiase J, Silverman H and Brandstein M 2001 *Microphone Arrays, Signal Processing Techniques and Applications* Springer, Chapter 8.

Dörbecker M 1998 *Mehrkanalige Signaverarbeitung zur Verbesserung akustisch gestörter Sprachsignale am Beispiel elektronischer Hörhilfen.* PhD Thesis, Technische Hochschule, Aachen.

Drews M 1999 *Mikrofonarrays und mehrkanalige Signalverabeitung zur Verbesserung gestörter Sprache.* PhD Thesis, Technische Universität, Berlin.

Ealey D, Kelleher H and Pearce D 1999 Harmonic tunneling: tracking non-stationary noises during speech. *EUROSPEECH, Aalborg,* pp. 437–440.

(ed. Brandstein M and Ward D) 2001 *Microphone Arrays.* Springer.

(ed. Gay SL and Benesty J) 2000 *Acoustic Signal Processing in Telecommunications.* Kluwer, Boston, Massachusetts, USA.

(ed. Kalouptsidis N and Theodoridis S) 1993 *Fast transversal RLS algorithms.* Prentice Hall, Englewood Cliffs.

Ephraim Y and Malah D 1984 Speech enhancement using a minimum mean-square error short-time spectral amplitude estimator. *IEEE Transactions on Acoustics, Speech, and Signal processing* **ASSP-32**(6), 1109–1121.

Ephraim Y and Malah D 1985 Speech enhancement using an minimum mean-square error log-spectral amplitude estimator. *IEEE Transactions on Acoustics, Speech and Signal Processing* **ASSP-32**(2), 443–445.

Fischer V, Gänsler T, Diethorn EJ and Benesty J 2001 A software stereo acoustic echo canceler for microsoft windows *Proceedings of the 7th IEEE International Workshop on Acoustic Echo and Noise Control*, pp. 87–90.

Gay SL and Travathia S 1995 The fast affine projection algorithm. In: *Acoustic Signal Processing for Telecommunication*, Kluwer: Boston, pp. 3023–3027.

Gilloire A, Moulines E Slock D and Duhamel P 1996 State of the art in acoustic echo cancellation. In Figueiras AR (ed.) *Digital Signal Processing in Telecommunications* **80**(7), 45–91.

Gilloire A, Scalart P, Lamblin C, Mokbel C and Proust S 2000 Innovative speech processing for mobile terminals: an annotated bibliography. *IEEE Signal Processing Magazine* **80**(7), 1149–1166.

Griffiths LJ and Jim CW 1982 An alternative approach to linearly constrained adaptive beamforming. *IEEE Transactions on Antenna Propagation* **30**, 27–34.

Herbordt W and Kellermann W 2001 Efficient frequency-domain realization of robust generalized sidelobe cancellers *Proceedings of the International Workshop on Acoustic Echo and Noise Control*, pp. 377–382.

Herbordt W and Kellermann W 2002 Frequency-domain integration of acoustic echo cancellation and a generalized sidelobe canceller with improved robustness. *European Transactions on Telecommunications (ETT)* **13**(2), 123–132.

Hänsler E 1992 The hands-free telephone problem: An annotated bibliography. *Signal Processing* **27**, 259–271.

Hänsler Eand Schmidt G 2004 *Acoustic Echo and Noise Control*. Wiley New York.

Huang Y, Benesty J and Elko GW 2001 An efficient linear-correction least-squares approach to source localization. *Applications of Signal Processing to Audio and Acoustics*, pp. 67–70.

Kim HG, Morea N and Sikora T 2004 *MPEG-7 Audio: Content Description and Retrieval*. Wiley.

Kim HG, Schwab M, Moreau N and Sikora T 2003 Speed enhancement of noisy speech using logspectral amplitude estimator and harmonic tunnelling. *Proceedings of the International Workshop on Acoustic Echo and Noise Control (IWAENC)*, pp. 119–122.

Knapp CH and Carter GC 1976 The generalized correlation method for estimation of time delay. *IEEE Transactions on Acoustics, Speech and Signal Processing* **24**(4), 320–327.

Krim H and Viberg M 1996 Two decades of array signal processing research. The parametric approach. *IEEE Signal Processing Magazine* **13**(4), 67–94.

Kuttruff H 1990 *Room Acoustics* 3rd edn. Applied Science Publishers.

Malah D, Cox R and Accardi A 1999 Tracking speech presence uncertainty to improve speech enhancement in non-stationary noise environments. *International Conference on Acoustics, Speech and Signal Processing (ICASSP)*, pp. 789–792.

Martin R 1994 Spectral subtraction based on minimum statistics. *Proceedings of the European Signal Processing Conference*, pp. 1182–1185.

Martin R 2001 Noise power spectral density estimation based on optimal smoothing and minimum statistics. *IEEE Transactions on Speech and Audio Processing* **9**(5), 504–512.

Martin R 2002 Speech enhancement using mmse short time spectral estimation with gamma distributed speech priors. *International Conference on Acoustics, Speech and Signal Processing (ICASSP)*, pp. 253–256.

McAulay R and Malpass M 1980 Speech enhancement using a soft-decision noise suppressing filter. *IEEE Transactions on Acoustics, Speech and Signal Processing* **28**, 137–145.

Noll P 1974 Adaptive quantizing in speech coding systems. *International Zürich Seminar in Digital Communications*, pp. B3.1–B3.6.

Schmidt RO 1986 Multiple emitter location and signal parameter estimation. *IEEE Transactions on Antenna Propagation*, **AP-34**, 276–280.

Silverman HF, Patterson WR and Flangan JL 1996 The huge microphone array (HMA). *Technical Report*, Laboratory for Engineering Man/Machine Systems (LEMS) Brown University.

Stoica P and Moses R 1997 *Introduction to Spectral Analysis*. Prentice Hall.

Wang H and Kaveh M 1985 Coherent signal-subspace processing for the detection and estimation of angles of arrival of multiple wide-band sources. *IEEE Transactions on Acoustics, Speech, and Signal Processing*, **ASSP-33**, 823–831.

11

Coding and Standardization

Aljoscha Smolic[1] and Thomas Sikora[2]

[1] Fraunhofer Institute for Telecommunications/Heinrich-Hertz-Institut, Berlin, Germany
[2] Technical University Berlin, Berlin, Germany

11.1 INTRODUCTION

Modern speech, audio, image and video compression techniques today offer the possibility to store or transmit the vast amount of data necessary to represent digital audiovisual data in an efficient and robust way. New audiovisual applications in the field of communication, multimedia and broadcasting became possible based on digital coding technologies developed over the past 25 years (Sikora 2005). As manifold as applications for audio and image coding are today, as manifold are the different approaches and algorithms, and these were the first hardware implementations and even systems in the commercial field, such as private teleconferencing systems. However, with the advances in VLSI and DSP technology it became possible to open more application fields to a larger number of users and therefore the necessity for coding standards arose. International coding standards made a dramatic impact in the field — both for commercial applications as well as for fostering research and development activities (Sikora 1997). Today the most successful and most well-known image, audio, and video coding standards are ISO-JPEG, ISO-JPEG 2000, ISO-MPEG-1, ISO-MPEG-2, ISO-MPEG-4 and ITU-H.263 and ITU-H.264 (Legall 1992; Haskell *et al.* 1997; Pereira and Ebrahimi 2002; Wallace 1991; Lee 2005; Luthra *et al.* 2003). While standard compression algorithms made a dramatic impact, proprietary industry solutions from Microsoft, Real Networks and others have also found widespread application, most notably for audio and video broadcast and video-on-demand streaming over the Internet.

The purpose of this chapter is to outline the basic coding strategies that enable today's successful audio, image and video applications. In the first part of the chapter we provide an

3D Videocommunication — Algorithms, concepts and real-time systems in human centred communication
Edited by O. Schreer, P. Kauff and T. Sikora © 2005 John Wiley & Sons, Ltd

overview on basics in image and video coding. Particular reference is made to international standards in the field.

The second part of the chapter focuses on audio and video coding using the MPEG-4 standard, which provides efficient and flexible specifications for data representation formats, coding, and transmission. We highlight specific elements that are useful to implement standard conform 3D video systems. A particular focus is given to recent and ongoing activities in the 3DAV (3D audiovisual) group of MPEG (ISO/IEC JTC1/SC29/WG11) that works on specific extensions for 3D media (Smolic and Kauff 2005).

11.2 BASIC STRATEGIES FOR CODING IMAGES AND VIDEO

Dependent on the application requirements we may envisage 'lossless' and 'lossy' coding of image and video data (Jayant and Noll 1984; Sikora 1997). The aim of 'lossless' coding is to reduce image or video data for storage and transmission while retaining the quality of the original images — the decoded image quality is required to be identical to the image quality prior to encoding. The lossless coding mode of JPEG is an example of a lossless coder strategy. In contrast the aim of lossy coding techniques — and this is relevant to the applications envisioned by lossy compression standards such as lossy JPEG, MPEG and ITU — is to meet a given target bit rate for storage and transmission. Important applications comprise image storage and transmission, both for low-quality Internet applications as well as for high-quality transmission of video over communications channels with constrained or low bandwidth and the efficient storage of video.

11.2.1 Predictive Coding of Images

The purpose of predictive coding strategies is to de-correlate adjacent pel information and to encode a prediction image rather than the original pels of the images. Image sequences usually contain significant correlation in spatial directions (Netravali and Haskell 1995). Video sequences in both spatial and temporal directions.

A predictive coding strategy is a very efficient operation for removing correlation between pels prior to coding (Jayant and Noll 1984). To this end a pel value to be coded is predicted from already coded and transmitted/stored adjacent pel values — and only the small prediction error value is coded. Note that predictive coding is suitable for both lossless and lossy coding.

Figure 11.1 depicts a predictive coding strategy for images which is in fact identical to the one employed by the lossless coding mode of the JPEG standard (Wallace 1991). A pel to be coded is predicted based on a weighted combination of values of the most adjacent neighbouring pels. A JPEG lossless coder has to decide on one of the predictor coefficient sets outlined in Figure 11.1. The predictor is stored/transmitted with the bitstream. Thus the decoder can perform the same prediction from the already decoded pels and reconstruct the new pel value based on the predicted value and decoded prediction error.

Figure 11.2(b–d) illustrate the prediction error images achievable using 1D as well as 2D predictors for test image 'Lenna'. It is apparent that the prediction error image that needs to be coded contains much less amplitude variance compared with the original – this results in significantly reduced bit rate (Jayant and Noll 1984). In a lossless predictive coder each pel magnitude of the prediction error image is assigned a binary codeword for storage or transmission.

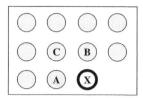

Figure 11.1 Predictive coding of image pixels. X: pixel to be predicted and coded. A, B, C: already coded and stored or transmitted pixels. JPEG lossless coding standard specifies the following possible predictors (mappers): $X = A, X = B, X = C, X = A + B - C, X = A + (C - B)/2, X = (A + C)/2, X = C + (A - B)/2$

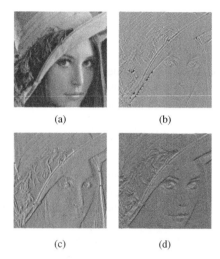

Figure 11.2 Predictive coding of test image 'Lenna'. (a) original image; (b) optimal two-dimensional prediction using coefficients $X = \alpha, B = \beta, C = \alpha\beta$; (c) one-dimensional prediction, $X = \alpha, B = C = 0$; (d) one-dimensional prediction, $B = \alpha, A = C = 0$. α, β are correlation coefficients in horizontal and vertical image dimension

11.2.2 Transform Domain Coding of Images and Video

Transform coding is a strategy that has been studied extensively during the last two decades and has become a very popular compression method for lossy image and video coding. The purpose of transform coding is to de-correlate adjacent pel information and to quantize and encode transform coefficients rather than the original pels of the images. As such the goal is identical to predictive coding strategies describe above. The most popular and well-established transform techniques are the celebrated discrete cosine transform (DCT) (Ahmed *et al.* 1984) used in the lossy JPEG, MPEG and ITU standards and the discrete wavelet transform (DWT) standardized in MPEG-4 and JPEG 2000. The DCT is applied in these standards strictly as a block-based approach on blocks of size 4×4, 8×8 or 16×16 pels. The DWT in contrast is implemented in JPEG 2000 and MPEG-4 as a frame-based approach applied to entire images (Pereira and Ebrahimi 2002; Lee 2005).

For video compression, transform coding strategies are usually combined with motion compensated prediction into a hybrid approach to achieve very efficient coding. These techniques are described in the following in more detail.

11.2.2.1 Discrete Cosine Transform (DCT)

For the block-based DCT Transform approach the input images are split into disjoint blocks of pels $\mathbf{I}$ (i.e., of size $N \times N$ pels) as indicated in Figure 11.3. In general, a linear, separable and unitary forward 2D-transformation strategy can be represented as a matrix operation on each block matrix $\mathbf{I}$ using a $N \times N$ transform matrix $\mathbf{A}$, to obtain the $N \times N$ transform coefficients $\mathbf{C}$

$$\mathbf{C} = \mathbf{A}\mathbf{I}\mathbf{A}^{\mathrm{T}} \qquad (11.1)$$

Here, $\mathbf{A}^{\mathrm{T}}$ denotes the transpose of the 1D-transformation matrix $\mathbf{A}^{\mathrm{T}}$ (Netravali and Haskel 1995). Note that the transformation is reversible, since the original $N \times N$ block of pels $\mathbf{I}$ can be reconstructed using a linear and separable inverse transformation (for a unitrary transform the inverse matrix $\mathbf{A}^{-1}$ is identical with the transposed matrix $\mathbf{A}^{\mathrm{T}}$, that is $\mathbf{A}^{-1} = \mathbf{A}^{\mathrm{T}}$):

$$\mathbf{I} = \mathbf{A}^{\mathrm{T}}\mathbf{C}^{*}\mathbf{A} = \sum_{k=1}^{N}\sum_{l=1}^{N} B(k, l)C^{*}(k, l) \qquad (11.2)$$

as long as the transform matrix is invertible and the transform coefficients are not quantized, $\mathbf{C}^{*} = CB(k, l)$ is a basis image with index (k, l) of size $N \times N$.

In a practical coding scheme the coefficients will be quantized ($\mathbf{C}^{*} \neq \mathbf{C}$) and the original image block $\mathbf{I}$ is approximated as a weighted superposition of basis-images $B(k, l)$, weighted with the quantized transform coefficients $C^{*}(k, l)$. The coding strategy thus results in a lossy reconstruction — the coarser the quantization the less bits required to store/transmit the image, the worse the reconstruction quality at the decoder.

Upon many possible alternatives the DCT applied to image blocks of usually 8×8 pels has become the most successful transform for still image and video coding so far. Figure 11.4

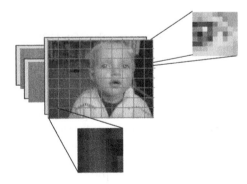

Figure 11.3 Decomposition of images into adjacent, non-overlapping blocks of $N \times N$ pels for transform coding (i.e., with the JPEG lossy coding standard). Colour information in images is usually separated into RGB or YUV color images, which are often subsampled and coded separately

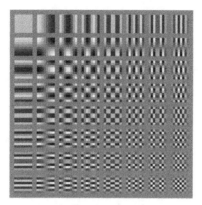

Figure 11.4 The 64 basis-images $B(k, l)$ of the 2D 8×8 DCT. A block-based DCT decoder, such as JPEG, reconstructs the stored or transmitted image blocks as a weighted superposition of these basis images, each weighted with the associated decoded DCT coefficient $C^*(k, l)$

depicts the basis images $B(k, l)$ of the 8×8 DCT that are used for reconstruction of the images according to Equation (11.2). In most standards coding schemes, i.e., lossy JPEG, each 8×8 block of coefficients is quantized and coded into variable or fixed length binary code words (Netravali and Haskel 1995).

Today, block-based DCT transform strategies are used in most image and video coding standards due to their high de-correlation performance and the availability of fast DCT algorithms suitable for real-time implementations.

The DCT is a so-called compact transform, because most signal energy is compacted into the lower-frequency coefficients, most higher coefficients are small or zero after quantization and small or zero-valued coefficients tend to be clustered together (Jayant and Noll 1984). Thus on average only a small number of quantized DCT coefficients need to be transmitted to the receiver to obtain a valuable approximate reconstruction of the image blocks based on the basis images in Figure 11.4 and Equation (11.2). At very low bit rates only few coefficients will be coded and the well-known DCT block artefacts will be visible (i.e., in low-quality JPEG or MPEG images and video).

11.2.2.2 Discrete Wavelet Transform (DWT)

The DWT provides the key technology for the JPEG 2000 and the MPEG-4 Still Image Coding standard and is applied to the entire image rather than to separate blocks of pels (Lee 2005).

A DWT can be implemented as a filter bank, as illustrated in Figure 11.5, enabling perfect reconstruction in the reverse process. The example filter bank decomposes the original image into horizontal (H), vertical (V), diagonal (D) and baseband (B) sub-band images, each being one-fourth the size of the original image. Multiple stages of decomposition can be cascaded together to recursively decompose the baseband — the sub-bands in this case are usually arranged in a pyramidal form as illustrated in Figure 11.6. Similar to the linear DCT approach most signal energy is compacted by the DWT into the lower-frequency sub-bands, most coefficients in higher sub-bands are small or zero after quantization and small or zero-valued

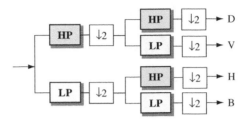

Figure 11.5 One stage of discrete wavelet transform (DWT) decomposition, composed of low-pass (LP) and high-pass (HP) filters with subsequent subsampling

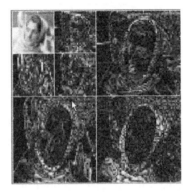

Figure 11.6 A three-scale DWT of test image 'Lenna' in a pyramidal arrangement of the sub-bands

coefficients tend to be clustered together. Also clusters of small or zero-valued coefficients tend to be located in the same relative spatial position in the sub-bands.

There is good evidence that DWT coding provides improved coding gains compared with DCT strategies. Most importantly, the DWT enables — alongside excellent compression efficiency — so-called fine-granularity embedded coding functionalities fully integrated into the coder. DWT-embedded coding enables to reconstruct images progressively in fine-granular stages using partial bitstream information, a capability in high demand in modern applications yet absent from prior standards. DWT image coders are essentially built upon three major innovative components: the DWT, successive approximation quantization, and significance-map encoding (i.e., using zero-trees, Shapiro 1993). It is also important to note that in JPEG 2000 the DWT is employed both in the lossless and lossy coding mode.

11.2.3 Predictive Coding of Video

For video sources it is assumed that pels in consecutive video images (frames of the sequence) are also correlated. Moving objects and part of background in a scene then appear in a number of consecutive video frames — even though possibly displaced in horizontal and vertical direction and somehow distorted when motion of objects or camera motion/projection is not purely translatory. Thus, the magnitude of a particular image pel can be predicted from nearby pels within the same frame, or even more efficiently from pels of a previously coded frame — so-called motion-compensated (MC) prediction (Netravali and Haskel 1995). This requires that one or more coded frames are stored at encoder and decoder.

The strategy for predicting motion of pels in video sequences is vital for achieving high compression gains. This subject has thus been focus of intense research over the last 25 years and continues to be one of the prime areas where most advances in compression efficiency may be seen in the future. The most established and implemented strategy is the so-called block-based motion compensation technique employed in all international MPEG or ITU-T video compression standards described below (Legall 1992). Other strategies employ global motion prediction (Pereira and Ebrahimi 2002), SPRITE motion prediction (Kauff *et al.* 1997; Smolic *et al.* 1999a), segmentation-based (Karczewicz *et al.* 1997) or object-based motion compensation (Smolic *et al.* 1999b). The block-based motion compensation strategy is based on motion vector estimation outlined in Figure 11.7 (Legall 1992). The images are separated into disjoint blocks of pels.

The motion of a block of pels between frames is estimated and described by only one motion vector x (MV). This reduces the bit rate overhead per pel significantly and assumes that all pels have the same displacement. The motion vectors x are quantized to pel or sub-pel accuracy prior to coding. The higher the precision of the MVs the better the prediction. The bit rate overhead for sending the MV information to the receiver needs to be well balanced with the gain achieved by the motion-compensated prediction. MPEG and ITU-T video coding standards employ $1/2$-pel to $1/8$-pel MV accuracy for MC blocks of sizes between 4×4 and 16×16 pels (non-quadratic blocks as 4×8 are also possible with H.264/AVC, see below). It is worth noting that motion estimation can be a very difficult and time-consuming task, since it is necessary to find one-to-one correspondence between pels in consecutive frames. Figure 11.8 depicts the impressive efficiency of motion compensated prediction for a TV size video sequence.

11.2.4   Hybrid MC/DCT Coding for Video Sequences

The combination of temporal block-based motion-compensated prediction and block-based DCT coding provides the key elements of the MPEG and ITU-T video coding standards (Sikora 1997; Legall 1992; Sikora 2005). For this reason the MPEG and ITU-T coding algorithms are usually referred to as hybrid block-based MC/DCT algorithms. DWT coding so far has not shown significant compression gains compared with DCT for video coding —

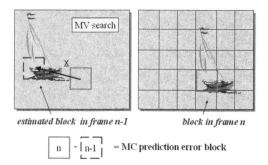

Figure 11.7   MPEG/ITU-type block matching approach for motion prediction and compensation: One motion vector x is estimated for each block in the actual frame n to be coded. The motion vector points to a reference block of same size in a previously coded frame $n - 1$. The motion compensated (MC) prediction error is calculated by subtracting each pel in a block with its motion shifted counterpart in the reference block of the previous frame

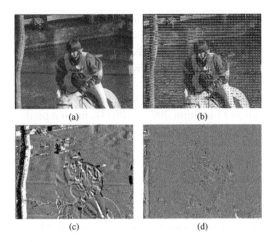

Figure 11.8 Efficiency of MPEG/ITU-type video coding using block-based motion prediction. Top left: frame to be coded. Top right: estimated motion vectors x for each 16×16 block used for motion compensated prediction. Bottom right: motion compensated prediction error image. Bottom left: prediction error using simple frame difference (all motion vectors are zero)

HYBRID DCT/DPCM CODING SCHEME

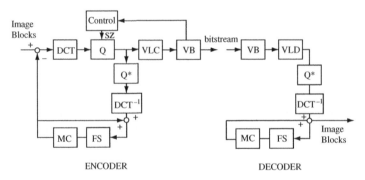

Figure 11.9 Block diagram of a basic hybrid MC/DCT encoder and decoder structure (as used in MPEG/ITU-type video coders)

probably due to the additional hybrid block-based motion compensation required to achieve high coding gains. The basic building blocks of such a hybrid video coder are depicted in Figure 11.9 — in a broad sense a JPEG coder plus block-based motion compensated prediction.

In basic MPEG and ITU-T video coding schemes the first frame in a video sequence (I-picture) is encoded in INTRA mode without reference to any past or future frames (Sikora 1997; Legall 1992; Netravali and Haskel 1995). At the encoder the DCT is applied to each $N \times N$-block and, after output of the DCT, each of the $N \times N$-DCT coefficients is uniformly quantized (Q) and coded with Huffman code words of variable lengths (VLC).

The decoder performs the reverse operations, first extracting and decoding (VLD) the variable length coded words from the bitstream to obtain locations and quantizer values of the non-zero DCT coefficients for each block. With the reconstruction (Q*) of all non-zero

DCT coefficients belonging to one block and subsequent inverse DCT (DCT^{-1}) the quantized block pixel values are obtained. By processing the entire bitstream all image blocks are decoded and reconstructed.

For motion predicted coding (P-pictures), the previously coded I- or P-picture frame $N - 1$ is stored in a frame store (FS) in both encoder and decoder. Motion compensation (MC) is performed — only one motion vector is estimated between frame N and frame $N - 1$ for a particular block to be encoded. These motion vectors are coded and transmitted to the receiver. The motion-compensated prediction error is calculated by subtracting each pel in a block with its motion-shifted counterpart in the previous frame. A $N \times N$-DCT is then applied to each of the $N \times N$ blocks contained in the block, followed by quantization (Q) of the DCT coefficients with entropy coding (VLC). A video buffer (VB) is needed to ensure that a constant target bit rate output is produced by the encoder. The quantization step size (sz) can be adjusted for each block in a frame to achieve a given target bit rate and to avoid buffer overflow and underflow. The decoder uses the reverse process to reproduce a block of frame N at the receiver.

11.2.5 Content-based Video Coding

The coding strategies outlined above are designed to provide the best possible quality of the reconstructed images at a given bit rate. At the heart of 'content-based' functionalities (i.e., ISO MPEG-4 standard) is the support for the separate encoding and decoding of content (e.g., physical objects in a scene, Pereira and Ebrahimi 2002). This extended functionality provides the most elementary mechanism for flexible presentation of single video objects in video scenes without the need for further segmentation or transcoding at the receiver. Figure 11.10 illustrates a virtual environment teleconference application example that makes extensive use of this functionality (Tanger *et al* 2004). Participants at diverse locations are coded as arbitrarily shaped objects in separate streams. The receiver can place the objects flexibly into a virtual environment, i.e., to provide the impression that all participants are gathered round a table as in a normal conference situation.

Figure 11.10 Immersive virtual environment video conference setup at Fraunhofer HHI, Berlin. In this scenario persons from different remote locations are arranged around a virtual table. Each participant can be compressed as an arbitrarily shaped video object using the MPEG-4 object-based coding modes (Tanger *et al.* 2004)

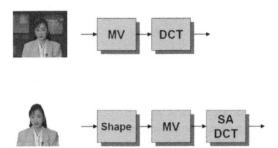

Figure 11.11 Standard video coding compared with the MPEG-4 object-based approach. The basic MC/DCT approach (coding of motion vectors MV and DCT coefficients) is extended towards the coding of the arbitrarily shaped video objects using shape coding and shape-adaptive DCT (SA-DCT)

The content-based approach can be implemented as an algorithmic extension of the conventional video coding approach towards image input sequences of arbitrary shape. In MPEG-4 this is achieved by means of advanced shape coding algorithms and a low-complexity shape-adaptive DCT approach, as outlined in Figure 11.11 (Ostermann *et al.* 1997; Sikora 1995).

11.3 CODING STANDARDS

11.3.1 JPEG and JPEG 2000

The JPEG still image coding standard is the most widely employed compression algorithm for still colour images today. JPEG finds applications in many diverse storage and transmission application domains, such as the Internet, digital professional and consumer photography and video. The standard enables both lossy as well as lossless coding of images in an efficient way and handles very small image sizes as well as huge size images. The lossless coding strategy employed by JPEG involves predictive coding, as outlined in Figure 11.1, using one of the standardized predictors per image as described with the figure. The well known lossy compression algorithm is based on DCT transform coding of image blocks of size 8×8, followed by quantization of the DCT coefficients and subsequent assignment of variable length binary code (VLBC) words to position and amplitudes of DCT coefficients.

The JPEG 2000 standard was approved in 2002 and commercial deployment is still in its early stages. Compared with JPEG it is probably its very efficient progressive coding functionality that sparks most interest in industry — a functionality that is not efficiently provided in JPEG. A compression performance comparison between JPEG and JPEG 2000 coding has been reported in (Santa-Cruz *et al.* 2002) and also indicates superior compression gains by the JPEG 2000 standard in terms of peak-SNR (PSNR).

11.3.2 Video Coding Standards

Standardization work in the video coding domain started around 1990 with the development of the ITU-T H.261 standard — targeted at transmission of digital video-telephone signals

Table 11.1 Coding parameters and options for various video standards

	MPEG-1 (1993)	MPEG-2 (1995)	MPEG-4 (2000)	H.261 (1993)	H.263 (1995)	H.264/MPEG-4 AVC (2002)
Transform	8×8 DCT	8×8 DCT	8×8 DCT	8×8 DCT	8×8 DCT	4×4
MC block size	16×16	16×16 8×16	8×8, 16×16	16×16	8×8, 16×16	16×16, 16×8, 8×8, 8×4, 4×4
MC accuracy	1/2-pel	1/2-pel	1/4-pel	1-pel	1/2-pel	1/8-pel
Additional motion prediction modes	B-frames	B-frames Interlace	B-frames Interlace GMC (global MC) SPRITE coding		B-frames	B-frames Long-term frame memory In-loop deblocking filter CAVLC/CABAC

over ISDN channels at data rates of $n \times 64$ kbit/s ($n = 1, 2, \ldots, 30$). Since then a number of international video coding standards have been released by ITU-T and ISO-MPEG standards bodies, targeted at diverse application domains. Table 11.1 summarizes the basic features of the standards available today. The H.263 standard was released by ITU-T in 1998 for transmission of videotelephone signals at low bit rates and H.264 (as a joint development with ISO-MPEG as MPEG-4 AVC) in 2002. H.264 essentially covers all application domains of H.263 while providing superior compression efficiency and additional functionalities over a broad range of bit rates and applications.

MPEG-1 was released by ISO-MPEG in 1993 and is now, together with MPEG-2, the most widely introduced video compression standard. Even though MPEG-1 was primarily developed for storage of compressed video on CD, MPEG-1 compressed video is widely used for distributing video over the Internet. MPEG-2 was developed for compression of digital TV signals at around 4–6 Mbit/s and has been instrumental in enabling and introducing commercial digital TV around the world. MPEG-4 was developed for coding video at very low bit rates and for providing additional object-based functionalities. MPEG-4 has found widespread application in Internet streaming, wireless video, digital consumer video cameras as well as in mobile phones and mobile palm computers.

All these standards build on the block-based motion-compensated DCT approach outlined in Figure 11.9. However, over the last 10 years various details of the basic approach have been refined and have resulted in more complex, but also more efficient compression standards. It appears that significant compression gains have been achieved over the years based on advanced motion vector accuracy and more sophisticated motion models; ITU-H.264 is presently the most advanced standard in terms of compression efficiency. In the H.264 (MPEG-4 AVC) standard also more precise motion compensation prediction using variable size MC blocks is employed along with context based arithmetic coding (Luthra *et al.* 2003). Another novelty is the introduction of long-term frame memories in H.264/AVC that allows the storage of multiple frames of the past for temporal prediction. Needless to say, most of the advanced techniques are implemented with much increased complexity at encoder and decoder. However, the last 10 years have also witnessed much improved processor speed and improvements in VLSI design to allow real time implementation of the algorithms. Future coding strategies will meet even less complexity constraints.

11.4 MPEG-4 — AN OVERVIEW

As outlined above, the MPEG-4 standard provides a rich, flexible and efficient toolbox for representation, coding and transmission of multimedia content. This includes natural audio and video as well as various types of synthetic content. Different types of interaction with the content can be supported. MPEG-4 allows implementation of a great variety of interoperable systems, content and applications.

The central concept of MPEG-4 is the audiovisual scene, as illustrated in Figure 11.12. Such a scene has three (or two) spatial dimensions and a temporal dimension. The scene is composed of different types of media objects, such as video objects, audio objects, synthetic objects or composed objects. The coded representations of media objects, so-called elementary streams (ES), are multiplexed in a systems layer along with scene composition information in a specific script language (BIFS, see Section 11.4.2).

At the receiving terminal the incoming stream (from a transport channel, file, etc.) is de-multiplexed, the scene graph is built, the media objects are decoded from ESs, the scene is composed and audio and visual presentations are rendered. The last step may take into account user interaction for instance to interactively change the viewpoint and direction (other types of interaction such as editing, selection, etc., are also possible), which is of special interest in the context of this book.

This flexible framework is particularly interesting for the 3D technologies described in this book. It provides a variety of suitable media object types to realize interoperable 3D audio and video systems and applications. Further extensions as investigated for instance in the 3DAV activity (see Section 11.5) will further enhance these capabilities.

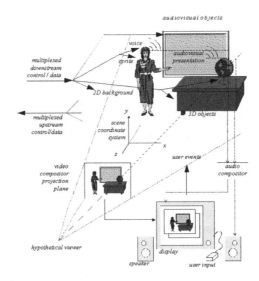

Figure 11.12 MPEG-4 scene composition and rendering

11.4.1 MPEG-4 Systems

MPEG-4 content can be carried over many different transport layers, and move from one transport to the other. However, for two important cases transport has been explicitly specified: MPEG-2 transport and IP. The specification of carriage of MPEG-4 content over MPEG-2 transport (MPEG-2 TS, systems part of MPEG-2) ensures that MPEG-4 can be used wherever MPEG-2 infrastructure is available, for instance in broadcast environments (DVB) and for storage applications (DVD). The specification of MPEG-4 carriage over Internet Protocol (IP) jointly developed with the Internet Engineering Task Force (IETF) opens the Internet for MPEG-4 data (ISO 2002a).

The so called delivery multimedia integration framework (DMIF) is an interface between the application and the transport which allows applications to be developed without taking transport into account. Further, the systems layer specifies de-multiplexing of ESs and other data, provides synchronization and timing, specifies a file format (MP4), provides mechanisms for intellectual property rights (IPR) management and protection, and specifies JAVA APIs (MPEG-J) to various aspects of the terminal and networks.

11.4.2 BIFS

MPEG-4 specifies a script language called 'binary format for scenes' (BIFS) for description of audiovisual scene composition (ISO 2002a). BIFS is specified as an extension of the ISO/IEC virtual reality modelling language (VRML, ISO/IEC 14772) thus BIFS inherits all elements that are available in VRML.

VRML was developed mainly for transmission of 3D graphics over the Internet, but can just as well be used for other applications. VRML data can be visualized with an appropriate player and the user can navigate through the virtual environments. Apart from the formats and attributes for object description (geometric primitives, 3D meshes, textures, appearance, etc.), VRML also contains other elements necessary to define interactive 3D worlds (e.g., light sources, collision, sensors, interpolators, viewpoint).

In addition to the functionality of VRML, BIFS provides, for instance, a better integration of natural audio and video, 2D graphics nodes, advanced 3D audio features, a timing model, an update mechanism to modify the scene in time, a script to animate the scene temporally, new graphics elements (e.g., face and body animation), and an efficient binary encoding for the scene description. The last point is particularly important for streaming and broadcast applications, since scene description files in text format might be insufficiently large and not well suited for real-time streaming.

As such MPEG-4 is much more than just another video and audio codec, although advanced audiovisual coding is again an important part of MPEG-4. In fact, it is a real multimedia standard, combining all types of media within an interoperable format. An example image of a rendered MPEG-4 BIFS scene is shown in Figure 11.13. The background is a still image coded in JPEG format. The foreground scene contains 3D graphics elements (box, pawn in the game) that the user can interact with (move, rotate, change of object properties such as shape, size, texture, opacity,). In addition, the interaction elements at the bottom allow a change of the image background and browsing from scene to scene. Such scene changes can be downloaded on demand and the scene graph is updated online while viewing the scene. Furthermore, a live video stream showing a person is decoded and mapped onto the surface

Figure 11.13   Example of a rendered MPEG-4 BIFS scene

of the 3D box. This simple example illustrates some basic features of MPEG-4 BIFS and its potential for content creators. Especially the possibility of online scene updates and the live streaming of audiovisual data and their seamless integration into virtual worlds represent a clear progress over VRML.

In a complete interactive multimedia application, the interactive video must be accompanied by corresponding interactive audio. If the user rotates the viewpoint, the associated sound should follow the interaction, by changing its direction of origin. If the user approaches a sound source, the associated sound should become louder. These functionalities are provided by MPEG-4 AudioBIFS (Scheirer 1999), which provides the means for setting up 3D audio scenes. Sound sources with various attributes and properties can be placed anywhere in a virtual 3D space. With a suitable 3D audio player the user can navigate arbitrarily within such a scene and corresponding audio is rendered for every position and orientation.

11.4.3 Natural Video

MPEG-4 provides a variety of tools to efficiently encode different types of natural video. These can be used for the colour data as well as for other types of data such as per pixel depth or disparity information. The combination of temporal block-based motion compensated prediction and block-based transform coding provides the key elements of the MPEG and ITU-T video coding standards (Luthra *et al.* 2003).

The latest video codec is the already mentioned MPEG-4 Part 10, Advanced Video Coding (AVC), jointly developed with the ITU (where it is called H.264, Luthra *et al.* 2003). H.264/AVC outperforms all other available standard codecs by at least a factor of 2 (i.e., half the bitrate at the same visual quality). Therefore it is expected that H.264/AVC will be adopted for a great variety of applications and systems in the near future and might replace MPEG-2 as main format for digital video. Several adoptions in different applications domains such as third-generation mobile networks, DVB, Blue-Ray-Disc, high-definition DVD, DMB, or mobile broadcast have already been announced by different organizations and authorities.

Due to its superior compression performance H.264/AVC is also highly attractive for different 3D video applications and systems, wherever standard rectangular video has to be transmitted. Additionally, the standard supports transmission of so-called alpha channels.

These are luminance-only signals that can be multiplexed with the video data (NAL units syntax); it is possible to attach additional properties to the video signal such as transparency, shape or, as most important for 3D, depth. H.264/AVC can be very efficiently used to compress per-pixel depth data, and with that 3D-TV systems and services can be built.

11.4.4 Natural Audio

MPEG-4 provides a rich tool set for representation and coding of natural and synthetic audio. Figure 11.14 outlines the various coders as a function of bit rate compared with quality.

Several dedicated speech codes are available, covering the whole range from very low bitrates to high-quality wideband speech. This includes a parametric coder and a CELP coder covering the range of about 1 kbits/s up to 16 kbits/s. A text-to-speech coder enables the transmission of text and the synthesis of speech from text at the decoder at extremely low bit rates.

For coding of natural audio in mono or stereo from lowest bit rates to highest quality extended versions of advanced audio coding (AAC) are available. A prior version of AAC was already specified in MPEG-2, which was itself an extension of the popular MP3 format (correctly MPEG-1 Audio Layer 3). The MPEG-4 AAC coder relies on the successful perceptual coding strategy that has already been standardized with MP3. These perceptual coders transform audio frames into a frequency domain using suitable filter banks. As illustrated in Figure 11.15 the perceptual impact of audio events is analysed in the FFT domain. The audio information that is below a masking threshold is not important to the human listener and is cancelled or diminished at the encoder using a dynamic parameter allocator and coder.

Also MPEG-2 AAC and MP3 can be used in a MPEG-4 framework. Suitable formats are also specified for multi-channel and surround sound, which enables for instance high-quality home cinema applications (e.g., via DVD or DVB). Also for highest-quality requirements, as in studios and production environments, MPEG-4 has defined an efficient compression format. This format, known as audio lossless coding (ALS), allows compressing of digital PCM audio signals by more than a factor of two without loosing a single bit when reconstructing the signal (Liebchen *et al.* 2004). In consequence MPEG-4 provides optimized

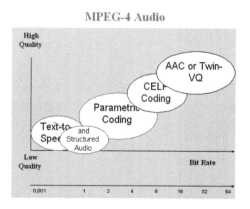

Figure 11.14 MPEG-4 coders for speech and audio

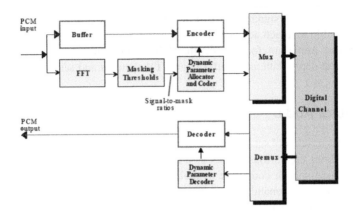

Figure 11.15 MPEG perceptual audio coding

standard solutions for any type of natural audio, ranging from low bit rate Internet and AM radio to highest studio quality.

Besides natural types of audio, MPEG-4 also addresses different kinds of synthetic sound. This includes text-to-speech synthesis as well as MIDI-like music and sound synthesis.

11.4.5 SNHC

SNHC is the abbreviation for synthetic natural hybrid coding. This covers all aspects related to computer graphics and rendering. One important SNHC tool for 3D applications is a very efficient and flexible compression tool for 3D polygonal meshes called 3D mesh coding (3DMC). 3DMC allows compression of 3D mesh data by a factor of 30–40 compared with the textual description. This feature is particularly important for large models with a high level of detail and realism. Further 3DMC provides level-of-detail (LOD) scalability that allows the decoder reconstruction a simplified version of the original mesh. Less powerful rendering engines can display the same object at a reduced quality. All these features make MPEG-4 3DMC suitable and efficient for various kinds of streaming and broadcast applications with dynamic and complex objects and scenes. VRML in contrast is mainly designed for Internet (i.e., download) scenarios. Also 2D animated meshes are supported. Figure 11.16 shows examples for 3DMC.

Figure 11.16 Original mesh (left, 48 bit per vertex) and 3DMC decoded meshes at 8 (middle) and 6 (right) bit per vertex

For 3D modelling and animation of humans and humanoids MPEG-4 provides a set of dedicated tools. 3D face models can be defined, coded and transmitted using facial definition parameters (FDPs) including 3D shape and optionally texture. The models can be animated using facial animation parameters (FAPs). These FAPs are derived from muscle models of the human face, and suitable combinations of them allow one to generate any human facial impression. Efficient compression schemes are available for the FAPs, resulting in extremely low bit rates when compared with transmission of corresponding video data. This makes the technology very attractive for animation-based videoconferencing, e.g., on mobile phones, as well as for virtual and virtualized character animation applications.

3D modelling and animation of human bodies is enabled in a similar way by body definition parameters (BDP) and body animation parameters (BAP). BDPs can be used for definition of custom bodies in relation to a generic virtual human body model (3D shape and optionally texture). BAPs are used in analogy to FAPs for animation. The physical model underlying the MPEG-4 BDP/BAP-framework is derived with simplification from the human skeleton.

11.4.6 AFX

Computer graphics research has of course continued successfully since the initial versions of MPEG-4 BIFS and SNHC were finalized. Some of these developments had been integrated into an extension of MPEG-4 called animation framework extension (AFX) (ISO 2002b). Three of the new tools are of specific interest in the scope of 3D video: depth image-based rendering (DIBR), point rendering, and view-dependent multi-texturing.

The AFX tool DIBR implements the concept of layered depth images (Shade *et al.* 1998). In this case a 3D object or scene is represented by a number of views with associated depth maps as shown in Figure 11.17 (Bayakovski *et al.* 2002). The depth maps define a depth value for every single pixel of the 2D images. Together with appropriate scaling and information

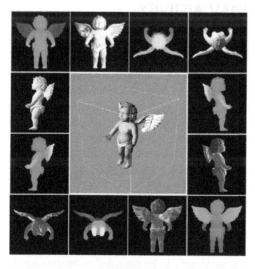

Figure 11.17 Example of AFX DIBR (Bayakovski *et al.* 2002)

from camera calibration it is possible to render virtual intermediate views as shown in the middle image in Figure 11.17. The quality of the rendered views and the possible range of navigation depend on the number of original views and the setting of the cameras. A special case of this method is stereo-vision, where two views are generated in accordance with the geometry of the human eyes basis. In this case the depth is often calculated using disparity estimation. Supposing that the capturing cameras are fully calibrated and their 3D geometry is known therefore, corresponding depth values can be recalculated one-to-one from the estimated disparity results. In the case of simple camera configurations (such as a conventional stereo rig or a multi-baseline video system) this disparity estimation can even be used for fully automatic real-time depth reconstruction in 3D video or 3D-TV applications. See also the Chapters 2 and 7 for details.

In a point-based representation shape, colour and other properties of 3D objects are defined by a point cloud in 3D (Würmlin *et al.* 2004). This can be regarded as an alternative to classical 3D mesh models with associated textures. Points are samples of surfaces in 3D. In contrast to 3D meshes such a point cloud representation does not require additional information such as connectivity, texture maps, etc., (note that color and other attributes are assigned to each point individually), which is an advantage over 3D mesh representations. Each point represents a small surface element or surfel, describing the surface in a small neighbourhood. A point-based representation as contained in MPEG-4 AFX allows for high-quality rendering of 3D objects at reasonable computational costs. It is especially attractive for applications where real-world objects are reconstructed from real camera views, for instance 3D video objects as described in the next section. A point cloud can be regarded as natural extension of 2D image pixels as they come from cameras into 3D. Since there is no need for connectivity the 3D reconstruction process is also simplified.

View-dependent multi-texturing on the other hand as supported by MPEG-4 AFX is attractive for classical computer graphics applications such as games or movie production, but also for reconstructed real world 3D video objects as described in the next section.

11.5 THE MPEG 3DAV ACTIVITY

This section outlines and discusses the most recent trend related to standardization for 3D video communications, MPEG 3DAV (3D audiovisual). This activity investigates possible technology for standardization and supports in particular interactivity. Interactivity in the sense of 3DAV is the ability to look around within an audiovisual scene by freely choosing viewpoints and viewing directions. Applications include omnidirectional video, interactive stereo video and free viewpoint video, where the last is further divided into model-reconstruction (also called 3D video objects) and ray-space methods (ISO 2003); (Smolic and McCutchen 2004). Interactive stereo video, including standard-compatible realization has been described in detail in Chapter 2. In the following we concentrate on omnidirectional and free-viewpoint video.

11.5.1 Omnidirectional Video

Onmidirectional or panoramic video can be regarded as an extension of the planar 2D image plane to a spherical or cylindrical image plane. Omnidirectional video can be displayed in a suitable player, with the key types of interaction being zoom and rotation to give the effect of looking around. However, in contrast to free-viewpoint video, the user is not able to change the position of the viewpoint interactively. The viewpoint might change, but that implies that the camera has been moved during the period of capturing. With the scene projected onto a dome, or with a head-mounted display, the user can get the impression of being part of the scene (Figure 11.18).

MPEG-4 BIFS can be used to represent and code such omnidirectional video in a standardized way. Thus the scene can be displayed interactively by any MPEG-4 3D player. For this purpose it is possible to define a 3D sphere and to map the omnidirectional video onto it, by analogy to the box with video texture in Figure 11.13. The user's viewpoint would be placed at the centre of the sphere.

More sophisticated solutions use knowledge from cartography for a more efficient representation of the spherical video texture (Smolic and McCutchen 2004). Due to mapping a sphere to a plane the image content is distorted. It is well known that no rectangular coordinate map can accurately depict a sphere. Different types of projections may preserve distances, shape or area while distorting the others. The most common projections for cartographic maps preserve distances. From the point of view of coding efficiency a projection-preserving area (i.e., areas on the sphere and on the map have the same size) is more desirable, because it keeps the original video resolution. Figure 11.19 shows an example of such an equal-area projection. The video texture is mapped onto a 3D mesh object, approximating a sphere. In this case the resolution of the mapped video texture fits well to the resolution of the original camera signals.

The flexible syntax of MPEG-4 BIFS also enables a variety of other solutions to represent and encode omnidirectional video. Among those, the group of polygonal mappings has received particular attention in 3DAV (ISO 2003). Here the 3D geometry is represented by a regular polyhedron such as a hexahedron, octahedron or icosahedron. The associated 2D video texture has the shape of the unwrapped surface and is therefore not rectangular. In terms of coding efficiency (i.e., the quality of rendered views at a given bit rate) of the different approaches, the icosahedron mapping provides the best performance (ISO 2003).

A practical problem arises from the dimensions of omnidirectional video. If good quality is required (e.g., similar to TV resolution) for a rendered view, for instance, a 60° field of view, the complete panorama must have an extremely large resolution. The minimum resolution to arrive at a good quality for each rendered view is 600×600 pixels. Thus a full spherical panorama would need a total resolution of at least 3600×1800 pixels

Figure 11.18 Immersive panoramic video by head-mounted display

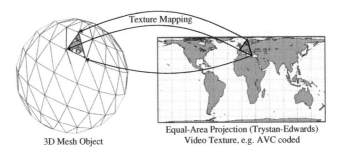

Texture Mapping

3D Mesh Object

Equal-Area Projection (Trystan-Edwards)
Video Texture, e.g. AVC coded

Figure 11.19 Representation of spherical video using a dense 3D mesh object and a rectangular video texture with equal-area projection

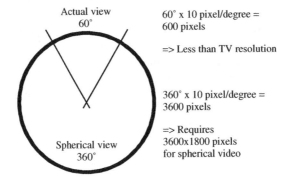

Actual view
60°

60° x 10 pixel/degree =
600 pixels

=> Less than TV resolution

360° x 10 pixel/degree =
3600 pixels

=> Requires
3600x1800 pixels
for spherical video

Spherical view
360°

Figure 11.20   Requirement for very-high-resolution panoramic video

(Figure 11.20). Transmitting or even processing such a large video is almost impossible. Furthermore, the usage of one large picture for the whole panorama is a heavy burden for the player as it has to keep the whole image in memory for rendering purposes.

The key for solving this problem lies in the interactivity itself. Only a small portion of the panorama is displayed at a certain time with an interactive player. The panorama can thus be partitioned into patches that are encoded separately. If the data representation allows spatial random access to these patches and partial decoding, the necessary load can be limited.

Figure 11.21 illustrates a suitable implementation using BIFS (Grünheit *et al.* 2002). Each patch (i.e., a video sequence) is encoded separately using a MPEG-4 video coding tool (e.g., H.264/AVC). Each of the patches is assigned to a visibility sensor that is slightly bigger than the patch. While the user navigates over the panorama, the temporarily visible patches are streamed, loaded and rendered. This procedure allows smooth rendering of even very-high-resolution panoramas.

11.5.2 Free-viewpoint Video

This is the most general and challenging application scenario investigated in 3DAV. A detailed description of algorithms and systems can be found in Chapter 4 and related chapters. Within the context of MPEG 3DAV free-viewpoint video has been divided into model-based and ray-space approaches (ISO 2003; Smolic and McCutchen 2004).

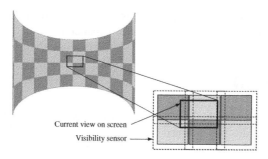

Figure 11.21   BIFS representation for high-resolution panoramic views

Figure 11.22   Left, right: original camera views; middle: virtual intermediate view of 3D video object integrated into a virtual environment (Würmlin *et al.* 2004)

If multiple synchronized views of a scene are available, it is also possible to reconstruct 3D video objects, as described in Chapter 8. MPEG-4 already supports different representation, coding and rendering formats for 3D video objects. Especially the new AFX tools as described in Section 11.4.6 are interesting in this context. Figure 11.22 shows an example of a 3D video object generated using visual hull reconstruction and represented and rendered using a point-based approach (Würmlin *et al.* 2004). The left and the right image show two original camera views captured in a multi-camera cave-like setting (Gross *et al.* 2003). The middle image shows the 3D video object rendered at a virtual viewpoint and integrated into a virtual environment. Such a 3D video object can be placed into a virtual 3D environment (such as a BIFS scene) and viewed from any direction while it is moving. It is worth noting that this application is not an animation of a virtual character or avatar as commonly used in 3D graphics, but that it depicts the natural motion of a real object captured by a number of cameras. More details about reconstruction of volumetric models can be found in Chapter 8.

Figure 11.23 shows a virtual camera fly around a 3D video object within a simple 3D scene. This example is represented, coded and rendered using 3D meshes and view-dependent

Figure 11.23   Virtual camera fly, rendered views at four different times from four different virtual viewpoints

texture mapping. Note that both examples in Figures 11.22 and 11.23 are fully compatible with MPEG-4 due to the new AFX tools, see also Chapter 4.

11.6 CONCLUSION

Coding and standardization is a highly dynamic process and it may be stated that the efficient coding of video and audio has reached its limits. Nevertheless additional improvements could be achieved in video coding due to the development of H.264. The latest work on standardization activities is now concentrated on three-dimensional audiovisual information and MPEG 3DAV is playing the key role in this context.

REFERENCES

Ahmed N, Natrajan T and Rao KR 1984 Discrete Cosine Transform. *IEEE Transactions on Computers*, **C-23**(1), 90–93.

Bayakovski Y, Levkovich-Maslyuk L, Ignatenko A, Konushin A, Timasov D, Zhirkov A, Han M and Park IK 2002 Depth image-based representations for static and animated 3D objects. *Proceedings of the International Conference on Image Processing*, Rochester, NY, USA, Vol. **3** pp. 25–28.

Chen CF and Pang KK 1993 The optimal transform of motion-compensated frame difference images in a hybrid coder. *IEEE Transactions on Circuits and Systems for Video Technology - II*: Analog and Digital Signal Processing, pp. 289–296.

Gross M, Würmlin S, Naef M, Lamboray E, Spagno C, Kunz A, Koller-Meier E, Svoboda T, Van Gool L, Lang S, Strehlke K, Vande MA, Staadt O (2003) Blue-c: a spatially immersive display and 3D video portal for telepresence. *Proceedings of ACM SIGGRAPH 2003*, 819–827.

Grünheit C, Smolic A and T. Wiegand 2002 Efficient representation and interactive streaming of high-resolution panoramic views. *Proceedings of the International Conference on Image Processing*, Rochester, NY, USA, pp. 209–212.

Haskell BG, Puri A and Netravali A.N 1997 Digital video: an introduction to MPEG-2. *Digital Multimedia Standards Series*. Chapman & Hall, 1997.

ISO/IEC JTC1/SC29/WG11 2002a MPEG-4 Overview. Doc. N4668, Jeju, Korea.

ISO/IEC JTC1/SC29/WG11 2002b Text of ISO/IEC 14496-16:2003/FDAM4. Doc. N5397, Awaji, Japan, December.

ISO/IEC JTC1/SC29/WG11 2003 Report on 3DAV Exploration. Doc. N5878, Trondheim, Norway, July.

Jayant NS and Noll P 1984 *Digital Coding of Waveforms*. Englewood Cliffs, NJ, Prentice Hall.

Karczewicz M, Niewglowski J and Haavisto P 1997 Video coding using motion compensation with polynomial motion vector fields. *Signal Processing; Image Communication*, **10**(1–3), 63–91.

Kauff P, Makai B, Rauthenberg S, Gölz U, DeLameillieure JLP and Sikora T 1997 Functional coding of video using a shape-adaptive DCT algorithm and object-based motion prediction toolbox. *IEEE Transactions on Circuits for Systems for Video Technology*, **7**(1), 181–196.

Lee D 2005 JPEG 2000: retrospective and new developments. *Proceedings of the IEEE*, **93**(1), 32–41.

Legall D 1992 The MPEG video compression algorithm. *Image Communication* **4**, 129–140.

Liebchen T, Reznik Y, Moriya T and Yang D 2004 MPEG-4 Audio lossless coding. *Proceedings of the 116th AES Convention*, Berlin.

Luthra A, Sullivan GJ and Wiegand T (ed.,) 2003 Special Issue on the H.264/AVC Video Coding Standard. *IEEE Transactions on Circuits Systems for Video Technology* **13**(7).

Netravali A and Haskel B 1995 *Digital Pictures: Representations, Compression, and Standards* (Applications of Communications Theory). Plenum.

Ostermann J, Jang ES, Shin JS and Chen T 1997 Coding the arbitrarily shaped video objects in MPEG-4. *Proceedings of the International Conference on Image Processing*, Santa Barbara, California, USA, 496–499.

Pereira F and Ebrahimi T 2002 *The MPEG-4 Book*. Prentice Hall.

Santa-Cruz D, Grosbois R and Ebrahimi T 2002 JPEG 2000 performance evaluation and assessment. *Signal Processing: Image Communication*, **17**(1), 113–130.

Scheirer E, Väänänen R and Huopaniemi J 1999 AudioBIFS: Describing audio scenes with the MPEG-4 multimedia standard. *IEEE Transactions on Multimedia*, **1**(3), 237–250.

Shapiro JM 1993 Embedded image coding using zerotrees of wavelet coefficients. *IEEE Transactions Signal Processing*, **41**(12), 3445–3460.

Shade J, Gortler S, He LW and Szeliski R 1998 Layered depth images. *Proceedings of ACM SIG-GRAPH*, Orlando, FL, USA, pp. 231–242.

Sikora T 1997 MPEG Digital Video-Coding Standards. *IEEE Signal Processing Magazine*, **14**(5), 82–100.

Sikora T 2005 Trends and perspectives in image and video coding. *Proceedings, of the IEEE* **93**(1), 6–17.

Sikora T 1995 Low complexity shape-adaptive DCT for coding of arbitrarily shaped image segments. *Signal Processing: Image Communication* **7**(4–6), 381–95.

Smolic A, Sikora T and Ohm JR 1999 Long-term global motion estimation and its application for sprite coding, content description and segmentation. *IEEE Transactions on Circuits Systems for Video Technology*, **9**(8), 1227–1242.

Smolic A, Makai B and Sikora T 1999 Real-time estimation of long-term 3D motion parameters for SNHC face animation and model-based coding applications. *IEEE Transactions on Circuits Systems for Video Technology* **9**(2), 255–263.

Smolic A and McCutchen D 2004 3DAV Exploration of video-based rendering technology in MPEG. *IEEE Transaction on Circuits and Systems for Video Technology* **14**(3), 348–356.

Smolic A and Kauff P 2005 Interactive 3D video representation and coding technologies. *Proceedings of the IEEE*, **93**(1), 98–110.

Tanger R, Kauff P and Schreer O 2004 Immersive meeting point. *Proceedings of the IEEE Pacific Rim Conference on Multimedia*, Tokyo, Japan.

Vetterli M and Kovacevi'c J 1995 *Wavelets and Subband Coding*. Prentice Hall, Part 1, 89–96. Englewood Cliffs, NJ.

Wallace GK 1991 JPEG Still picture compression standard. *Communications of the ACM*, **34**(4), 30–44.

Würmlin S, Lamboray E and Gross M 2004 3D video fragments: dynamic point samples for real-time free-viewpoint video. *Computers and Graphics* **28**(1), p. 3–14.

Section III
3D Reproduction

12

Human Factors of 3D Displays

Wijnand A. IJsselsteijn, Pieter J.H. Seuntiëns and Lydia M.J. Meesters

Eindhoven University of Technology, Eindhoven, The Netherlands

12.1 INTRODUCTION

Despite considerable advances in technology, stereoscopic television has not yet delivered on its promise of providing a high-quality broadcast 3D TV service that guarantees an optimal and strain-free viewing experience for multiple viewers under normal viewing circumstances. The general public's awareness of stereoscopic displays still tends to centre around its gimmicky side. The boundaries of common experience are defined by children's toys (e.g., the ViewMaster stereoscope), stereoscopic arcade games, Pulfrich and anaglyph television experiments, and 3D cinemas at theme parks and carnivals. An interesting exception is 3D IMAX, which has been commercially relatively successful since its introduction in 1986, and launches a limited number of specialized stereoscopic productions each year. The most promising applications of 3D displays, however, have been outside the field of 3D cinema and broadcasting — in niche applications where the true benefits of stereoscopy are needed, and users are willing to put up with some of its drawbacks. Examples include simulation systems (e.g., flight simulators), medical systems (e.g., endoscopy), telerobotics (e.g., remote inspection of hazardous environments), computer-aided design (e.g., car interior design), data visualization (e.g., molecular or chemical modelling, oil and gas exploration, weather forecasting), and telecommunication systems (e.g., videoconferencing). In these applications, the capacity of stereoscopic imaging to provide an accurate representation of structured layout, distance and shape can be utilized for precise perception and manipulation of objects and environments.

Some of the factors limiting the introduction of a successful stereoscopic broadcast service can be traced back to human factors. Such issues can originate at various stages of

3D Videocommunication — Algorithms, concepts and real-time systems in human centred communication
Edited by O. Schreer, P. Kauff and T. Sikora © 2005 John Wiley & Sons, Ltd

the stereoscopic broadcast chain. Some will be related to the way stereoscopic images are produced, such as keystone distortions due to a converging camera setup or left – right camera misalignment (e.g., rotational error, vertical offsets, luminance differences). Others are transmission-related errors (e.g., compression artifacts) or an unwanted side-effect of the specific display technology used to address the left and right eyes separately (e.g., cross-talk, loss of resolution). Each of the current stereoscopic display alternatives has its own inherent limitations and drawbacks (see Chapter 13). In addition to the human factors issues associated with stereoscopic TV, there is also the issue of downward compatibility (enabling a smooth transition from existing 2D systems) and the availability of a large enough range of interesting and up-to-date broadcast content to be of interest to the general viewing public. New data formats are currently being explored in relation to (pseudo)stereoscopic TV in order to both address the issue of comfortable viewing as well as to develop tools that will enable content producers to more easily generate new 3D material or convert existing 2D material into 3D. For a detailed account of such a new data format in the service of stereoscopic TV, see Chapter 2.

The human viewer is, of course, the final judge of the quality of any 3D system and, eventually, its widespread acceptance and commercial success. Understanding the physiological, perceptual, cognitive, and emotional processes underlying the viewer's experience and judgement is essential to informing the development of technology that aims to be usable in its interaction, pleasurable to watch, and fun to own. In this chapter, we will highlight human factors issues that have been associated with viewing stereoscopic displays. First, we will discuss the basics of human depth perception, with particular relevance to stereoscopic displays. Next, we will briefly describe the basic principles of stereoscopic image production and display. We will then discuss some of the most pertinent issues that are detrimental to stereoscopic image quality and viewer comfort, and will conclude with some observations regarding our current understanding of stereoscopic image quality.

12.2 HUMAN DEPTH PERCEPTION

12.2.1 Binocular Disparity and Stereopsis

Complex, natural scenes contain a wide variety of visual cues to depth. Our visual system utilizes monocularly available information (or cues) such as accommodation, occlusion, linear and aerial perspective, relative size, relative density, and motion parallax to construct a perception of depth. The effectiveness of monocular cues is illustrated by the fact that we can close one eye and still have a considerable appreciation of depth. These cues are already available in monoscopic displays, such as traditional 2D television. The binocular cues, stereopsis and vergence, require both eyes to work in concert. Random-dot stereograms or Julesz patterns (Julesz 1971) demonstrate that in the absence of consistent monocular information, stereopsis alone provides the visual system with enough information to extract a fairly accurate estimate of depth.

Generally speaking, the more *consistent* cues, the better an accurate depth percept can be established. Cutting and Vishton (1995), based on Nagata (1993), provide a thorough discussion of the relative information potency of depth cues at various distances. Each cue on its own is an ambiguous indicator of distance, layout and surface structure. However, by combining several sources of information, this ambiguity may be reduced, even to the point of near-metric accuracy. A detailed treatment of all depth cues is beyond the purpose of this chapter, but interested readers are referred to Cutting and Vishton (1995), Palmer (1999) and Sedgwick (2001).

Stereopsis is available because the human eyes are horizontally separated (on average by approximately 6.3 cm) which provides each eye with a unique viewpoint on the world. This horizontal separation causes an interocular difference in the relative projections of monocular images onto the left and right retina, i.e., *binocular disparity*. When points from one eye's view are matched to corresponding points in the other eye's view, the retinal point-to-point disparity variation across the image provides information about the relative distances of objects to the observer and the depth structure of objects and environments.

The line that can be drawn through all points in space that stimulate corresponding retinal points for a given degree of convergence is called the *horopter* (Figure 12.1). The theoretical horopter is a circle, known as the *Vieth–Müller circle,* which passes through the fixation point and the nodal points of both eyes. Experimental work has shown that the empirical horopter lies slightly behind the theoretical horopter, although no satisfactory explanation has yet been provided to account for this fact. However, differences between the theoretical and the empirically measured horopter are small and can usually be ignored for practical purposes (Palmer 1999).

Points which are not on the horopter will have a retinal disparity. Disparities in front of the horopter are said to be *crossed* and disparities behind the horopter *uncrossed.* As long as

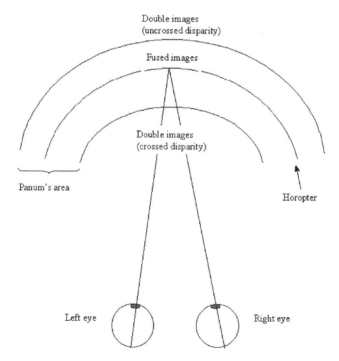

Figure 12.1 Panum's fusional area. Points within Panum's area are fused perceptually into a single image. Points that are closer or farther produce double images of crossed or uncrossed disparity. Adapted from Palmer (1999)

the disparities do not exceed a certain magnitude, the two separate viewpoints are merged into a single percept (*fusion*). The small region around the horopter within which disparities are fused is called *Panum's fusional area* (Figure 12.1).

If disparity is large the images will not fuse and double images will be seen, a phenomenon that is known as *diplopia*. The largest disparity at which fusion can occur is dependent on a number of factors. Duwaer and van den Brink (1981) showed that the diplopia threshold is dependent on the subject tested (i.e., individual variations), the amount of training a subject received, the criterion used for diplopia (unequivocal 'singleness' of vision compared with unequivocal 'doubleness' of vision), and the conspicuousness of the disparity.

Although limits vary somewhat across studies, some representative disparity limits for binocular fusion can be given. For small stimuli (i.e. smaller than 15 minutes of arc) the range of disparity for the foveal area is about ±10 minutes of arc, while at 6 degrees eccentricity the range increases to around ±20 minutes of arc. For large stimuli (1.0–6.6 degrees) the range for the foveal region is about ±20 minutes of arc, i.e., twice the range for smaller stimuli (Patterson and Martin 1992). However, most individuals have an appreciation of depth beyond the diplopia threshold, i.e., the region where single vision has been lost. Up to 2 degrees of overall disparity between two images is tolerated before the sensation of depth is lost (Howard and Rogers 1995), yet such a level of disparity is not comfortable for the viewer.

Stereoscopic vision greatly enhances our ability to discriminate differences in depth, particularly at close distances. Stereoscopic sensitivity is remarkably good at regions close to the horopter: people can detect a disparity of around 2 seconds of arc. Better stereo-acuity performance is reported for crossed than for uncrossed disparities (Woo and Sillanpaa 1979). The only monocular cue that provides a similar degree of depth resolution is movement parallax created by side-to-side head motion (Howard and Rogers 1995). In fact, monocular movement parallax and binocular disparity are closely related, since temporally separated successive views can in principle provide the same information to the visual system as spatially separated views (Rogers and Graham 1982). Estimates of the effective range of stereopsis vary across the literature, but it is clear that stereoscopic information becomes less effective as distance increases and retinal disparities become smaller. For distances beyond 30 m disparities become negligible.

12.2.2 Accommodation and Vergence

In normal vision, the oculomotor mechanisms, accommodation and vergence, work in concert with stereopsis. Thus, when viewing an object the eyes converge on it, so that the disparities providing depth information about the object and its environment fall within Panum's fusional area. The eye's lens automatically focuses (accommodation) on the currently fixated object, making it stand out against its surroundings. Thus, double images in front or behind the plane of fixation tend to be out of focus and will 'disappear' in increasing optical blur.

Accommodation and convergence operate in a closely coupled fashion, that is, accommodation may produce vergence movements (i.e., accommodative vergence) and vergence may produce accommodation (i.e., vergence accommodation). Under natural viewing conditions all objects which are in focus are also fixated upon, bringing them within the limits of fusion. However, most of the currently available stereoscopic display techniques do not support the linkage between accommodation and convergence, forcing the observer to focus at a fixed

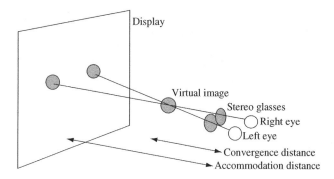

Figure 12.2 Difference between accommodation distance and vergence distance when looking at a stereoscopic image. Reproduced by permission of Stereographics

distance (i.e., on the screen plane where the image is sharpest) irrespective of the fixation distance (i.e., where the object is located in depth according to the disparity information), as illustrated in Figure 12.2. This implies that observers' retinal images may be blurred, especially for stereoscopic images containing large image parallaxes. As a consequence, the accommodative system (implemented by the circular ciliary muscles that surround the lens of the eye) will attempt to bring the blurred image back into focus, which may lead to a constant oscillation of the visual system between fixating on the object of interest, and adjusting focus back towards the sharpest image. This decoupling of accommodation and vergence when viewing stereoscopic displays has often been suggested as a potential cause of visual strain (Hiruma and Fukuda 1993; Inoue and Ohzu 1990; Ohzu and Habara 1996).

12.2.3 Asymmetrical Binocular Combination

A mechanism that is of particular interest to (asymmetrical) stereoscopic image coding is how the different inputs to the two eyes are combined to form a single percept. As was discussed earlier, with similar images presented to each eye, fusion will occur, as long as the disparities fall within certain boundaries. There is said to be *binocular summation* when visual detection or discrimination is performed better with two eyes than with one eye. However, if the stimuli presented to the two eyes are very different (e.g., large differences in pattern direction, contour, contrast, illumination, etc.), the most common percepts are binocular mixture, binocular rivalry and suppression, and binocular lustre.

Mixture typically occurs, for instance, when a uniform field in one eye is combined with a detailed stimulus in the corresponding part of the other eye. When corresponding parts of the two retina's receive very different high-contrast images, then *binocular rivalry* may occur. In a way, this is the opposite of fusion, as the two monocular images will alternate repetitively, in whole or in part, with the unseen portion somehow suppressed. The stimulus seen at any given time is called the *dominant* stimulus. *Binocular lustre* occurs in areas of uniform illumination in which the luminance or colour is different in the two eyes. In this response, the images are stable and fused, yet appear to be shimmering or lustrous and cannot be properly localized in depth.

The relevance for asymmetrical image coding lies in the fact that the combination of a high-quality image presented to one eye and a low-quality (highly compressed) image to the other may still result, through binocular mixture, in a fairly high-quality overall stereoscopic percept, particularly if the high-quality image is presented to the dominant eye (in the majority of cases the right one). As was noted, the perceptual effect of binocular combination strongly depends on the individual stimulus patterns presented to each eye. Therefore, it is likely that the nature of the monocular deterioration (e.g., blockiness or blur) will be a principal determining factor in the resulting overall stereoscopic image quality.

12.2.4 Individual Differences

The mechanism that underlies stereopsis is extremely precise, but also somewhat fragile. Visual disorders in early childhood, even if only temporary, may result in *stereoblindness,* which is estimated to affect 5–10% of the population. Richards (1970) performed a survey of 150 members of the student community at MIT and found that 4% of the population was unable to use the depth cue offered by disparity and 10% had great difficulty deciding the direction of a hidden Julesz figure (i.e., a random-dot stereogram) relative to the background.

More recent studies have shown that the performance of observers on tests of stereoanomaly depends to a large extent on the duration the observer is allowed to look at the test figures. The proportion of the population unable to perform a depth discrimination task decreases with increasing display duration (Patterson and Fox 1984; Tam and Stelmach 1998). Tam and Stelmach (1998) systematically varied display duration until viewers were able to perform a depth discrimination task with 75% accuracy. Their results showed that the proportion of observers unable to perform the depth discrimination task decreased with increasing display duration. At the shortest display duration (20 ms) approximately half the observers ($N = 100$) were unable to perform the stereoscopic depth discrimination task at criterion level. However, at the maximum display duration of 1000 ms, failure rate dropped to about 5%. One important implication of these findings for stereoscopic televisual services is that about 95% of the population is expected to experience an enhanced sense of depth when viewing a stereoscopic programme, since most scene durations typically exceed 1 s between cuts or fades, which is a sufficiently long duration for most people.

The single most common cause of stereoblindness is *strabismus,* the misalignment of the two eyes. If this condition is surgically corrected at an early age, stereoscopic vision may develop normally. Stereopsis also fails to develop when children have good vision in only one eye, due to monocular nearsightedness or a cataract. Obviously, stereoblind individuals can still perceive depth, but are restricted to information sources other than binocular disparity.

It is well known that visual abilities change with age. The major age differences in visual functioning are the result of two types of changes in the structure of the eye. The first type of change is related to the transmissiveness and accommodative power of the lens that begins to manifest itself between the ages of 35 and 45. This affects binocular depth perception, sensitivity to glare, and color sensitivity. The second type of change occurs in the retina and the nervous system, and usually starts to occur between 55 and 65 years of age. This affects the size of the visual field, sensitivity to low light levels, and sensitivity to flicker (Hayslip and Panek 1989). A study by Norman *et al.* (2000) demonstrated that older adults were less sensitive than younger adults to perceiving stereoscopic depth, in particular when image disparity was higher. Overall however, older adults performed to 75% of their expected depth intervals, demonstrating that their stereoscopic abilities have largely been preserved during the process of aging.

12.3   PRINCIPLES OF STEREOSCOPIC IMAGE PRODUCTION AND DISPLAY

> Like many others who came before me, I had no idea how terribly difficult it is to make such [stereoscopic] equipment perform properly. The naive position is that since one is simply using two of this and two of that, stereoscopic film-making ought to be fairly simple. I had no notion that stereoscopic photography was in many ways substantially different from conventional, or planar, photography. . . . The danger with stereoscopic film-making is that if it is improperly done, the result can be discomfort. Yet, when properly executed, stereoscopic films are beautiful and easy on the eyes. Lenny Lipton (1982)

The earliest stereoscopic photographs date from about 1841 and were taken using a single-lens plate camera mounted on a tripod with a slide bar. After the first picture was taken, a new photographic plate would be inserted, the camera would be moved by about 6.5 cm, and a second picture was taken. Of course, with one mono-camera, only stereo exposures of non-moving objects could be made. From approximately 1856 onwards, twin-lens cameras came in use, and by the end of the 19th century, dedicated stereoscopic cameras had been developed for the amateur photographer.

Stereoscopic motion pictures are usually produced by using a stereoscopic camera, consisting of a coplanar configuration of two separate, monoscopic cameras, each corresponding to one eye's view. Usually, the aim of stereoscopic video is to capture real-world scenes in such a way that viewing them stereoscopically accurately simulates the information the viewer would receive in the real life situation. To achieve this, the stereoscopic filming process must be carefully planned and executed. Stereoscopic parameters need to be calculated and adjusted such that they accurately capture the disparities a viewer would receive from the actual scene and do not cause any visual discomfort.

The mathematical basis for stereoscopic image formation was first described by Rule (1941). In the 1950s, the seminal paper by Spottiswoode *et al.* (1952) extended this work considerably, introducing new concepts, such as the *nearness factor,* which facilitated the communication between the film director and the stereo-technician. Various accounts of the geometry of stereoscopic camera and display systems have been published since (Gonzalez and Woods 1992; Woods *et al.* 1993), both for the parallel camera configuration, where the optical axes of both monoscopic cameras run parallel, and for the converging camera configuration, where the optical axes of the cameras intersect at a convergence point. The equations by which the horizontal and vertical disparities can be calculated for both types of camera configurations are discussed in Woods *et al.* (1993) and IJsselsteijn (2004); see also Chapter 2.

For parallel camera configurations, a horizontal translation between right and left cameras results in horizontal disparities. For the converging camera setup there is also a vertical disparity component, which can be calculated separately. This vertical component is equal to zero in the plane defined by $X = 0$ as well as for the plane defined by $Y = 0$. It is largest at the corners of the image. Vertical parallax causes what is known as *keystone distortion,* which will be discussed in Section 12.4. When properly calibrated, the parallel camera setup does not produce vertical disparities, and as a consequence no keystone distortion will occur. Based on this consideration, Woods *et al.* (1993) recommend using a parallel camera configuration.

The resulting three-dimensional scene must be free from visual errors. Particular care needs to be taken to assure that the two stereoscopic images are well calibrated, both during filming and subsequent display. Various binocular asymmetries, either based on asymmetrical optical geometry (e.g., image shift, rotation, magnification) or filter characteristics (e.g., luminance, colour, contrast, cross-talk) may significantly diminish the image quality and visual comfort experienced by the viewer.

After image pickup, the camera sensor coordinates are transformed to two sets of screen coordinates (i.e., the x and y positions of the left and right images on the stereoscopic display) by multiplying them by the screen magnification factor M, which is the ratio of the horizontal screen width to the camera sensor width. The resulting disparities for the observer are a function of: (i) the screen parallax P, which is the horizontal linear distance between common (homologous) points on the display surface; (ii) the distance of the observer to the screen V; and (iii) the observer's eye separation E, also known as interocular or interpupillary distance (IPD). Variations in IPD for adult males range from 5.77 to 6.96 cm, with a median of 6.32 cm (Woodson 1981). Although 6.3 cm is usually taken as the average adult IPD, it is advisable to produce and display stereoscopic video in such a way that people with smaller eye separations (e.g., children) can still comfortably fuse the stereoscopic scene.

Recent developments in sensor technologies, computing power, and image processing, have allowed for new (pseudo)stereoscopic data formats to emerge (see Chapter 2). Camera systems have been developed that are based on registering one conventional RGB view, and an accompanying *depth map,* containing the corresponding depth or z-value for each image point. Examples of cameras based on this format are the *AXI-Vision* camera developed by NHK (Kawakita *et al.* 2000; 2002) and the *Zcam* developed by 3DV Systems (Gvili *et al.* 2003; Iddan and Yahav 2001). The RGB+depth format has the advantage of being very flexible in post-production and rendering, and requiring less bandwidth than two separate left–right RGB images would need. Moreover, it also allows for a higher degree of user control over the final rendered image through, for example, interactively adjusting the amount of depth in an image, much like a volume control for adjusting the sound level. In such a way, viewers are enabled to adjust the image parallax to a level that is comfortable for them to watch, a privilege that was previously reserved only for the stereo-technician, and, to a lesser degree, the operator of the projection system.

12.4   SOURCES OF VISUAL DISCOMFORT IN VIEWING STEREOSCOPIC DISPLAYS

Several types of stereoscopic distortions have been described in the literature, and can be experienced firsthand when interacting with stereoscopic systems. Potential imperfections can be related to image pickup (e.g., choice of camera configuration), transmission (e.g., compression), rendering (as in the case of stereoscopic depth reconstruction from an RGB+depth signal), and display. Also, various factors normally under the viewer's control (viewing position, ambient light levels, etc.) will influence the perceived quality, as is also the case in conventional TV viewing.

The relative perceptual importance (weighting) of each distortion will need to be determined by studying the effects of distortions in isolation as well as in various combinations, particularly as some of the interactions are likely to be non-linear. We will not provide a comprehensive overview of all possible sources of viewer discomfort here, but will highlight some of the most pertinent stereoscopic impairments. These include: keystone distortion, depth plane curvature, cardboard effect, puppet theatre effect, shear distortion, cross-talk, picket fence effect and image flipping. For a recent, more in-depth treatment of stereoscopic image quality issues and methods of quality assessment, we refer to Meesters *et al.* (2004).

12.4.1 Keystone Distortion and Depth Plane Curvature

Woods *et al.* (1993) reviewed several stereoscopic distortions including *keystone distortion* and *depth plane curvature.* These are typically introduced in a stereoscopic image captured with a converging camera setup where the left and the right camera are positioned at an angle towards each other. In this case, the imaging sensors of the two cameras are directed towards slightly different image planes. This results in a trapezoidal picture shape in opposite directions for the left- and right-eye camera recordings (Figure 12.3).

In a stereoscopic image these oppositely oriented trapezoidal picture shapes may induce incorrect vertical and horizontal parallax. Incorrectly introduced vertical parallax is the source of *keystone distortion.* Keystone distortion is most noticeable in the image corners and increases with increasing camera base distance, decreasing convergence distance, and decreasing lens focal length. Incorrectly introduced horizontal parallax is the source of *depth plane curvature,* whereby objects at the corner of the image appear further away from the observer compared with objects in the middle of the image.

Perceptually, keystone distortion and depth plane curvature may have a negative effect on appreciation-oriented assessments (IJsselsteijn *et al.* 2000). Furthermore, Woods *et al.* (1993) reported that subjects experienced eyestrain at higher vertical parallax values. More recently, Stelmach *et al.* (2003) demonstrated that, whereas extreme levels of camera convergence produce visual discomfort, moderate vertical parallax, introduced with camera convergence distances in the range 60–240 cm, hardly affects visual comfort ratings. Thus, although perceptual studies generally show a negative effect of keystoning, its effect appears minimal as long as reasonably small convergence angles are being employed.

Other geometric distortions are often caused by misalignment of the left- and right-eye cameras. For example, Yamanoue (1998) discusses rotation error, size inconsistency and

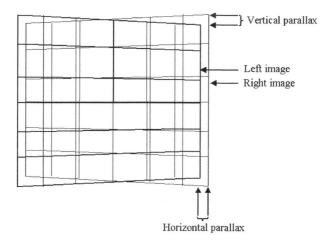

Figure 12.3 Unwanted horizontal and vertical parallax due to a converging camera setup (exaggerated for illustration purposes)

vertical shift. These distortions were introduced independently in static images and the visibility and acceptability limits for each geometric distortion were measured. Overall the measured detection limits were approximately half the tolerance limits. The author showed that rotation error and vertical shift had a similar effect on perceived impairment while the effect of size inconsistency was relatively weak.

12.4.2 Magnification and Miniaturization Effects

Visual size distortions occur if the angular retinal size of a displayed object and its perceived distance do not covary as in real world conditions. In the real-world a change in angular size corresponds to a change in distance. In stereoscopic displays however, various depth cues may give conflicting estimates of depth (e.g., depth from stereo versus depth from size estimations), such that the resulting percept (usually some kind of weighted average of the various sources of depth information) may actually contain some unnatural magnification or miniaturization effects.

One such effect is known as the *cardboard effect* and refers to an unnatural flattening of objects and people in stereoscopic images, as if they were cardboard cut-outs. It makes a stereoscopic scene appear as if it is divided into discrete depth planes. As an explanation of this frequently encountered phenomenon, Howard and Rogers (1995) suggest that it is a consequence of the fact that size is inversely proportional to the distance $1/d$, whereas disparity is inversely proportional to the square of the distance $1/d^2$. Thus, an object's perceived depth obtained from disparity is too small compared with its perceived size and as a result of this, objects may appear flattened. Howard and Rogers (1995) comment that size and depth scaling often are inappropriate in stereoscopic displays as vergence and vertical disparity signal a closer viewing distance than in the original scene.

Another perceptual effect of the mismatch between depth and size magnification is known as the *puppet theatre effect*. This refers to an annoying miniaturization effect, for instance making people appear as if they are small puppets (Pastoor 1995; Yamanoue 1997). Yamanoue (1997) showed that orthostereoscopic parallel shooting and display conditions (i.e., simulating human viewing angles, magnification, and convergence in the most natural way possible) do not cause the puppet theatre effect. Hopf (2000) describes a display technique which reduces the puppet theatre effect. An autostereoscopic display with collimation optics enables a large volume of depth, and thus allows a larger image to be presented at a greater distance behind the screen so that the puppet theatre effect is not perceptible. Novel display techniques seem promising in this regard, contributing to a more natural appearance of the 3D image.

More recently, with the introduction of the RGB+depth format for generating stereoscopic scenes, the application of compression algorithms on the depth map may result in a coarse quantization of disparity or depth values, which, after rendering, may also result in a cardboard-like effect (Schertz 1992). In this case, it can also manifest itself also as torn-up objects such that a coherent object is represented in several depth planes and perceived as a disjointed object. Schertz (1992) showed that scenes with torn objects were judged as more annoying than if the objects are reunited by reducing the depth resolution. Therefore, choosing a suitable number of depth layers is a subtle issue, where higher depth resolution does not always imply better quality. Obviously, the cost of introducing depth (in this case, a single object becoming disjointed across various depth layers) should never outweigh the perceived benefit.

12.4.3 Shear Distortion

Shear distortion is typically experienced with stereoscopic displays that allow only one correct viewing position (Postoor 1993; Woods *et al.* 1993). For most stereoscopic displays a stereoscopic image can only be viewed correctly from one particular viewpoint or 'sweet spot'. If the observer changes his viewing position the image seems to follow the observer and therefore appears perspectively distorted. Objects with positive parallax move in the opposite direction to the observer's movement and objects with negative parallax appear to move in the same direction.

Shear distortion can be avoided by using head-tracked displays where the correct stereoscopic image is presented according to the viewpoint of the observer. For instance, a solution proposed by Surman *et al.* (2003) involves an autostereoscopic display with steerable exit pupils that can be used to present either a single stereo pair correctly when viewed from different positions or a correct unique stereoscopic view onto the 3D scene for different viewing positions. The latter approach requires a large data set containing multiple stereo image pairs of the same scene. A more efficient approach in this case is based on the IBR techniques discussed in this volume, such that the appropriate view can be reconstructed from an RGB+depth image format. Of course, occlusion problems will need to be taken into account as only one RGB image will form the basis of multiple (pseudo)stereoscopic perspectives.

12.4.4 Cross-talk

Cross-talk or image ghosting in stereoscopic displays is caused primarily by: (i) phosphor persistence of a CRT-display and (ii) imperfect image separation techniques by which the left-eye view leaks through to the right-eye view and vice versa. Cross-talk is perceived as a ghost, shadow or double contours and is probably one of the main display-related perceptual factors detrimental to image quality and visual comfort. Moreover, the visibility of cross-talk increases with increasing screen parallax, thus limiting the degree of depth that can be introduced while maintaining a high 3D image quality. Even a relatively small amount of cross-talk can lead to problems (Pastoor 1995).

Cross-talk due to phosphor persistence may occur in time-sequential displays when the left and the right view are displayed alternately and the image intensity leaks into the subsequent view. Thus, the right eye also receives a small proportion of the left-eye view. Additionally, for linear polarization techniques an incorrect head position of an observer (e.g., tilted head) also causes annoying image ghosting. This can be avoided by using circular polarization techniques. Pastoor (1995) demonstrated that the perceptibility of cross-talk increases with increasing contrast and disparity values. The author suggests that the cross-talk of a display should not go beyond a threshold of 0.3%. In an overview that discusses the sources of image ghosting in time-sequential stereoscopic video displays, Woods and Tan (2002) introduced a cross-talk model which incorporates phosphor afterglow and shutter leakage.

Autostereoscopic displays also suffer from cross-talk, sometimes even significantly so. This can have a variety of reasons, depending on the specific eye addressing technique used. One problem in tracked displays, for instance, is the latency in the tracking system and/or in the steering mechanism of the directional lenses supporting movement parallax of the observer(s).

As a potential solution, Konrad *et al.* (2000) proposed an algorithm to compensate for cross-talk in stereoscopic images. The authors show that suppression of cross-talk enhances visual comfort. The main problem of cross-talk, however, is that solutions need to be system-specific, that is, an algorithm to compensate for it needs to be adapted specifically to each type of display and/or glasses.

12.4.5 Picket Fence Effect and Image Flipping

Typical artifacts found in multi-view autostereoscopic displays are the *picket fence effect* and *image flipping*. Both artifacts are perceived when observers move their head laterally in front of the display. The picket fence effect is the appearance of vertical banding in an image due to the black mask between columns of pixels in the LCD (see Chapter 13 for an explanation of the principles behind lenticular displays). Image flipping refers to the noticeable transition between viewing zones which results in discrete views and is experienced as unnatural compared with the continuous parallax experienced in the real world (Sexton and Surman 1999). Display techniques can be improved so that both artifacts are less visible. For instance, van Berkel and Clarke (1997) placed a tilted lenticular sheet in front of the LCD, which had the effect of making a constant amount of the black mask always visible. Owing to habituation an observer no longer perceives the picket fence effect and image flipping is also softened. Progress with user-tracking in autostereoscopic displays is another promising approach to significantly reduce image flipping.

12.5 UNDERSTANDING STEREOSCOPIC IMAGE QUALITY

Image quality is a standard psychological criterion whereby imaging systems are evaluated. It is a subjective preference judgement which is widely used to evaluate the perceptual effects of compression algorithms, image processing techniques, and display parameters. Image quality is regarded as a multidimensional psychological construct, based on several attributes (Ahumada and Null 1993; Meesters 2002). Engeldrum (2000) describes an image quality circle, illustrating how, through careful experimentation, a relation can be mapped between various technological parameters, their objectively measurable attributes (image physics), and their associated human perceptual effects. Eventually, the goal is to reliably *predict* subjective image quality through instrumental image quality models so that one does not need to go through the laborious and expensive process of subjective testing for each iteration in product development. Moreover, competing systems can then be compared by standardized assessment factors.

In the case of stereo images, no comprehensive stereoscopic image quality model has been formulated to date, yet it is likely that a diverse set of image attributes contribute to the overall perceived quality of stereoscopic TV images. Some attributes will have a positive contribution to the overall image quality (e.g., increased sharpness, or increased depth sensation), others may have a limiting or negative effect (e.g., diplopic images due to exaggerated disparities or lack of sharpness due to compression artifacts). A good stereoscopic image quality model will account for both contributing and attenuating factors, allowing for a weighting of the attributes based on perceptual importance, and for interactions that may occur as a consequence of (potentially asymmetric) binocular combinations.

For example, cross-talk becomes more noticeable with an increase in left–right image separation, a manipulation that also increases perceived depth. In such a case the perceptual 'benefit' of increased depth can be nullified by the perceptual 'cost' of increased cross-talk. The interactions between such positive and negative contributions, and their relative weighting deserve further study, in order to arrive at a more complete understanding of stereoscopic image quality. First steps in this direction have been taken (Meegan *et al.* 2001; Seuntiëns *et al.* 2004).

The added value of depth needs to be defined explicitly in a stereoscopic image quality model, especially when 2D picture quality is used as reference. The positive contribution of depth to the perceived image quality has been demonstrated for stereoscopic uncompressed and blur-degraded images (Berthold 1997; IJsselsteijn *et al.* 2000). In addition, Schertz (1992) demonstrated that human observers seem to prefer DCT-coded stereoscopic images over the monoscopic originals, even though the perceived impairment was rated as perceptible and slightly annoying. Other research has shown, however, that when observers were asked to rate the perceived image quality of images compressed with MPEG-2 and JPEG, the image quality results are mainly determined by the impairments introduced and not so much by depth (Seuntiëns *et al.* 2003; Tam and Stelmach 1998).

It is unclear at the moment whether the variable results in the literature and in our own studies regarding the presumed added value of depth are due a variability in stimulus characteristics (e.g., blur versus blockiness artifacts), or whether it is an effect of the specific scaling paradigms that are being used across various studies. Subjective measurements of image quality are known to be sensitive to contextual effects based on the composition of the stimulus set as well as judgement strategies of the observers (de Ridder 2001). This is a focus of our future endeavours, as we are working towards a stereoscopic image quality model that takes both perceptual *costs* and *benefits* of the various stereoscopic video production, coding, and display factors into account.

REFERENCES

Ahumada A and Null C 1993 Image quality: A multidimensional problem. In *Digital Images and Human Vision* (Watson A ed.) MIT Press Cambridge, MA pp. 141–148.

Berthold A 1997 The influence of blur on the perceived quality and sensation of depth of 2D and stereo images. *Technical Report,* ATR Human Information Processing Research Laboratories.

Cutting J and Vishton P 1995 Perceiving layout and knowing distances: The integration relative potency, and contextual use of different information about depth. In: *Perception of Space and Motion* (Epstein W and Rogers S ed) Academic Press San Diego, CA pp. 69–117.

de Ridder H 2001 Cognitive issues in image quality measurement. *Journal of Electronic Imaging* **10** 47–55.

Duwaer A and van den Brink G 1981 What is the diplopia threshold?. *Perception and Psychophysics* **29,** 295–309.

Engeldrum P 2000 *Psychometric Scaling.* Imcotek Press, Winchester, Massachusetts, USA.

Gonzalez R and Woods R 1992 *Digital Image Processing.* Addison Wesley.

Gvili R, Kaplan A, Ofek E and Yahav G 2003 Depth keying. *Proceedings of the SPIE* **5006,** 554–563.

Hayslip B and Panek P 1989 *Adult Development and Aging.* Harper & Row, New York, NY.

Hiruma N and Fukuda T 1993 Accommodation response to binocular stereoscopic TV images and their viewing conditions. *SMPTE Journal* **102,** 1137–1144.

Hopf K 2000 An autostereoscopic display providing comfortable viewing conditions and a high degree of telepresence. *IEEE Transaction on Circuits and Systems for Video Technology* **10,** 359–365.

Howard IP and Rogers BJ 1995 *Binocular Vision and Stereopsis.* Oxford University Press, New York.

Iddan G and Yahav G 2001 3D imaging in the studio (and elsewhere . . .). *Proceedings of the SPIE* **4298,** 48–55.

IJsselsteijn W 2004 *Presence in Depth*. Eindhoven University of Technology, Eindhoven, The Netherlands.

IJsselsteijn W, de Ridder H and Vliegen J 2000 Subjective evaluation of stereoscopic images: Effects of camera parameters and display duration. *IEEE Transactions on Circuits and Systems for Video Technology* **10**, 225–233.

Inoue T and Ohzu H 1990 Accomodation and convergence when looking at binocular 3D images In: *Human Factors in Organizational Design and Management – III* (Noro K and Brown O ed.) Elsevier New York, NY pp. 249–252.

Julesz B 1971 *Foundations of Cyclopean Perceptions*. University of Chicago Press, Chicago, IL.

Kawakita M, Kurita T, Kikurchi H and Inoue S 2002 HDTV AXI-vision camera. *Proceedings of the IBC* pp. 397–404.

Kawakita M, Lizuka K, Aida T, Kikuchi H, Fujikake H, Yonai J and Takizawa K 2000 AXI-vision camera (real-time distance-mapping camera). *Applied Optics* **39**, 3931–3939.

Konrad J, Lacotte B and Dubois E 2000 Cancellation of image crosstalk in time-sequential displays of stereoscopic video. *IEEE Transactions on Image Processing* **9**, 897–908.

Lipton L 1982 *Foundations of the Stereoscopic Cinema - A Study in Depth*. Van Nostrand Reinhold, New York, NY.

Meegan DV, Stelmach LB and Tam WJ 2001 Unequal weighting of monocular inputs in binocular combinations: Implications for the compression of stereoscopic imagery. *Journal of Experimental Psychology: Applied* **7**, 143–153.

Meesters L 2002 *Predicted and Perceived Quality of Bit-Reduced Gray-Scale Still Images*. Eindhoven University of Technology, Eindhoven, The Netherlands.

Meesters L, IJsselsteijn W and Seuntiëns P 2004 A survey of perceptual evaluations and requirements of three-dimensional television. *IEEE Transactions on Circuits and Systems for Video Technology* **14**, 381–391.

Nagata S 1993 How to reinforce perception of depth in single two-dimensional pictures In *Pictorial Communication in Virtual and Real Environments* (Ellis S ed.) Taylor & Francis: London, pp. 527–545.

Norman J, Dawson T and Butler A 2000 The effects of age upon the perception of depth and 3-D shape from differential motion and binocular disparity. *Perception* **29**, 1335–1359.

Ohzu H and Habara K 1996 Behind the scenes of virtual reality: Vision and motion. *Proceedings of the IEEE* **84**, 782–798.

Palmer S 1999 *Vision Science: Photons to Phenomenology*. MIT Press, Cambridge, MA.

Pastoor S 1993 Human factors of 3D displays in advanced image communications. *Displays* **14**, 150–157.

Pastoor S 1995 Human factors of 3D imaging: Results of recent research at Heinrich-Hertz-Institut Berlin. *ASIA Display'95 Conference*, pp. 66–72.

Patterson R and Fox R 1984 The effect of testing method on stereoanomaly. *Vision Research* **24**, 403–408.

Patterson R and Martin WL 1992 Human stereopsis. *Human Factors* **34**, 669–692.

Richards W 1970 Stereopsis and stereoblindness. *Experimental Brain Research* **10**, 380–388.

Rogers B and Graham M 1982 Similarities between motion parallax and stereopsis in human depth perception. *Vision Research* **22**, 261–270.

Rule J 1941 The geometry of stereoscopic pictures. *Journal of the Optical Society of America* **31**, 325–334.

Schertz A 1992 Source coding of stereoscopic television pictures. *IEE International Conference on Image Processing and its Applications*, pp. 462–464.

Sedgwick H 2001 Visual space perception In *Blackwell Handbook of Perception* (Goldstein E ed.) Blackwell: Oxford pp. 128–167.

Seuntiëns P, Meesters L and IJsselsteijn W 2003 Perceptual evaluation of JPEG coded stereoscopic images. *Proceedings of the SPIE* **5006**, 215–226.

Seuntiëns P, Meesters L and IJsselsteijn W 2004 Perceptual attributes of crosstalk in 3-D TV. *Perception* **33**, 98.

Sexton I and Surman P 1999 Stereoscopic and autostereoscopic display systems. *IEEE Signal Processing Magazine*, pp. 85–99.

Spottiswoode R, Spottiswoode N and Smith C 1952 3-D photography - Basic principles of the three-dimensional film. *Journal of the SMPTE* **59**, 249–286.

Stelmach L, Tam W, Speranza F, Renaud R and Martin T 2003 Improving the visual comfort of stereoscopic images. *Proceedings of the SPIE* **5006,** 269–282.

Surman P, Sexton I, Bates R, Lee WK, Craven M and Yow KC 2003 Beyond 3D television: The multi-modal, multi-viewer TV system of the future. *SID/SPIE FLOWERS 2003,* pp. 208–210.

Tam W and Stelmach L 1998 Display duration and stereoscopic depth perception. *Canadian Journal of Experimental Psychology* **52,** 56–61.

van Berkel C and Clarke J 1997 Characterization and optimization of 3D-LCD module design. *Proceedings of the SPIE* **3012,** 179–186.

Woo G and Sillanpaa V 1979 Absolute stereoscopic threshold as measured by crossed and uncrossed disparities. *American Journal of Optometry and Physiological Optics* **56,** 350–355.

Woods A and Tan S 2002 Characterising sources of ghosting in time-sequential stereoscopic video displays. *Proceedings of the SPIE* **4660,** 66–77.

Woods A, Docherty T and Koch R 1993 Image distortions in stereoscopic video systems. *Proceedings of the SPIE* **1915,** 36–48.

Woodson W 1981 *Human Factors Design Handbook.* McGraw-Hill, New York, NY.

Yamanoue H 1997 The relation between size distortion and shooting conditions for stereoscopic images. *Journal of the SMPTE* pp. 225–232.

Yamanoue H 1998 Tolerance for geometrical distortions between L/R images in 3D-HDTV. *Systems and Computers in Japan* **29**(5), 37–48.

13
3D Displays

Siegmund Pastoor

Fraunhofer Institute for Telecommunications/Heinrich-Hertz-Institut, Berlin, Germany

13.1 INTRODUCTION

Stereoscopic 3D displays provide different perspective views to the left and right eye. As in natural vision, the differences in perspective are immediately used by the visual system to create a vivid, compelling and efficient sensation of depth in natural and computer-generated scenes. Thus, stereoscopic displays not only add a unique sense of naturalness to any video communications system; they are an absolute must if applications require seamless and accurate mixture of video output with real 3D environments (see Chapter 14). Various human factor experiments with advanced display technologies (including large-screen HDTV and 3DTV) have demonstrated that stereoscopy significantly enhances the spatial impression, telepresence and attractiveness of videoconferencing systems. It was concluded that 3D can be considered *the* promising feature of future videocommunication (Suwita *et al.* 1997).

Apart from videocommunication, 3D displays have proven to be particularly useful in applications where immediate and accurate depth perception is vital. Remote control of vehicles and robotic arms in hazardous environments as well as handling and coordination of instruments in endoscopic (minimally invasive) surgery are the most obvious examples. Various medical applications, including diagnosis, surgical planning, training and telemedicine benefit from the visualization of imaging data, such as MRI/CT scans, in full 3D. Room-sized virtual 3D environments such as the CAVE system provide total immersion into the world of scientific data. Such surround displays are efficiently used in education, engineering, and scientific computing — both as stand-alone devices and as networked systems for telecooperation. 3D displays are increasingly employed in engineering applications such as computer-assisted design (CAD) and manufacturing (CAM), genetic and pharmaceutical engineering (modelling of complex protein molecules, structure-based drug design), virtual design review of new products, virtual prototyping, and architecture. Consequently, most renowned 3D software tools already come with special plug-ins for the different stereo formats of today's display hardware. Emerging 3D applications in the military and security fields vary from battlefield simulation to surveillance, intelligence analysis, and baggage scanning. Due to the availability of affordable displays and production technologies the use of 3D displays is spreading out into the domains of entertainment and infotainment (3DTV, 3D film, 3D video, arcade games), trade (electronic show rooms, point-of-sale product presentation with interactive 3D kiosk systems) and education (museums, simulator-based training, teaching).

There are two basic approaches in the underlying 3D display technologies. One relies on special user-worn devices, such as stereo glasses or head-mounted miniature displays, in order to optically channel the left- and right-eye views to the appropriate eye. Such *aided-viewing* systems have been firmly established in many applications. Another approach integrates the optical elements needed for selective addressing of the two eyes in a remote display device, hence allowing free 3D viewing with the naked eye. In general, *free-viewing* 3D displays are more comfortable to the viewer. Since the eye region is not occluded by technical equipment, these displays are particularly well suited for human-to-human video communication. The respective technologies are challenging. Most of the numerous concepts presented in recent years are still under development. Based on an earlier survey (Pastoor and Wöpking 1997) this chapter reviews the state-of-the-art of both, aided-viewing and free-viewing techniques.

The next section will briefly introduce the relevant perceptual processes and definitions needed to understand the basic display concepts. This is followed by a classification of 3D techniques. The main part reviews the basic forms of display implementation. Interested readers may find detailed treatises on the foundations of 3D displays and spatial vision in the works of Valyus (1966) Okoshi (1976) and Howard and Rogers (1995).

13.2 SPATIAL VISION

Spatial vision relies on monocular and binocular depth cues. Monocular cues are visible with one eye and are used to create an illusion of volume and depth on flat image surfaces. Common examples are linear and aerial perspective, texture gradients, overlap of objects, and the distribution of light and shadows.

Binocular depth perception is based on local displacements between the projections of a scene onto the left and right retina (disparity). These displacements depend on the (relative) distances of the objects in the scene observed. As long as they do not exceed a certain magnitude, the two projections are merged into a single percept with structured depth (fusion). This condition is safeguarded by the operation of several interacting mechanisms:

Depth of field is an integral part of spatial vision. Automatic focusing of the eye's lens (accommodation) makes the currently fixated object stand out against its surroundings; details in the near foreground and in the far background 'disappear' in increasing optical blur. This way, disparity is kept within physiologically tolerated limits. In order to maintain a moving object within the zone of sharpest vision, the eyeballs perform rotational movements (convergence). Thus, accommodation and convergence are used to carrying out coordinated changes. As a consequence, all objects which are in focus are also within the limits of fusion — independent of their absolute distance from the observer.

Most of the currently feasible 3D display techniques neither support the gliding depth of field nor the link between accommodation and convergence (Figure 13.1). This failure

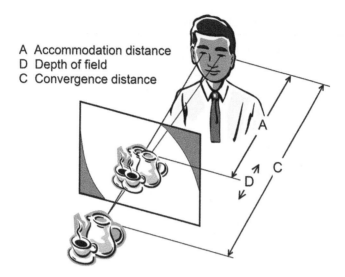

A Accommodation distance
D Depth of field
C Convergence distance

Figure 13.1 3D displays usually force the observer to 'freeze' ocular accommodation on the screen plane, irrespective of the fixated object's distance. Consequently, all details of the scene appear within unvarying depth of field. Normally, accommodation distance *A* corresponds to convergence distance *C* and the depth of field range *D* is centred at the fixated object

is a potential cause of visual strain, and generally limits the comfortably viewable depth volume (Wöpking 1995). 'Freezing' accommodation to a fixed display distance is a particular problem when users want to directly manipulate a virtual object floating stereoscopically in front of the display with the hand or a real tool. Looking at the real thing will make the virtual one appear blurred, and vice versa. The accommodation conflict may hamper seamless mixture and interactions in mixed reality applications (Pastoor and Liu 2002).

13.3 TAXONOMY OF 3D DISPLAYS

Stereoscopic 3D displays can be categorized by the technique used to channel the left- and right-eye views to the appropriate eye (Table 13.1): some require optical devices close to the eyes, while others have the eye-addressing techniques completely integrated into the display itself. Such free-viewing 3D displays are called autostereoscopic and are technically much more demanding than the type with viewing aids.

With *aided-viewing displays* two different perspective views are generated (quasi-) simultaneously. The rays of light entering the eyes may have their origin on a fixed image plane, a steerable eye-gaze-controlled plane, or at steerable pixel-specific locations. Various multiplexing methods have been proposed to carry the optical signals to the appropriate eye. It is possible to adapt the image content to the current head position (motion parallax), e.g., via corresponding camera movements or signal processing.

Most *free-viewing display* techniques produce a limited number of distinct views (at least two). In this case, the only exploitable constraint for selective eye-addressing is the fact that the eyes occupy different locations in space. The waves of light forming the 3D image may originate from fixed or gaze-controlled image planes. In both cases, direction multiplex is the

Table 13.1   Classification of 3D display techniques

Eye-addressing method	Multiplex method	Origin of waves	Number of views	Motion parallax
Aided-viewing	Colour	Fixed image plane	Two	Optional (for a single head-tracked viewer)
	Polarization Time	Eye-gaze-related image plane		
	Location	Pixel-specific location		
Free viewing (autostereoscopic)	Direction (e.g. by diffraction, refraction, reflection, occlusion)	Fixed image plane	Two	Optional (for a single head-tracked viewer)
		Eye-gaze-related image plane	Multiple	Inherent (for a small group of viewers)
	Volumetric display	Multiple fixed image planes	Unlimited	Inherent (for a small group of viewers)
	Electroholography	Entire space		

only way to channel the views into the appropriate eyes. Two-view autostereoscopic displays can be combined with a head tracker, in order to optically address the eyes of a moving observer and to render motion parallax. It is also possible to multiplex more than two views at a time. Thus, individual head position-dependent views can be delivered without the aid of a head tracker, and each viewer will perceive motion parallax effects when moving.

Volumetric and electroholographic approaches produce autostereoscopic 3D images where the effective origin of the waves entering the viewer's eyes match the apparent spatial positions of the corresponding image points. Thus, the fundamental mechanisms of spatial vision are perfectly supported (basically, there is no difference to natural viewing conditions). Current approaches to 3D displays are described in detail in the following sections, starting with the aided-viewing methods.

13.4   AIDED-VIEWING 3D DISPLAY TECHNOLOGIES

13.4.1   Colour-multiplexed (Anaglyph) Displays

In anaglyph displays the left- and right-eye images are filtered with near-complementary colours (red and green, red and cyan or green and magenta) and the viewer wears respective colour-filter glasses for separation. Combining the red component of one eye's view with the green and blue components of the other view allows some limited colour rendition (binocular colour mixture). The Infitec system developed by DaimlerChrysler and Barco uses three-colour wavelength-sensitive notch filters with peak transmissions in the red, green and blue spectral areas (Jorke and Fritz 2003). The transmissions are centred about slightly different peak frequencies for the left- and right-eye images, different enough to separate the views, but close enough to avoid binocular rivalry. The Infitec system provides full colour rendition.

13.4.2 Polarization-multiplexed Displays

The basic setup consists of two monitors arranged at a right-angle and with screens covered by orthogonally oriented filter sheets (linear or circular polarization). The two views are combined by a beam combiner (half-silvered mirror) and the user wears polarized glasses. This technique provides full colour rendition at full resolution and very little cross-talk in the stereo pair (less than 0.1% with linear filters). However, over 60% of the light emitted is lost through the filters, and the remaining light flux is halved by the mirror.

Polarization techniques are very well suited for stereoscopic projection. When using CRT projectors with separate optical systems for the primary colours, the left- and right-view colour beams should be arranged in identical order to avoid binocular colour rivalry. The light flux in LC projectors is polarized by the light-valves. Commercial LC projectors can be fitted for 3D display by twisting the original polarization direction via half-wave retardation sheets to achieve, e.g., the prevalent V-formation. Highest light efficiencies are achieved with single-panel LC projectors as well as with three-panel systems with coincident polarization axes of the primary colours.

Stereoscopic projection screens must preserve polarization. Optimum results have been reported for aluminized surfaces and for translucent opalized acrylic screens. Typical TV rear-projection screens (sandwiched Fresnel lens and lenticular raster sheet) depolarize the passing light. LCD-based, direct-view 3D displays and overhead panels have been developed (Faris 1994). Their front sheet consists of pixel-sized micro-polarizers, tuned in precise register with the raster of the LCD. The stereoscopic images are electronically interlaced line-by-line and separated through a line-by-line change of polarization.

13.4.3 Time-multiplexed Displays

The human visual system is capable of merging stereoscopic images across a time-lag of up to 50 ms. This 'memory effect' is exploited by time-multiplexed displays. The left- and right-eye views are shown in rapid alternation and synchronized with an LC shutter, which opens in turns for one eye, while occluding the other eye. The shutter system is usually integrated in a pair of spectacles and controlled via an infrared link (Figure 13.2). In the open (close) state transmittance is about 30% (0.1%). Time-multiplexed displays are fully compatible for 2D presentations. Both constituent images are reproduced at full spatial resolution by a single monitor or projector, thus avoiding geometrical and colour differences.

Currently, only cathode ray tubes (CRT displays) and some special digital mirror device (DMD) projectors provide the required frame frequencies of more than 100 Hz. Impairing cross-talk may result from the persistence of CRT phosphors, particularly of the green one. Ideally, image extinction should be completed within the blanking interval. With the standard phosphor P22, cross-talk can amount to over 20%, while the perception threshold is at 0.3%

Figure 13.2 The CrystalEyes3 eyewear is the current standard for engineers and scientists, who develop, view and manipulate 3D computer graphics models (Reproduced by permission of Stereo-Graphics Corporation)

(Pastoor 1995). Phosphor decay limits CRT based systems to about 180 Hz. Since the two views are displayed with a time delay, they should be generated with exactly this delay. Otherwise, moving objects will appear at incorrect positions in depth.

Latest DMD projectors operate in stereo mode at frame frequencies of 2×55 Hz at 9000 ANSI Lumen without noticeable crosstalk (McDowall *et al.* 2001; Bolas *et al.* 2004). Some can be used with active shutter glasses or with passive polarized eyewear. Alternatively, standard-frequency dual-projectors, continually projecting the left and right images are combined with high-speed LC shutters in front of the lenses, switching in synchronism with the stereo glasses (Lipton and Dorworth 2001).

Tektronix has developed a 3D monitor which combines the time and polarization-multiplex techniques. The monitor's faceplate is covered with a modulator, consisting of a linear polarizer, a liquid-crystal π-cell and a quarter-wave retardation sheet (to turn linear polarization into circular polarization). The π-cell switches polarization in synchronism with the change of the left- and right-eye images. Circular polarizing glasses serve for de-multiplexing. This approach offers three advantages over systems with active shutter-glasses: First, polarizing glasses are inexpensive and light-weight. Second, the π-cell can be constructed from several segments which operate independently on the active portions of the screen, hence ensuring that each eye is only reached by the intended image contents. This way, cross-talk can be greatly reduced (down to 3.6% in a system with five segments and the P22 standard phosphor (Bos 1991). Third, multiple display arrays can be operated without any extra synchronizing circuitry.

13.4.4 Location-multiplexed Displays

Location-multiplex means that the two views are created at separate places and relayed to the appropriate eye through separate channels (e.g., by means of lenses, mirrors and optical fibres). In the following examples, the image plane appears either in a fixed (but adjustable) accommodation distance (HMD and BOOM displays) or in a variable, gaze-controlled position (3DDAC) or in pixel specific distances (VRD).

With head-mounted displays (HMD), the perceived images may subtend a large field of view of up to 140° (horizontal) by 90° (vertical). As the natural surroundings are occluded from sight (with optional see-through mode realized by electro-optical effects or a movable visor), HMDs are apt to convey a feeling of immersion in the scene displayed. Attached head-tracking devices are used to create viewpoint, dependent changes in perspective when the user moves. Accommodation distance is usually below 2 m, or can be adjusted in see-through displays to the distance of the fixated real-world objects. Meanwhile, numerous manufacturers offer a variety of systems with different performances.

HMDs allow free head movement without losing screen contact, thus avoiding musculoskeletal problems. There is no need to dim ambient light (visor closed), and peripheral vision must not necessarily be obstructed (visor open). On the other hand, latencies and tracking errors tend to provoke odd, uneasy sensations through conflicting visual stimulation and postural feedback, and adaptation can lead to reciprocal aftereffects. A disadvantage of see-through HMDs is that the viewer must direct his or her regard towards a dim background in order to perceive a clear HMD image.

The Binocular Omni-Orientation Monitor (BOOM, Bolas *et al.* 1994) was designed to release the user from the encumbrance of wearing an HMD. Two miniature displays are accommodated in a case, which is mechanically supported by a counterweighted, six-degrees-of-freedom arm. The user regards the stereo images as if looking through binoculars. Tracking is implemented by optical shaft encoders at the joints of the boom. The monitor case is moved either with hand-held handles or with the head (like an HMD).

ATR Labs (Kyoto) have developed an HMD concept with movable relay lenses interposed between the screen and the eyepiece lenses (Omura *et al.* 1996). The current convergence distance of the eyes is sensed by a gaze-tracker. The position of the relay lenses is constantly adjusted, so that the screen surface appears in the same distance as the convergence point of the gaze-lines. Thus, the 3D display with accommodative compensation (3DDAC) supports the link between accommodation and convergence.

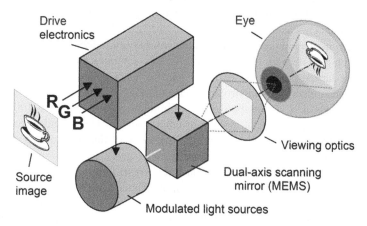

Figure 13.3 Basically, the VRD consists of a modulated light source, a dual-axis scanner, and viewing optics to focus the light beam on the retina. The display's exit pupil must be aligned with the pupil of the viewer's eye. In case of a sufficiently small exit pupil and beam diameter, no ocular accommodation is required (Ando and Shimizu 2001)

Unlike conventional imaging techniques which use a screen outside the eye, a virtual retinal display (VRD, invented at the University of Washington) directly 'paints' an image onto the retina with an intensity-modulated light beam (Figure 13.3). The beam (or three primary colour beams) is rapidly moving in a raster pattern. The display's resolution is only limited by diffraction and aberrations in the scanning beam and not by the physical size of pixels on a display panel. High brightness and low power consumption are additional advantages of the VRD concept. Moreover, it is possible to vary the image focus distance on a pixel-by-pixel basis allowing variable accommodation to the different object distances in the displayed image. This is because the VRD generates pixels sequentially, at different points in time. The challenge for the implementation of pixel-specific wavefronts is to devise a variable-focal-length lens (Kollin and Tidwell 1995) or deformable membrane mirror (Schowengerdt *et al.* 2003) with switching rates of the order of the pixel clock rate.

13.5   FREE-VIEWING 3D DISPLAY TECHNOLOGIES

The present situation regarding autostereoscopic displays is still marked by a variety of competitive approaches. Therefore, this survey mainly focuses on fundamental trends. The range of currently recognized techniques is presented in the following order: electroholography, volumetric displays and direction-multiplexed displays.

13.5.1   Electroholography

Holographic techniques can record and reproduce the properties of light waves — amplitude (luminance), wavelength (chroma) and phase differences — almost to perfection, making them a very close approximation of an ideal free-viewing 3D technique. Recording requires coherent light to illuminate both the scene and the camera target (used without front lenses). For replay, the recorded interference pattern is again illuminated with coherent light. Diffraction (amplitude hologram) or phase modulation (phase hologram) will create an exact reproduction of the original wavefront.

Electroholographic techniques are still in their infancy, although they have received much attention over the past decade. Organized by the Telecommunications Advancement Organization, several Japanese research institutes tried to adapt the principle of holography to an LCD-based video electronics environment (Honda 1995). However, the spatial resolution of today's LC panels is a serious bottleneck (minimum requirements are 1000 lp/mm). A possible solution is to partition the hologram among several panels and to reassemble the image with optical beam combiners. Currently, the scope of this approach is still limited to very small and coarse monochromatic holograms.

Holograms cannot be recorded with natural (incoherent) lighting — a decisive shortcoming. Therefore, they will remain confined to applications where the scene is available in the form of computer-generated models. An approach based on computer-generated holograms has been pursued since the late 1980s at MIT's Media Lab (Benton 1993). The MIT approach makes intense use of data reduction techniques (elimination of vertical parallaxes, sub-sampling of horizontal parallaxes). Yet, the pixel rate for a monochrome display with a diameter of 15 cm, a depth of 20 cm, and a viewing zone of 30° amounts to 2 Gigapixel/s (at 30 Hz frame rate). The hologram is displayed with acousto-optical modulators (tellurium

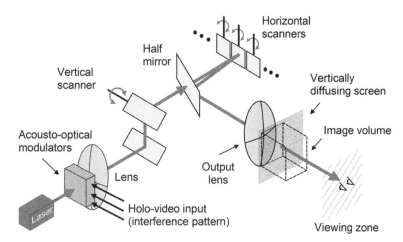

Figure 13.4 MIT electroholographic display. Light modulation is effected by frequency-modulated soundwaves which propagate through the acousto-optical modulator and cause local changes of the refractive index due to pressure variations. Phase shifts in the lightwaves cause local retardations used to create a phase hologram

dioxide crystals transversed by laser light, Figure 13.4). At any one instant, the information of nearly 5000 pixels travels through a crystal. For optical stabilization, synchronized oscillating mirrors are interposed between the modulator and the observer's eyes. Vertical sweep is accomplished with a nodding mirror. A recent approach at KIST (Korea) used pulsed lasers and additional acousto-optical deflectors, in order to do without mechanically moved mirror surfaces.

13.5.2 Volumetric Displays

Volumetric displays produce volume-filling imagery. 3D image points (voxels) are projected to definite loci in a physical volume of space where they appear either on a real surface, or in translucent (aerial) images forming a stack of distinct depth planes. With the first type of system, a self-luminous or light reflecting medium is used which either occupies the volume permanently or sweeps it out periodically. Technical solutions range from the utilization of fluorescent gas with external excitation through intersecting rays of infrared light (Langhans *et al.* 2003) over rotating or linearly moved LED panels to specially shaped rotating projection screens. Rotating screens have been implemented, e.g., in the form of a disc, an Archimedian spiral or a helix, winding around the vertical axis.

Volumetric displays of the first type differ in the scanning technique applied. Raster scanned approaches allow very high voxel densities of up to 100% of the addressable resolution, compared with only 1% density achievable with vector scanning (due to modulation and scanning constraints). Among the *vector-scanned* displays with a real collecting surface, the helical design has reached maturity. The most elaborate equipment uses a double-helix filling a 91-cm-diameter by 46-cm-high volume at 10 revolutions per second with a maximum of 40 000 colour pixels per frame (Soltan *et al.* 1995). Actuality Systems' *raster-scanned* display (Favalora *et al.* 2001) provides 100 million voxel resolution. The volume is comprised

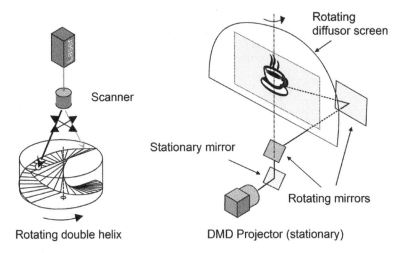

Figure 13.5 Volumetric displays based on vector (left side) and raster scanning (right side)

of 198 2D slices created by a rotating screen, with 768×768 colour-pixel slice-resolution at 30 Hz volume refresh rate. A fast DMD projection illuminates the rotating projection surface via relay mirrors, which rotate with the screen (Figure 13.5). With both displays observers can walk around the display and see the image volume from different angles.

The second type of volumetric display creates aerial images in free space which the viewer perceives as serial cross-sections of a scene, lined up one behind the other (multiplanar or slice-stacking displays). The images belonging to different depth layers are, for example, written time-sequentially to a stationary CRT. The viewer looks at the screen via a spherical mirror with a varying focal length (membrane mirror, McAllister 1993). Rendition of the depth layers is synchronized with changes of the mirror's shape, so that *virtual* images of the 3D volume are successively created in planes of varying distance behind the oscillating mirror. With the aid of large-diameter lenses and beam combiners it is possible to create two or more *real* aerial images floating in front of the display (Dolgoff 1997). Another approach employs pixel-sized spherical miniature lenses with different fixed lengths (Grasnick 2000) or steerable focal lengths (multifocal lens array) to project images of the individual pixels at fixed or variable distances from the physical display panel. In a different implementation two (or more) overlapping LCD panels showing physically separated image planes are stacked one behind the other to form a multilayer display (Deep Video Imaging 2004). In volumetric displays the portrayed foreground objects appear transparent, since the light energy addressed to points in space cannot be absorbed or blocked by (active) foreground pixels. Practical applications seem to be limited to fields where the objects of interest are easily iconized or represented by wireframe models.

13.5.3 Direction-multiplexed Displays

Direction-multiplexed displays apply optical effects such as diffraction, refraction, reflection and occlusion in order to direct the light emitted by pixels of distinct perspective views exclusively to the appropriate eye. Various technical approaches make use of this concept.

13.5.3.1 Diffraction-based Approaches

DOE approach. With the diffractive-optical-elements (DOE) approach, corresponding pixels of adjacent perspective views are grouped in arrays of 'partial pixels' (Kulick *et al.* 2001; Toda *et al.* 1995). Small diffraction gratings placed in front of each partial pixel direct the incident light to the respective image's viewing area (first-order diffraction). Current prototypes yield images of less than 1.5 inches in diameter.

HOE approach. Holographic optical elements (HOE) model the properties of conventional optical elements (e.g., lenses) by holographic methods. Thus, an HOE contains no image information, but serves to diffract rays of light modulated elsewhere. The HOE can be made an integral part of the back light of an LCD. In the approach of Trayner and Orr (1997) it consists of a holographic diffuser which is rastered so as to direct the light of alternating lines to specified viewing zones. Outside the stereo zone both eyes receive a 2D view. The stereo zone can be made to follow the viewer's head movement by moving the light source. Creating this HOE takes two exposures of a hologram with the same illumination setup. Between the exposures, the object, which in this case is a diffuse plane, is shifted horizontally by its own width. For each exposure, parts of the hologram corresponding to the odd or even image lines are occluded.

The 'moving aperture' multiview system developed at KIST (Korea) employs a special holographic screen operating as a field lens (Son and Bobrinev 2001). The different views are shown sequentially in rapid alternation on a CRT display. An LC shutter placed in front of the tube limits the aperture to a vertical slit which is moved stepwise across the tube, in synchronism with the change of perspectives. The holographic screen forms multiple images of the slit apertures, thus creating several viewing zones for multiple viewers.

The Holografika display (Balogh 2001) uses a direction-selective holographic rear projection screen which emits the beams of light at well-defined emitting angles, depending on the incident angle. As opposed to traditional multiview systems, the basic 2D images rear-projected onto the holographic screen are composite images, comprising details to be emitted into different deflecting directions from the screen points.

13.5.3.2 Refraction-based Approaches

Numerous display concepts have been proposed based on conventional, refractive optical elements (such as picture-sized large lenses or small lenslets) to deliver a unique image for each eye observing the 3D scene. These concepts are discussed in the following.

Integral imaging. With integral imaging, the spatial image is composed of multiple tiny 2D images of the scene, captured with a very large number of small convex lenslets (fly's-eye lens sheet). Each lenslet captures the scene from a slightly different perspective. Usually, a lens sheet of the same kind is used for display. Between capture and replay, the image must be inverted for orthoscopic depth rendition (Okoshi 1976). As the image plane is positioned into the focal plane of the lenslets, the light from each image point is emitted into the viewing zone as a beam of parallel rays at a specific direction. Therefore, the viewer perceives different compositions of image points at different viewpoints. The individual lenslets must be very small, since each pixel is spread to the lens diameter at replay, and the elemental images formed behind each lenslet should be as complete and detailed as possible. As a consequence, the display must provide an extremely high spatial resolution. Moreover, optical barriers between the lenses of the pickup device must avoid optical crosstalk when capturing the images.

NHK's (Japan) experimental integral 3DTV system (Okano *et al.* 2002) uses an array of gradient-index (GRIN) lenses, each measuring about 1 mm in diameter, for image capture. A GRIN lens is a special optical fibre with a refractive index decreasing continuously from the centre of the fibre (optical axis) to the border. Hence, the optical path within the fibre curves periodically. Depending on the length of the fibre, the GRIN lens can invert the elemental image as required. Moreover, the fibres efficiently prevent optical crosstalk. For pick-up, a large-aperture convex lens creates a real intermediate image of the scene close to the lens array. This "depth-control lens" is required to focus the pick-up device on different object distances. The output plane showing the numerous elemental images is finally recorded with a high-definition camera and reproduced on an LCD display equipped with an array of convex micro lenslets. The effective image resolution (number of the elemental images) of the current experimental system is 160(H) × 118(V) pixel (Arai *et al.* 2003). Liao *et al.* (2004) use multiple projectors to create a high-resolution integral image.

The integral imaging approach allows variable accommodation, corresponding with the apparent stereoscopic distances. This requires that at least two rays of light, having their origins in corresponding points of the same object in adjacent views, enter the eye pupil. Since the eye attempts to avoid double images produced by the two rays hitting the retina, it will focus on the stereoscopic distance of the object. In this case, the two rays passing through the virtual object point are imaged at the same location on the retina (Figure 13.6).

This concept requires a very large number of elemental images with extremely small exit pupils. The 'focused light-source array display' proposed by Honda *et al.* (2003) employs a multitude of small semiconductor light sources which are densely arranged in an arc and focused on the centre of the arc. The focal point is regarded as a multiview pixel: the light beams virtually emitted by the multiview pixel are modulated by image signals, depending on the beam direction. The entire 3D image is created by raster scanning the beams. Hence,

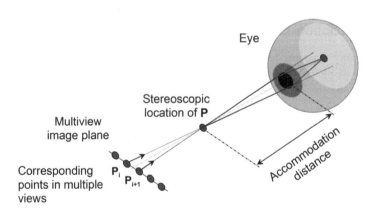

Figure 13.6   Accommodation distance with the focused light-source array super-multiview display. P1 and P2 are two direction-selectively emitted pixels of an object, representing the same object point, but viewed from two slightly different angles. If the two points are close enough the eye lens will focus on the apparent stereoscopic position P, in order to avoid double images on the retina

each scanned image point represents numerous different elemental image points being visible from distinct viewing angles.

Lenticular imaging.   Lenticular techniques use cylindrical lenslets and can be regarded as a one-dimensional version of integral imaging. Due to the vertical orientation of the lenslets, the rays of light from each image point are emitted at specific directions in the horizontal plane, but non-selectively in the vertical plane. Various direct view and projection type 3D displays have been implemented (Börner 1993; Börner *et al.* 2000).

The working principle of a direct-view display is shown in Figure 13.7. The two stereo views are presented simultaneously, with two columns of pixels (one for the left-eye and one for the right-eye image) behind a lenslet. There is one diamond-shaped main viewing zone with optimal stereo separation and a number of adjacent side lobe zones. In the side lobes depth may be inverted if the left and right images are not seen with the proper eye. When traversing from a viewing zone to the adjacent one, the display appears more or less darkened (depending on the viewing distance) since the spaces between the pixel cells are magnified to black spots between the individual views. In order to avoid such interferences, the observer's head position should be constantly sensed and the lenticular sheet should be shifted to track lateral and frontal movements.

Because of the horizontal selectivity of the lens array, the colour-filter stripes of the display panel must be aligned one above the other. Otherwise, the viewer would see separated colour components from different viewpoints. Unfortunately, in available flat-panel displays the RGB sub-pixels are arranged horizontally in a line. The 'simple' solution is to operate the display in portrait mode. The landscape mode requires a more complex sub-pixel multiplexing scheme (c.f. Figure 13.16) and a significant reduction of the lens pitch. Moreover, it must be taken into account that the usable horizontal aperture of the sub-pixel cells of an RGB triplet is less than one-third of the aperture in the vertical direction. Overall, these unfavourable conditions put very high demands on the production and alignment tolerances of a lenticular raster display operating in landscape mode.

Head-tracking can also be achieved by moving the entire display to and fro and rotating it around its vertical axis. Alternatively, a large number of pixel columns belonging to different perspective views have to be displayed simultaneously behind each lenslet, forming

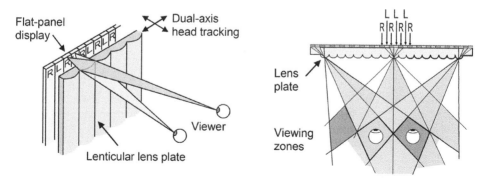

Figure 13.7 Principle of operation of a direct-view lenticular lens display with head-tracking. L and R denote corresponding columns of the left- and right-eye images. The lens pitch is slightly less than the horizontal pitch of the pixels (depends on the design viewing distance)

a multiview display (Uematsu and Hamasaki 1992). The principle of operation of a purely electronic head tracking display is shown in Figure 13.8 (Hentschke 1996).

Figure 13.9 shows the principle of a rear-projection 3D display using a stereo projector and a dual lenticular screen (Takemori *et al.* 1995). The rear screen focuses the projected images in the form of vertical stripes onto a translucent diffuser. The lenses of the front screen map the vertical image stripes to specific loci in the viewing zone – quasi-mirroring the initial path of light during rear-projection. Since the image stripes on the diffuser also pass through a set of adjacent lenslets, there are several adjacent viewing zones. Head-tracking is possible

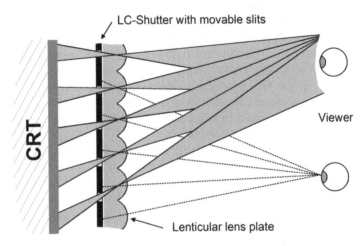

Figure 13.8   Direct-view 3D display with purely electronic head tracking. An LC shutter with movable vertical slits is placed in the focal plane of the lens sheet, and the left and right-eye views are time-sequentially displayed on the CRT screen. The slit position is switched in synchronism with the change of views and shifted laterally according to the viewer's head position

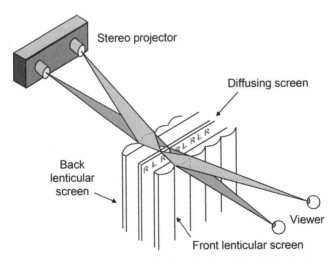

Figure 13.9   Principle of a rear-projection display with dual lenticular screen

by corresponding shifts of the front lens sheet (same principle as in Figure 13.7), or by moving the stereo projection unit on a mirror-inverted path.

In the latter case, multiple user access systems would need to operate several independent pairs of projectors (Omura *et al.* 1995). Lenticular 3D displays with head-tracking were not constructed until very recently. In the past, multiview capability was approached by a large number of stationary projectors (up to 40, see Montes and Campoy 1995). Manufacturing dual lenticular screens is a rather delicate process, since the two raster sheets must be brought into precise register and the whole surface must be of perfect homogeneity. These problems are somewhat attenuated with front-projection systems (single raster sheet with reflective rear coating), since identical optical elements are used to write and read the image. On the other hand, front-projection screens require special optical treatment (e.g., micro structures) to reduce reflections on the front surface of the lenslets.

If a single projector is used to rear-project rastered image stripes directly onto the diffuser behind the frontal lenslets, head-tracking can be achieved by small horizontal shifts of the rastered image (e.g., by shifting the projector's front lens). Frontal movements are followed up by adjusting the magnification of the rastered projection image. Again, an alternative for tracking is simultaneous projection of multiple rastered views, with a respective number of image stripes accommodated behind each lenslet (Isono *et al.* 1993).

An obvious shortcoming of the direct-view 3D approach shown in Figure 13.7 is that just a fraction of the native horizontal resolution of the basic display panel is available for the stereo images while nothing happens to the vertical resolution. The unsymmetrical loss of resolution is particularly serious in the case of multiview displays requiring multiple pixel columns behind each lenslet. This problem is neatly solved by setting the lenticular sheet at an angle in relation to the display panel (Figure 13.10). Depending on the number of views and the multiplex scheme applied, this concept gives developers a way to control the resolution trade-offs in the horizontal and vertical direction. Moreover, when the eye moves to the adjacent perspective there is no such abrupt change (flipping) as observed in the traditional design, where the lenses are vertically aligned with the columns of the

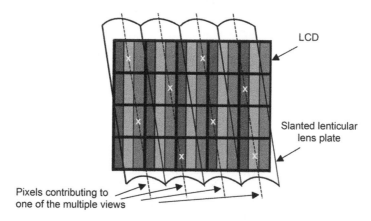

Figure 13.10 Principle of the slanted-raster multiview approach (seven-view display, after van Berkel and Clarke 2000). From a particular viewing angle the eye will see points of the display panel sampled along the dashed lines. Colour sub-pixels marked 'x' belong to the same perspective. When the eyes move from one perspective to the adjacent one, the overall brightness of the display is maintained and a gradual fading between the views is observed

display panel. On the other hand, with a slanted raster it is not possible to separate adjacent perspectives perfectly. The partial visibility of pixels belonging to adjacent views produces crosstalk limiting the usable depth range and contrast of the display.

Typically, lenticular techniques use micro-optical lenslets no wider than the intended pixel size; i.e., the lens pitch defines and limits the resolution of the 3D display. Another class of lenticular raster techniques use macro-optical cylindrical lenses with a width of 10 mm and more. In Großmann (1999) two horizontally compressed image stripes of the stereo pair are accommodated behind each lens. As opposed to the traditional approach, the liquid-crystal layer of the LCD panel does not lie in the focal plane of the lenses but within the focal distance. Hence, the viewer sees horizontally magnified (i.e., de-compressed) virtual images of the image stripes. From a particular viewing angle, the magnified stripes seamlessly line up to form the complete virtual image, which is perceived at a distance behind the lens plate. The position and width of the image stripes is easily adapted to the viewer's head position (purely electronic head-tracking). In an approach by Würz (1999), a single image stripe showing a complete, but horizontally compressed view is placed behind a lens (i.e., a seven-view display requires an array of seven lenses). The eyes look at different parts of the views which – due to the horizontal magnification – add up to form two complete stereo images across the display. The change in perspective is continuous when the viewer moves and head-tracking is not required. On the other hand, a large number of views are required to avoid visible disruptions between the partial images.

Field-lens displays. A field lens is placed at the locus of a real (aerial) image in order to collimate the rays of light passing through that image, without affecting its geometrical properties. Various 3D display concepts use a field lens to project the exit pupils of the left- and right-image illumination systems into the proper eye of the observer. The result is that the right-view image appears dark to the left eye, and vice versa. This approach avoids all the difficulties resulting from small registration tolerances of pixel-sized optical elements. The principle of operation is shown in Figure 13.11 (Ezra *et al.* 1995).

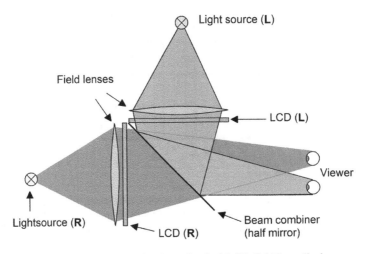

Figure 13.11 Schematic view of a dual-LCD field lens display

The left and right views displayed on distinct LC panels are superimposed by a beam combiner. Field lenses placed close to the LCDs direct the illumination beams into the proper eye. For head-tracking, the position of the light sources must be steerable (e.g., implemented by monochrome CRTs displaying high-contrast (binary) camera images of the left and right halves of the viewer's face (Omori *et al.* 1995). Multiple viewer access is possible with multiple independent illuminators. A very compact display setup can be achieved when the large-format field lens and single light source is replaced by an array of lenticular lenses in combination with an array of light strips (Woodgate *et al.* 1998).

A different solution developed at Dresden University (Germany) uses a single panel and light source in connection with a prism mask (Fig. 13.12). Alternating pixel columns (RGB triplets) are composed of corresponding columns of the left and right image. The prisms serve to deflect the rays of light to separated viewing zones. It is possible to integrate the function of the field lens, the collimator optics (in the form of pixel-sized micro lenses) and the prism mask in a single raster (Heidrich and Schwerdtner 1999).

Various recent approaches combine the field lens concept with polarization multiplexing. Typically, a very bright twisted-nematic LCD panel with the front polarizer removed is used as a light source. Hence, it is possible to rotate the polarization axis of any selected part of the panel by 90° (Benton *et al.* 1999). Alternatively, an array of switchable white LEDs is combined with an array of small orthogonally arranged filter elements (Hattori *et al.* 1999). This way, two switchable orthogonally polarized light sources are created.

As shown in Figure 13.13 a large field lens is used to focus separated images of the two polarized light sources on the user's left and right eye. The side of the imaging LCD facing the light sources is covered with an array of small strips of birefringent material having a retardance of one-half wavelength. (Birefringent material divides light into two components travelling at different speeds. Hence, after passing through the material there is a phase difference between the components, resulting in a rotation of the electric and magnetic field vectors of the combined beam. The rotation angle depends on the relative retardance between the components.) The $\lambda/2$ retardation strips rotate the polarization axis of the light beams passing through by 90°. The strips cover every second row of the imaging LCD, alternating with strips of non-birefringent material. The result is that the rays of one of the light sources will enter the odd numbered rows of the LCD, while beams of the other source will enter the even ones. The left- and right-eye images are shown on alternating rows. The filter array achieves that the left (right) eye will receive only light beams that have traversed the odd (even) numbered rows of the display panel. When the user moves, the polarized light sources are shifted so that their real images (as focused by the field lens) fall at the eye positions.

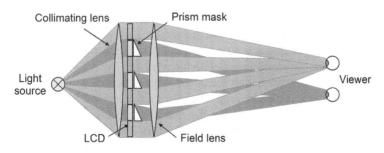

Figure 13.12 Underlying principle of a single-LCD field lens display

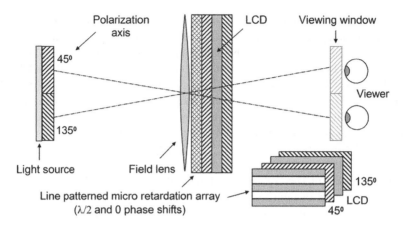

Figure 13.13 Schematic diagram of an autostereoscopic display combining the field-lens concept with polarization multiplexing

According to this basic principle it is possible to implement electronic head-tracking for multiple viewers. Again, the large-format field lens can be replaced by a micro-optical lens array in combination with a rastered light-source LCD (Tsai *et al.* 2001).

Based on the field lens concept it is also possible to realize a multiview display with motion parallax for multiple viewers. In order to achieve a high density of views, the multiple images are presented on a single projection LCD panel, where they are arranged on, e.g., five rows with an offset of one-fifth of the image width (for a 21-view display, Hines 1997). The images are rear-projected by individual projection lenses to the back of a weak diffuser screen. A large Fresnel-type lens close to the screen creates real images of the exit pupils at a comfortable viewing distance. An additional lenticular screen with the lenslets oriented horizontally is used to scatter light vertically, thus changing the circular form of the imaged exit pupils into vertical stripes. Hence, the viewing space is covered by a sequence of small, laterally spaced viewing windows for the different perspectives, formed by the images of the projectors' exit pupils.

A field-lens-based display concept with a steerable, gaze-controlled image plane was developed at the Fraunhofer Heinrich-Hertz-Institut (Boerger and Pastoor 2003). As shown in Figure 14.11 in Chapter 14, the stereo images are not projected onto a physical display screen, but appear in the form of aerial images floating in front of or behind a large Fresnel-type field lens. The position of the image plane is steerable by motorized focus adjustments of the projection optics. A gaze tracker senses the viewer's point of fixation, and the aerial image plane is moved to this position by focus adjustments. As the viewer focuses on the aerial-image plane, accommodation and convergence are coupled like in natural vision. Additionally, the display includes a natural depth-of-field effect by depth-selective spatial low-pass filtering of the projected images (Blohm *et al.* 1997).

Combining the field-lens projection technique with polarization multiplexing results in additional possibilities for eye-addressing when the viewer moves. Respective display concepts creating floating 3D images within the user's reach are described by Kakeya and Yoshiki (2002). In the basic configuration a polarization-multiplexed stereo pair is projected onto a large front-projection screen. The polarizing filters required for image separation are mounted on a horizontally movable support which is placed between the screen and a large

Fresnel-type field lens. The task of the Fresnel lens is twofold. First, it creates a real (but significantly reduced) aerial image of the projection screen floating in front of the viewer. Second, the Fresnel lens creates images of the two polarizing filters at the location of the viewer's eyes. The effect is comparable to looking directly through the physically existing filters: i.e., the polarized stereo images are separated. A head-tracker is used to shift the movable filter stage when the viewer moves.

In the traditional approaches a single large-field lens serves to image the exit pupils of the display's backlighting system at the viewer's eye location. A display concept developed at De Montfort University (UK) substitutes the function of the large lens and light source by an array of multiple small optical elements and illumination sources (Surman *et al.* 2003). The steering optics allow one to illuminate the entire screen area and to direct the emitted light beams to the viewer's eye position. Several illumination sources can be lit simultaneously, producing multiple exit pupils for multiple viewers.

13.5.3.3 Reflection-based Approaches

Retro-reflective techniques. This approach uses flat or curved mirrors as well as retro-reflective screens for direction multiplexing. Retro-reflective means that the incident rays of light are reflected only in their original direction (Okoshi 1976). In a prototype display (Harman 1997) dual video projectors are mounted on a laterally movable stage (Figure 13.14). The screen reflects the two images through a large half-mirror to the viewer's eyes. The system locates the current head position and adjusts the position of the projectors and the angle of the half-mirror accordingly. A simplified setup using a curved retro-reflective screen was developed by Ohshima *et al.* (1997).

Concave-mirror technique. The 'monocentric autostereoscopic immersive display' (Cobb *et al.* 2003) developed by Kodak (USA) applies a large concave spherical mirror as a functional replacement for the field lens. The exit pupils of the stereo projection system are placed close to the centre of curvature of the mirror so that they are imaged directly back at

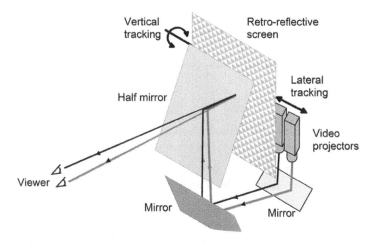

Figure 13.14 Retro-reflective display (after Harman 1997)

unit magnification. The viewer's eyes are also placed at the centre of curvature, physically separated from the projection device by a beam splitter (half-mirror), so that the pupils of the projection lenses are imaged on the left and right eye. The projection system employs 54-mm-diameter ball lenses; it creates intermediate images of the left- and right-eye views that lie close to the focal surface of the mirror. Hence, the viewer sees the virtual 3D image at infinity. The advantage of this design compared with the use of Fresnel-type field lenses is the absence of moiré distortions and chromatic aberrations. A similar approach, the IRIS-3D display, has been developed at Strathclyde University (UK) by McKay *et al.* (2000).

13.5.3.4 Occlusion-based Approaches

These approaches have one thing in common: due to parallax effects, parts of the image are hidden to the one eye but visible to the other eye. The technical solutions differ in the number of viewing slits (ranging from a dense grid to a single vertical slit), in presentation mode (time-sequential or stationary) and in whether the opaque barriers are placed in front of or behind the image screen (parallax barrier or parallax illumination techniques). Instead of employing regular structures for the partial occlusion of the image, another approach uses fragments of the constituent images, which are stacked one behind the other, for the viewpoint dependent optical combination of the stereo views.

Basically, any of the lenticular display designs can be implemented with parallax effects, replacing the raster of cylindrical lenslets by a raster of slit openings (Figures 13.7 and 13.16). Hence, there are numerous direct-view and projection-type displays (using single and dual parallax barriers in vertical as well as slanted orientation) operating completely analogous to the methods described in Section 13.5.3.2. Parallax barriers are easier to fabricate than lenticular screens (e.g., by printing or electro-optical methods). The barriers may be designed as opaque dark stripes or steps (Mashitani *et al.* 2004), orthogonally polarized filter stripes, and as colour filter stripes being opaque for the complementary colour (Schmidt and Grasnick 2002; Figure 13.15). They can be moved electronically for head-tracking and time-sequential presentation (moving slit techniques), and they can be switched on and off for 2D/3D compatibility. Moreover, the dark barriers increase the maximal contrast under ambient

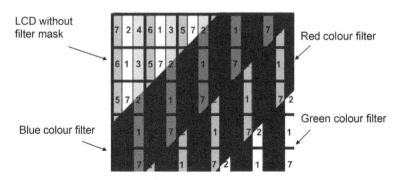

Figure 13.15 Seven-view parallax barrier display with slanted colour filter stripes. The filters are tuned to the primary colours. For example, a red filter stripe operates as a slit for the red pixels and as a barrier for the blue and green ones. Numbers 1–7 denote the membership of a sub-pixel to views number 1–7. In this example, we see predominantly pixels belonging to view number 1

illumination (black mask effect). On the other hand, the light efficiency of occlusion-based approaches is generally lower than the competing approaches based on imaging techniques. The following paragraphs list some recent developments of parallax displays.

Barrier-grid displays.   Sanyo has optimized the barrier-grid design for LCD-based direct-view displays (Hamagishi *et al.* 1995). One barrier is in front of the LCD and an additional one is placed between the LCD panel and the backlight case. The additional barrier has a reflective coating facing the backlight. It serves to exploit the light which would otherwise be blocked by the black mask of the LCD panel. This way, lighting efficiency was improved by a factor of 1.4.

As the rays of light pass through several adjacent slits, they produce a number of additional viewing windows for the left and right views. This can lead to the disturbing effect that the left eye sees the right view and vice versa (i.e., the stereo-depth is inverted). Ideally, a head-tracker should be used to switch the pixel positions of the multiplexed left- and right-eye images as soon as the observer enters a 'wrong' viewing window (Shiwa *et al.* 1994). In a prototype of NHK Labs (Japan), the barrier grid is generated by an LC panel (Isono *et al.* 1992). This way, it can be limited to software-selected areas of the panel to create scalable '3D windows'. The grid is switched off in the 2D area, making the full resolution of the imaging panel available. Another method for a 2D/3D compatible parallax-barrier display is presented by Sanyo (Japan). This display uses a special 'image splitter' LC panel consisting of a lattice of fixed opaque stripes with additional switchable barrier regions to the left and right of the fixed ones (Hamagishi *et al.* 2001). The fixed barriers are narrow enough to block only a small amount of light, which is not sufficient for image separation; in the 3D mode, the variable stripes expand the barriers as required. Depending on the viewer's head position, either the additional barrier regions to the left or right of the fixed ones are switched on, in order to direct the visible pixels to the intended eye.

If the native imaging panel emits polarized light (what is the case with LCDs) the parallax barriers can be made of differently polarized filter stripes. Sharp Laboratories Europe (UK) have developed a latent parallax barrier, made of an array of optical half-wave retarders that are patterned in a geometrical arrangement identical to a conventional absorbing barrier (Woodgate *et al.* 2000). The barrier consists of slit and non-slit regions which modify the output polarization vector of the LCD differently. This has almost no visible effect (full resolution 2D mode). In the 3D mode, an additional polarising filter sheet covering the entire display screen is added to the output of the latent barriers. The polarization axis of the filter sheet corresponds to the slit regions and blocks the orthogonally polarized light passing through the non-slit regions, thus 'developing' the function of the latent barriers.

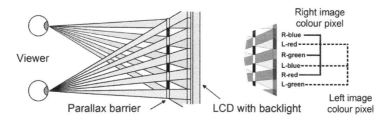

Figure 13.16 Principle of a two-view parallax-barrier 3D display and grid–pixel arrangement for standard direct-view panels with vertical colour-filter stripes

A full resolution 3D display based on the combination of polarized parallax-barriers with a time-multiplexed polarized image source is presented by Mantinband *et al.* (2002).

Parallax-illumination displays.   Dimension Technologies (USA) create the parallax effect by a lattice of very thin vertical lines of light, placed at a distance behind a standard LC panel (Eichenlaub 1994). As indicated in Figure 13.17, parallax causes each eye to perceive light passing through alternate image columns only. The illumination lattice is generated by means of a lenticular lens sheet which focuses the light from a small number of fluorescent lamps into a large number of light bars on the surface of a translucent diffuser. For head-tracking, multiple sets of lamps and a large field lens are incorporated into the display. The position of the light source is changed by switching between sets of laterally displaced lamps. It is also possible to create two or more viewing zones for multiple viewers by simultaneous operation of multiple light sources. Alternatively, switchable vertical light bars can be generated by an additional binary LC panel used as a backlight.

Two sets of blinking light sources, which generate two laterally displaced blinking lines behind each single column of the LCD, have been proposed to obtain full spatial resolution for both views. The sequential display of the left and right views is synchronized with the blinking light sources. An even more advanced version generates multiple sets of flashing light lines to create a stationary multiview display with look-around effect (Eichenlaub *et al.* 1995). Morishima *et al.* (1998) propose a parallax illumination display using a chequered illumination pattern with the spatially multiplexed views arranged on alternate pixel rows instead of columns.

Moving-slit displays.   With this technique, a single vertical slit opening serves to channel different perspective views to an array of adjacent viewing zones. The slit is generated by a fast-switching binary LC panel mounted on the front of the imaging display, and it passes the screen periodically at a rate of about 60 Hz. Image output of the display is synchronized with the slit position. Depending on the specific approach, the display shows either complete images or partial images composed of selected image columns of multiple views.

A 50-inch full-colour 3D display with 15 views displayed at VGA resolution and 30 Hz refresh rate has been developed at Cambridge University (UK) by Dodgson *et al.* (2000).

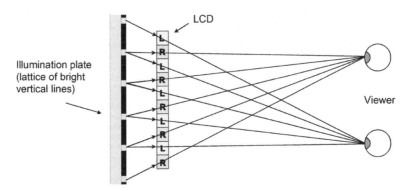

Figure 13.17 Basic concept of a parallax illumination 3D display. The left- and right-eye views are spatially multiplexed on alternate pixel columns

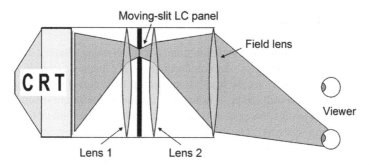

Figure 13.18 Optical setup of a moving-slit 3D display (Cambridge). Lenses 1 and 2 project the CRT image onto the field lens which in turn projects the slit aperture into viewing space

Complete views are presented on the RGB CRT screens for any given slit position and relayed to a viewing window by means of a large spherical mirror. Figure 13.18 shows the basic principle of operation using a Fresnel lens in place of the mirror. As the slit traverses the screen, adjacent windows for the neighbouring views line up.

The so-called Holotron is a moving-slit display using N simultaneously open slits without additional lenses in front of the CRT (Sombrowsky 1992). At each slit position, N partial image stripes composed of corresponding columns of different views are displayed side-by-side behind the slit aperture. As the slits move laterally, a new set of multiplexed image columns is displayed on the CRT. In this way, during a single movement of the slits in the range of $1/N$ of the width of the CRT screen, a complete multiview image set is generated. The Holotron employs a special CRT with vertical deflection to reduce cross-talk.

A general disadvantage of parallax-barrier displays according to the basic principle illustrated in Figure 13.16 is the fact that only one-half of the spatial resolution of the native panel is available for each of the two views. This problem can be overcome by time-sequential display of the two views. For example, in the first phase only the odd-numbered pixel columns of the two views are sent to the imaging display (where they are presented on alternate columns), followed by the even-numbered columns in the second phase. The left and right-view pixel columns are shifted by one pixel position when switching between the phases, and the parallax-barriers are inverted accordingly, in order to direct the pixels to the intended eye. An approach developed at the New York University (USA) by Perlin *et al.* (2001) suggests applying three phases. This allows separating the visible image stripes by black spaces which allows for some registration errors in the system. Hence, this concept reduces the required precision of head-tracking.

Parallax-combination displays.   This approach developed by Neurok Optics (USA and Russia) is based on the idea that different perspective views can be combined from fragmented images by parallax effects when viewed from the left- or right-eye positions (Neurok 2004). The display consists of two LC panels stacked one behind the other, with a special mask placed between them. A dedicated computer program is required to calculate which fragments of the 3D image should be shown on the rear panel and which on the front one, in order to achieve correct optical summation of image fragments at different angles. 2D compatibility is achieved by switching the rear LC panel to 'transparent' (i.e., showing a blank image) and presenting the 2D image on the front one only.

13.6 CONCLUSIONS

Recent developments in information and communication technology have stimulated a growing demand for mature 3D displays. As a consequence, 3D display technology has rapidly
gained momentum in terms of research and commercial activities. On the one hand, these
projects concentrate on the utilization of novel base technologies in conjunction with rather
conventional 3D concepts for the custom design of marketable equipment for special applications (aided-viewing systems). On the other hand, there is a rising interest in universal
purpose systems that do not fall behind familiar 2D displays in terms of image quality and
viewing comfort (free-viewing systems). Overall, the research scene is characterized by an
extremely large variety of competing concepts, each having particular advantages and flaws.
There are still enormous challenges waiting for future R&D work.

REFERENCES

Ando T and Shimizu E 2001 Head mounted display using holographic optical element; In *Three-
Dimensional Video and Display: Devices and Systems* (Javidi B and Okano F eds) *SPIE* Optical
Engineering Press, Bellingham.
Arai J, Okui M, Kobayashi M, Sugawara M, Mitani K, Shimamoto H and Okano F 2003 Integral
three-dimensional television based on superhigh-definition video system *SPIE* **5006**, 49–57.
Balogh T 2001 Method and apparatus for displaying 3D images. Patent WO 01/88598.
Benton SA 1993 The second generation of the MIT holographic video system. *TAO First International
Symposium 1993*, S-3-1-3–S-3-1-6.
Benton SA, Slowe TE, Kropp AB and Smith SL 1999 Micropolarizer-based multiple-viewer autostereo-
scopic display *SPIE* **3639**, 76–83.
Boerger G and Pastoor S 2003 Autostereoscopic Display. Patent DE 195 37 499.
Börner R 1993 Autostereoscopic 3D-imaging by front and rear projection and on flat panel displays.
Displays, **14**(1), 39–46.
Börner R, Duckstein B, Machui O, Röder H, Sinnig and Sikora T 2000 A family of single-user
autostereoscopic displays with head-tracking capabilities, *IEEE Transactions Circuits and Systems
for Video Technology*, **10**(2), 234–43.
Bolas MT, Lorimer ER, McDowall IE, Mead RX 1994 Proliferation of counterbalanced, CRT-based
stereoscopic displays for virtual environment viewing and control *SPIE* **2177**, 325–34.
Bolas MT, McDowall IE, Corr D 2004 New Research and Explorations into Multiuser Immersive
Display Systems, *IEEE Computer Graphics and Applications*, **24**(1), 18–21.
Bos PJ 1991 Time sequential stereoscopic displays: The contribution of phosphor persistence to the
"ghost" image intensity. *Proceedings 1991 ITE Annual Convention*, Tokyo, 603–60.
Blohm W, Beldie IP, Schenke K, Fazel K and Pastoor S 1997 Stereoscopic Image Representation with
Synthetic Depth of Field. *Journal of the SID*, **5**(33), 7–31.
Cobb JM, Kessler D, Agostinelli JA and Waldmann M 2003 High-resolution autostereoscopic immer-
sive imaging display using a monocentric optical system *SPIE* **5006**, 92–8.
Deep Video Imaging 2004 http://www.deepvideo.com/products/
Dodgson NA, Moore JR, Lang SR, Martin G and Canepa P 2000 A 50″ time-multiplexed autostereo-
scopic display *SPIE* **3957**, 177–83.
Dolgoff G 1997 A New No-Glasses 3-D Imaging Technology. *SID 97 Digest,* 269–72.
Eichenlaub J 1994 An autostereoscopic display with high brightness and power efficiency *SPIE* **2177**,
4–15.
Eichenlaub J, Hollands D, Hutchins J 1995 A prototype flat panel hologram-like display that produces
multiple perspective views at full resolution *SPIE* **2409**, 102–12.
Ezra D, Woodgate GJ, Omar BA, Holliman NS, Harrold J and Shapiro LS 1995 New autostereoscopic
display system *SPIE* **2409**, 31–40.
Faris S 1994 Novel 3-D stereoscopic imaging technology *SPIE* **2177**, 180–95.

Favalora G, Dorval RK, Hall DM, Giovinco M 2001 Volumetric three-dimensional display system with rasterization hardware *SPIE* **4297**, 227–35.

Grasnick A 2000 Anordnung zur dreidimensionalen Darstellung. Patent DE 198 25 950.

Großmann C 1999 Verfahren und Vorrichtung zur Autostereoskopie. Patent WO 99/67956.

Hamagishi G, Sakata M, Yamashita A, Mashitani K, Nakayama E, Kishimoto S, Kanatani K 1995 New stereoscopic LC displays without special glasses. *Asia Display '95*, 791–4.

Hamagishi G, Sakata M, Yamashita A, Mashitani K, Inoue M and Shimizu E 2001 15″ high-resolution non-glasses 3-D display with head tracking system *Transactions IEEJ*, **21-C**(5), 921–7.

Harman PV 1997 Retroreflective screens and their application to autostereoscopic displays *SPIE* **3012**, 145–53.

Hattori T, Ishigaki T, Shimamoto K, Sawaki A, Ishguchi T and Kobayashi H 1999 Advanced autostereoscopic display for G-7 pilot project *SPIE* **3639**, 66–75.

Heidrich H and Schwerdtner A 1999 Anordnung zur dreidimensionalen Darstellung von Informationen. Patent DE 198 22 342.

Hentschke S 1996 Personenadaptiver autostereoskoper Monitor (PAAS) - eine Option für den Fernseher? *Fernseh- und Kinotechnik*, **50** (5), 242–8.

Hines SP 1997 Autostereoscopic video display with motion parallax *SPIE* **3012**, 208–19.

Honda T 1995 Dynamic holographic 3D display using LCD. *Asia Display '95*, 777–80.

Honda T, Shimomatsu M, Imai H, Kobayashi S, Nate N and Iwane H 2003 Second version of 3D display system by fan-like array of projection optics *SPIE* **5006**, 118–27.

Howard IP and Rogers BJ 1995 Binocular Vision and Stereopsis. Oxford University Press, New York, Oxford.

Isono H, Yasuda M and Sasazawa H 1992 Autostereoscopic 3-D LCD display using LCD-generated parallax barrier. *Japan Display '92*, 303–6.

Isono H, Yasuda M, Takemori D, Kanayama H, Yamada C, Chiba K 1993 Autostereoscopic 3-D TV display system with wide viewing angles. *Euro Display '93 (SID)* 407–10.

Jorke H and Fritz M 2003 INFITEC – A new stereoscopic visualisation tool by wavelength multiplexing imaging. *Electronic Displays 2003* (see http://www.infitec.net/infitec_english.pdf).

Kakeya H and Yoshiki A 2002 Autostereoscopic display with real-image virtual screen and light filters *SPIE* **4660**, 349–57.

Kollin JS and Tidwell M 1995 Optical engineering challenges of the virtual retinal display *SPIE* **2537**, 48–60.

Kulick J, Nordin G, Kowel S and Lindquist R 2001 ICVision real-time autostereoscopic stereogram display, in *Three-Dimensional Video and Display: Devices and Systems* (Javidi B and Okano F eds) *SPIE* Optical Engineering Press, Bellingham.

Langhans K, Guill C, Rieper E, Oltmann K and Bahr D 2003 Solid Felix: a static volume 3D-laser display *SPIE* **5006**, 162–74.

Liao H, Iwahara M, Hata N, Dohi T 2004 High-quality integral videography using multiprojector. OSA **12**(6) Optics Express, 1067–76.

Lipton L and Dorworth B 2001 Eclipse Projection *SPIE,* **4297**, 1–4.

Mantinband Y, Goldberg H, Kleinberger I and Kleinberger P 2002 Autostereoscopic, field-sequential display with full freedom of movement *SPIE* **4660,** 246–53.

Mashitani K, Hamagishi G, Higashino M, Ando T and Takemoto S 2004 Stepbarrier system multi-view glass-less 3-D display *SPIE* **5291**, 265–72.

McAllister D 1993 Stereo computer graphics and other true 3D technologies. Princeton University Press, Princeton.

McKay S, Mair GM and Mason S 2000 Membrane-mirror-based autostereoscopic display for tele-operation and telepresence applications *SPIE* **3975**, 198–207.

McDowall I, Bolas M, Corr D and Schmidt T 2001 Single and Multiple Viewer Stereo with DLP Projectors *SPIE,* **4297**, 418–25.

Montes JD, Campoy P 1995 A new three-dimensional visualization system based on angular image differentiation *SPIE* **2409**, 125–32.

Morishima H, Nose H, Taniguchi N and Matsumura S 1998 An Eyeglass-Free Rear-Cross-Lenticular 3-D Display *SID Digest* **29**, 923–6.

Neurok (2004) see http://www.neurokoptics.com/products/technology.shtml

Okano F, Hoshino H, Arai J, Yamada M and Yuyama I 2002 Three-Dimensional Television System Based on Integral Photography, in *Three-Dimensional Television, Video, and Display Technologies* (Javidi B and Okano F eds), Springer, Berlin.

Okoshi T 1976 Three-Dimensional Imaging Techniques. Academic Press, New York.

Omori S, Suzuki J and Sakuma S 1995 Stereoscopic display system using backlight distribution. *SID 95 Digest*, 855–8.

Omura K, Shiwa S and Kishino F 1995 Development of lenticular stereoscopic display systems: Multiple images for multiple viewers. *SID 95 Digest*, 761–3.

Omura K, Shiwa S and Kishino F 1996 3-D Display with accommodative compensation (3DDAC) employing real-time gaze detection. *SID 96 Digest*, 889–92.

Ohshima T, Komoda O, Kaneko Y and Arimoto A 1997 A Stereoscopic Projection Display Using Curved Directional Reflection Screen *SPIE* **3012**, 140–4.

Pastoor S 1995 Human factors of 3D imaging: Results of recent research at Heinrich-Hertz-Institut Berlin. *2nd Internal Display Workshop*, Hamamatsu, 69–72.

Pastoor S and Wöpking M 1997 3-D displays: A review of current technologies. *Displays* **17**, 100–10.

Pastoor S and Liu J 2002 3-D Display and Interaction Technologies for Desktop Computing, in *Three-Dimensional Television, Video, and Display Technologies* (Javidi B and Okano F eds), Springer, Berlin.

Perlin K, Poultney C, Kollin JS, Kristjansson DT and Paxia S 2001 Recent advances in the NYU autostereoscopic display *SPIE* **4297**, 196–203.

Schmidt A and Grasnick A 2002 Multiviewpoint autostereoscopic displays from 4D-Vision GmbH *SPIE* **4660**, 212–21.

Schowengerdt BT, Seibel EJ, Kelly JP, Silvermann NL and Furness TA 2003 Binocular retinal scanning laser display with integrated focus cues for accommodation *SPIE* **5006**, 1–9.

Shiwa S, Tetsutani N, Akiyama K, Ichinose S, Komatsu T 1994 Development of direct-view 3D display for videophones using 15 inch LCD and lenticular sheet. *IEICE Transactions Information and Systems* **E77-D**(9), 940–8.

Soltan P, Trias J, Dahlke W, Lasher M and McDonald M 1995 Laser-based 3D volumetric display system - Second generation, in *Interactive Technology and the New Paradigm for Healthcare* (Morgan K eds) IOS Press and Ohmsha, 349–57.

Sombrowsky R 1992 Verfahren und Vorrichtung zur Erzeugung stereoskopischer Darstellungen. Patent DE 42 28 111.

Son JY and Bobrinev VI 2001 Autostereoscopic imaging system based on holographic screen; In *Three-Dimensional Video and Display: Devices and Systems* (Javidi B and Okano F eds) *SPIE* Optical Engineering Press, Bellingham.

Surman P, Sexton I, Bates R, Lee WK, Craven M and Yow KC 2003 Beyond 3D television: The multi-modal, multi-viewer, TV system of the future. *SID/SPIE FLOWERS 2003*, 208–10.

Suwita A, Böcker M, Mühlbach L and Runde D 1997 Overcoming human factors deficiencies of videocommunications systems by means of advanced image technologies. *Displays* **17**, 75–88.

Takemori D, Kanatani K, Kishimoto S, Yoshii S and Kanayama H 1995 3-D Display with large double lenticular lens screens. *SID 95 Digest*, 55–8.

Toda T, Takahashi S, Iwata F 1995 3D video system using Grating Image *SPIE* **2406** 191–8.

Trayner DJ and Orr E 1997 Developments in autostereoscopic displays using holographic optical elements *SPIE* **3012**, 167–174.

Tsai C, Hsueh W and Lee C 2001 Flat Panel Autostereoscopic Display *SPIE* **4297**, 165–74.

Uematsu S, Hamasaki J 1992 Clinical application of lenticular stereographic image system. *Internal Symposium 3D Image Technology and Arts (University of Tokyo, Seiken Symposium)* **8**, 227–34.

Valyus NA 1966 Stereoscopy. The Focal Press, London and New York.

Van Berkel C and Clarke JA 2000 Autostereoscopic Display Apparatus, U.S. Patent 6,064,424.

Wöpking M 1995 Viewing comfort with stereoscopic pictures: An experimental study on the subjective effects of disparity magnitude and depth of focus. *Journal of the SID*, **3**(3), 101–3.

Woodgate GJ, Ezra D, Harrold J, Holliman NS *et al.* (1998) Autostereoscopic 3D display systems with observer tracking. *Signal Processing: Image Communication*, **14**(1–2), 131–45.

Woodgate GJ, Harrold J, Jacobs AM, Moseley RR and Ezra D 2000 Flat-panel autostereoscopic displays: characterisation and enhancement *SPIE* **3957**, 153–64.

Würz H 1999 Verfahren und Anordnung zum Herstellen eines räumlich wiedergebbaren Bildes Patent WO 99/09449.

14
Mixed Reality Displays

Siegmund Pastoor and Christos Conomis

Fraunhofer Institute for Telecommunications/Heinrich-Hertz-Institut, Berlin, Germany

14.1 INTRODUCTION

The term 'mixed reality' (MR) was coined by Paul Milgram about a decade ago. It defines a system where both real and virtual (computer-generated) things appear to coexist in the same space. The potential value of MR systems is increasingly recognized and appreciated in a wide variety of fields, including simulation, medicine, architecture, driving, maintenance, industrial design and entertainment (Tamura 2002). A great potential lies in the capability of not just mimicking or enhancing properties of the real world, but of exceeding the physical laws governing reality, such as the possibility of stepping back and forth in time, performing an 'undo' function, and rendering hidden things visible.

The automotive industry is one of the early adapters of MR technologies. Emerging applications include assembly and disassembly of vehicles in production, maintenance and repair tasks, electrical troubleshooting, *in situ* visualization of calculated stress distribution and deformation in a crash test (Friedrich and Wohlgemuth 2002), as well as exterior and interior design. For example, Ohshima et al. (2003) applied MR technologies in order to evaluate design concepts using a partly physical and partly virtual car mock-up. The real components of the mock-up included the driver's seat, parts of the dashboard, the steering wheel and some basic devices for controlling the audio, air conditioner and navigation systems. The doors, ceiling, pillars, as well as the entire instrument panel with meters and displays were virtually overlaid onto the driver's field of view. The real control devices provided haptic feedback when touched, while their actual appearance was easily changeable using graphical overlays superimposed into vision. Hence, different designs could be visualized, interactively altered and evaluated from a usability point of view. In another application a new car model was visualized almost entirely in the digital domain (Tamura 2002). Interested customers could walk around the virtual car model and even get into it by 'opening' a door. A real seat was prepared to confirm the sight from the driver's position through the virtual windows.

3D Videocommunication — Algorithms, concepts and real-time systems in human centred communication
Edited by O. Schreer, P. Kauff and T. Sikora © 2005 John Wiley & Sons, Ltd

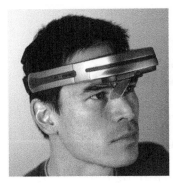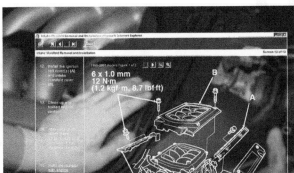

Figure 14.1   By superimposing test and repair data into vision, the Nomad Augmented Vision System allows reading detailed service information and following complex instructions directly at the point of task (Reproduced by permission of Microvision Incorporated)

Figure 14.1 illustrates how an advanced MR headset can be used in a car repair task. The transparent mirror in front of the eye superimposes computer-generated information upon the technician's field of view. Instructions for assembling parts of the engine, including the tightening torques of the various screws, are available without having to leave the vehicle. It is even possible to directly communicate with the spare parts inventory in order to see whether a required replacement part is in stock.

Apart from the automotive sector, MR systems are increasingly used in other industrial applications, particularly in order to give assistance and feedback in real-time processes. For example, Aiteanu *et al.* (2003) developed a welding system where the traditional protection helmet was replaced by a special helmet-mounted display. The display provides a better view of the working area; if required, an online agent suggests corrections of the welding gun's position, by analysing the electrical welding parameters, and points on welding errors. Pettersen *et al.* (2003) developed an easy to use MR system for interactively programming waypoints and specified actions of an industrial robot. During the programming sequence, the system presents visual feedback of results for the operator, allowing inspection of the process result before the robot has performed the actual task.

Various applications of MR technologies are emerging in the areas of city planning and architecture (e.g., Kato *et al.* 2003; ARTHUR 2003) since it is much easier to set up and modify virtual models instead of physical ones. Usually, the virtual models of buildings and facilities are complemented by simple real objects representing the virtual ones. The real objects operate as placeholders; they can easily and intuitively be grasped and moved on the planning desk, hence, providing a kind of tangible interface. When looking through the MR display, the user sees the virtual buildings at the locations of the placeholders.

In a similar way, antique ruins and monuments can be reconstructed virtually by computer software and visually superimposed onto the real location. Hence, visitors to archaeological sites can see temples, statues and buildings of old cities virtually placed on their original fundaments in their original surroundings (ARCHEOGUIDE 2002).

In medicine, MR imaging has been used to merge real-time video images (e.g., by an endoscope) with an underlying computer model derived from preoperative MRI or CT scans. In another approach, surfaces extracted from scan data were projected onto patients during surgery to provide the surgeon with 'X-ray' vision (Damini *et al.* 2000).

Regarding applications for entertainment, MR technologies allow one to create novel video games, where players can exercise their ingenuity in both real and virtual spaces — making full use of their hands and feet as interactive devices, instead of being limited to game pads, steering wheels, pedals, and the like (Tamura 2002; Stapleton *et al.* 2003). Broll *et al.* (2004) developed the mixed reality stage, an interactive pre-production environment which can be employed to rapidly plan, arrange and visualize scenes for music shows and theatre presentations. Virtual models and characters are projected into a real, but down-scaled model of the stage with tracked generic modules representing actors and decoration. Lighting conditions, moveable stage components and actors' paths can be planned and modified intuitively.

The overlay of virtual images on the real world is increasingly used to enhance computer-supported collaboration between people. Kato *et al.* (1999) developed a MR conferencing application where the remote collaborators are presented on virtual monitors which can be freely positioned in space. Users can view and interact with virtual objects using a shared virtual whiteboard. Further applications in the context of video communications are discussed in detail in Chapter 5.

MR systems require specialized displays allowing users to interact with mixed worlds in a natural and familiar way. Up to now, the available displays are a limiting factor, impeding proliferation. Most of the emerging applications impose very strong challenges in terms of performance and usability demands on the display devices that cannot yet be satisfied.

In this chapter we will briefly outline the challenges imposed on MR display systems and provide a survey of recent advances in the development of head-worn and free-viewing devices. With head-worn displays, viewers observe the mixed world through a headset, like the one shown in Figure 14.1. Free-viewing refers to monitor or projection-type displays whose special optics eliminate the need of headgear. Such displays are less intrusive but, in general, they are also more restrictive and limited to special applications.

In order to achieve a seamless mixture, precise registration of virtual objects with regard to the real world is indispensable. In addition to the general treatise of this topic in Chapter 17, a special section presents two generic MR systems recently implemented by the authors — one developed for desktop applications and the other one for mobile applications with a 2D or 3D handheld device. Interested readers may find valuable information about the foundations of MR systems in the first comprehensive book on this subject edited by Ohta and Tamura (1999) as well as in the surveys of Azuma (1997) and Azuma *et al.* (2001).

14.2 CHALLENGES FOR MR TECHNOLOGIES

According to Milgram, mixed reality environments are located somewhere along the reality — virtuality continuum which connects completely real environments to completely virtual ones (Milgram and Kishino 1994). Augmented virtuality (AV) and augmented reality (AR) are parts of this continuum (Milgram and Colquhoun, 1999).

Augmented virtuality means that the primary environment is *virtual* — enhanced by adding information-rich real objects. This may be done by mapping raw video data from natural scenes onto the surface of computer-generated solid objects, or by adding real objects as input tools, providing a sense of touch. In augmented reality the *real* environment dominates — enhanced, or augmented, by computer-generated imagery. It is interesting to note that the concept of augmented reality includes not only the *adding of virtual objects,* but also the apparent *removal of real objects* from an environment.

The definition of MR is not restricted to the visual sense. It includes other modalities such as hearing and touch. Hence, if we break down the creation of MR devices to engineering tasks we need to develop systems for visualization, sonification, haptification, etc., in order to support the relevant human sensory channels in a given application.

In this context, visualization is the most important field. Over 70% of the information humans collect from their environment is through the visual sense. Hence, the first implementations of MR displays were the head-up displays (HUD) used in military aircraft since the 1960s. These were employed in order to superimpose navigational and weapon-aiming information upon the pilot's field of view when looking through the windshield. Typically, a HUD includes a cathode-ray tube, collimating optics and a (holographic) beam combiner positioned between the pilot's eyes and the windshield (Wood 1992). The display creates virtual images focused at infinity, thus appearing merged with the distant external world. Since the early developments of HUDs most of the work done in the field of MR systems has concentrated on visual displays. The optical see-through displays discussed later in this chapter may be regarded as head-worn implementations of the HUD principle.

Milgram defines mixed reality as a technology that seamlessly integrates both real and virtual worlds. The notion of a *seamless* mixture is most important, because any break between the two worlds would immediately destroy the illusion of a joint environment — and our senses are very critical in this regard (Furmanski *et al.* 2002). Hence, we are faced with various challenges.

- *Spatio-temporal registration* must bring the virtual objects in precise alignment with the real ones, so that there is no noticeable offset in size, position and orientation over time. This requires adequate technologies for calibration and sensor fusion, indoor/outdoor object tracking and ego-motion estimation. Without precise registration, the feeling that the mixed objects coexist in the same environment is severely compromised (Holloway 1997).
- A comprehensive *photometric and geometric model* of the mixed environment is required to create appropriate illumination and occlusion effects. The system must measure the real lighting environment, model it in the computer, and calculate the shades and highlights of the virtual objects accordingly. Occlusion management needs knowledge of the 3D geometry of the real world in order to decide which of the virtual objects are visible from the user's current viewpoint and which are hidden behind real objects in the scene.
- Finally, MR *displays* are required that seamlessly blend the virtual objects and the real surroundings and, at the same time, support the relevant mechanisms of perception.

14.3   HUMAN SPATIAL VISION AND MR DISPLAYS

Since MR environments include contributions of the real, *three-dimensional* world, MR displays should adequately portray the visual cues required to perceive a sense of space. Some of the relevant perceptional mechanisms are very difficult to implement. These 'hard' implementation problems refer to:(1) the connection between the oculomotor processes of accommodation and convergence; (2) the limited, fixation-dependent depth of field of the eyes; and (3) potential accommodation conflicts when combining virtual and real things in the user's field of view (Drascic and Milgram 1996).

1. Normally, the lines of sight of both eyes converge at the fixated object, and the accommodation process, i.e., the focal adjustment of the eye lens, provides a clear retinal image.

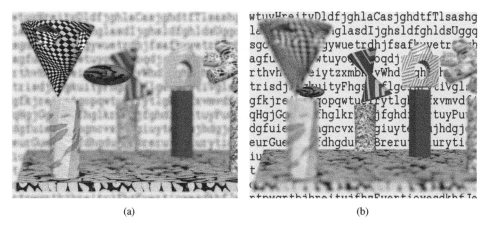

(a)   (b)

Figure 14.2 Effect of the limited depth-of-field of the human eye when looking at objects close to oneself (a) or far away (b)

Hence, eye movements cooperate and are reflexively linked with accommodation. Conventional 3D displays, however, require viewers to decouple these processes and focus the eyes on a screen distance to perceive a clear image, irrespective of the stereoscopic distance of the viewing target. Hence, the strange situation occurs that accommodation must be hold at a fixed distance, while convergence varies dynamically with the distance of the target object. The mismatching invokes a physical workload which may cause visual fatigue and deficits of binocular vision (Mon-Williams and Wann 1998). Moreover, conflicting depth information of the two processes is reported to the visual cortex.

2. When fixating on a real object we will see a clear image of it. Objects closer than the target or farther away will appear blurred (Figure 14.2). This blurring is caused by the limited depth of field of the eye, and it is a *wanted* artifact: it is necessary to suppress large retinal disparities related to the depth of non-fixated objects. With conventional 3D displays everything is shown in full focus, putting stress on the visual system (Blohm *et al.* 1997).

3. Another hard problem is to avoid accommodation conflicts between real and virtual objects, happening, e.g., when a virtual object is manipulated directly with a real tool. The eyes can focus at only one distance at a time. Hence, when looking at the real tool the eyes will focus on it — and see a blurred image of the virtual object, which stereoscopically appears to be at the same distance. Looking at the virtual object makes the eyes focus on the screen, and the tool appears blurred — although it is seen at the same distance. Again a paradoxical situation! Let us see what display technology can do to relieve these problems.

14.4 VISUAL INTEGRATION OF NATURAL AND SYNTHETIC WORLDS

14.4.1 Free-form Surface-prism HMD

First, we will have a look at head-mounted displays (HMD) which are currently the key device in MR applications. One of the most elaborate approaches is shown in Figure 14.3.

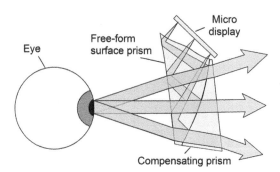

Figure 14.3 Free-form surface-prism optical see-through display (after Yamazaki *et al.* 1999)

This compact display uses a wedge-shaped prismatic module with three axially asymmetric curved optical surfaces. Images from the micro-display passing the transmitting surface are totally reflected at the front side, then reflected by a mirror surface and finally focused on the viewer's eye. The free-form surface-prism performs like a magnifying glass – the viewer sees a virtual image of the micro display as much as 10 times enlarged at a distance of typically 2 m.

When the mirror surface is made semi-transparent, the viewer can partially look through the display at the real environment. However, this view would be extremely distorted. Hence, it is necessary to compensate the optical distortions introduced by the first prism by a second one. The combined device operates like a plane parallel glass plate without noticeable distortions.

Major shortcomings of this concept are the fixed accommodation distance and the semitransparent (ghost-like) appearance of the superimposed virtual images. Moreover, registration of the virtual objects with regard to the real world is critical, requiring precise tracking. In spite of the elaborate design of the prisms and the use of light plastic material the weight of the HMD is still about 120 g. Hence, wearing the display tends to be uncomfortable after a while. Recent developments use OLEDs for the micro-display in order to reduce the weight and power consumption.

14.4.2 Waveguide Holographic HMD

Ideally, head-worn displays should not be larger and heavier than a pair of spectacles. (Azuma *et al.* 2001). Kasai *et al.* (2000) of Minolta Co. have developed a very light display, weighing only 25 g, which can be clipped onto an eyeglass frame.

As shown in Figure 14.4, the rays of light travel through a small acrylic waveguide only 3.4 mm thick with multiple total internal reflections. A holographic concave mirror forms an exit pupil of the display covering the pupil of the user's eye. Since the holographic element is tuned to the wavelength of the LED, it sends the emitted light to the viewer's eye and has almost perfect transmission for the rest of the spectrum. Hence, the waveguide appears completely transparent with no indication for others that the display is on. Another approach using a waveguide and holographic optical elements for a wide-field of view optical see-through HMD is under development at Physical Optics Corporation (POC 2004). In their approach the high-index polycarbonate waveguide is integrated into a visor. Two holograms,

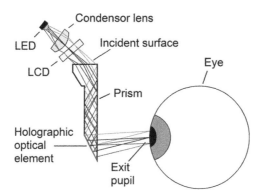

Figure 14.4   Ray paths in a waveguide holographic near-eye display (after Kasai *et al.* 2000)

one at the incident surface and another one at the surface in front of the viewer's eye, in connection with the curved shape of the visor, form the display's exit pupil.

14.4.3   Virtual Retinal Display

The virtual retinal display (VRD, also known as scanned-beam display) can create very bright, daylight-visible images and a very large depth of field so that almost no accommodation effort is required. The Nomad display shown in Figure 14.1 is a commercial implementation of a VRD; Figure 13.3 in Chapter 13 shows its principle of operation.

Unlike conventional displays which must create an image on an intermediate screen before they can be viewed, the VRD raster scans the visual information directly onto the retina, using low-power lasers or LEDs (Lewis 2004). It is possible to design the VRD with a small exit pupil of less than 1 mm and a small beam diameter, creating a large depth of focus, virtually 'bypassing' the optical effect of the eye lens. Hence, the advantage of this concept is that everything superimposed by the display will appear in full focus, irrespective of the accommodation distance of the real objects the user is looking at. However, in this case the exact eye-pupil position must be known and optically tracked, taking into account that the eye has a range of rotational motion that covers some 10–15 mm. Since the pixels are created sequentially it is also possible to include a deformable membrane mirror system, in order to modulate the beam divergence angle and to present images of the pixels at different focal distances (Schowengerdt *et al.* 2003). The VRD is commercialized by MicroVision (Bothell, USA).

14.4.4   Variable-accommodation HMD

A different approach to match the accommodation distance to the apparent distance of a virtual object has been developed at ATR (Sugihara and Miyasato 1998; see Figure 14.5). This concept uses a gaze detector to find the current point of fixation. The accommodation distance is adjusted to the fixation distance by real-time screen movements.

For practical use, the physical display screen is observed through a magnifying eyepiece lens. Hence, small movements of the screen result in large movements of the virtual image

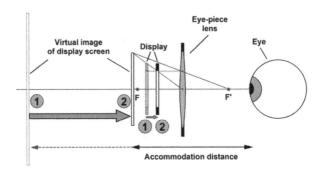

Figure 14.5   The variable accommodation display concept developed at ATR (Japan). Moving the display panel between positions 1 and 2 will change the accommodation distance accordingly

plane. This way, the system can move the displayed image from a distance of 25 cm to infinity at less than 0.3 s. Alternatively, a multi-planar miniature volumetric display has been proposed by Rolland *et al.* (1999). Human factor experiments show that multi-focal displays may produce less visual fatigue than conventional head-mounted displays.

14.4.5   Occlusion Handling HMD

In conventional optical see-through displays, virtual objects appear transparent, as ghost images through which the real world can be seen. Kiyokawa *et al.* (2001) of the Communications Research Labs (Japan) have proposed a display concept that makes it possible to block the visibility of the real environment at certain regions where virtual objects are superimposed. If, for example, a planned building is to be visualized at a certain location where a real one exists, the display can occlude the current building by an opaque mask and overlay the synthesized building upon that region (Figure 14.6). The viewer would perceive the mixed scene without a ghost image.

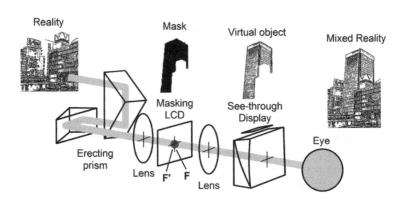

Figure 14.6   Principle of operation of an occlusion handling HMD (after Kiyokawa *et al.* 2001)

The occlusion handling display uses a masking LCD panel to opacify selected regions. The mask is placed at the focal point between a pair of convex lenses — the first one to focus the eye on the mask and the second one to re-focus on the real world. The lenses invert the optical path left/right and top/down. Hence, an erecting prism (known as Porro prism) is needed to put the image of the real world to its proper orientation. Possible shortcomings of this concept may result from the bulky and heavy setup due to the various lenses and prisms as well as from the reduced optical transmission of the masking LCD panel.

14.4.6   Video See-through HMD

Another class of HMDs stays completely in the electronic domain. Such video see-through displays use cameras mounted in front of the user's eyes (Figure 14.7). The camera images showing the real environment are electronically mixed with computer-generated objects.

The complete system, including the two cameras and displays, must be aligned with the viewer's eyes, so that corresponding points in the two virtual images are seen with parallel or inward-converging lines of sight (for close-range tasks, see Takagi *et al.* 2000). Video see-through allows precise registration of virtual and real-world objects as well as occlusion management via image processing. On the other hand, details of the real world are lost due to the limited resolution of today's cameras and displays. Moreover, everything appears at a fixed focal distance, and the cameras significantly add to the weight of the headset.

14.4.7   Head-mounted Projective Display

Conventional head-mounted displays create virtual images of a screen floating in space in front of the viewer. In an alternative approach real images are projected at objects in the real world by head-worn miniature projectors (Hua *et al.* 2001). As shown in Figure 14.8, a semi-transparent mirror in front of the eyes is used to overlay the projected images parallax-free with the user's line of sight. The real 'screen' for the projected images is made of retro-reflective material covering the surface of real objects. The microstructures of the screen reflect the light back to its source along the path of incidence. Hence, the left/right eye will see the image of the left/right projector superimposed upon the real environment.

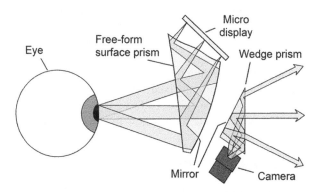

Figure 14.7   Concept of a video see-through HMD. The micro-display combines the video inputs generated by the computer and the camera (after Takagi *et al.* 2000)

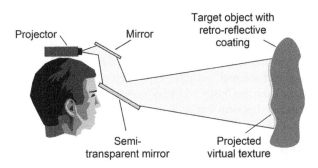

Figure 14.8 Head-mounted projective display concept

The projected images can be used for augmentation. A great advantage is that natural accommodation and convergence is supported when looking at the augmented scene. Different viewers can see different (personalized) information projected on the same object, and any real object not covered with reflecting material can occlude the virtual ones behind it. Moreover, if the projected textures are taken from the surroundings, it is possible to camouflage real world objects (Inami *et al.* 2000). Compared with the limited viewing angle of other HMDs, there is no such 'keyhole' effect narrowing the field of view, and optical distortions are negligible, even in a wide field of view. On the other hand, projective HMDs are limited to tasks where reflective screens are readily set up in real environments.

14.4.8 Towards Free-viewing MR Displays

The following sections are focused on free-viewing MR displays – e.g., placed on a desktop or workbench or projected onto reflective walls, respectively. Such displays are supposed to be better for long-term use, because the user does not experience the fatigue and discomfort of wearing a heavy head-mounted device. On the other hand, free-viewing displays are generally limited by a restricted working volume and potential space constraints.

Current approaches to non-head-mounted virtual reality displays use shutter glasses or polarizing filters worn by the user, in connection with a CRT monitor or a video projection system. Since the users need to wear stereo glasses, such approaches are not completely contact-free. More recent approaches work without the need of glasses. Here the optical elements required to separate the stereo views are included in the monitor. Chapter 13 surveys various approaches to glasses-free autostereoscopic 3D displays. However, most of the practical free-viewing display technologies have the typical drawbacks of conventional displays: the accommodation distance is fixed, irrespective of where the virtual objects appear, and there is an accommodation conflict when a user tries to touch the virtual objects with his/her hand or a real tool. The question is how to solve these problems.

14.4.8.1 Super-multiview Display

One current approach is the super-multiview concept (Honda *et al.* 2003) derived from the multiview techniques discussed in Chapter 13. Multiview means that the display provides multiple corresponding image points of an object related to different viewing angles. The super-multiview system differs from conventional approaches by producing hundreds of very densely arranged corresponding points. With this approach we can make the eye focus on the stereoscopic location of the point, and not on the screen distance.

How is that possible? Imagine the eye is focused on the screen distance. If two corresponding points in adjacent views are sufficiently close to each other their rays of light will enter the pupil at the same time, casting two images on the retina. Hence, the eye will see the same point twice (see Figure 13.6 in Chapter 13). The visual system tries to avoid such double images by focusing the eye lens on the stereoscopic location. In this case the two rays of light are focused on a single point on the retina, thus creating a single percept. The super-multiview approach supports the link between accommodation and convergence and avoids the accommodation conflict. However, the technological requirements of a 3D system with hundreds of perspective views are immense.

14.4.8.2 Virtual Image Overlay Display

Fortunately, in applications where only a single, primary stereoscopic distance is given, more practical solutions are available. An example is the Virtual Image Overlay system developed at the Carnegie Mellon University (Pittsburgh, USA) by Blackwell *et al.* (1998). As shown in Figure 14.9, the user views the real object through a semi-transparent mirror. A 3D or 2D video display device is positioned above the mirror, showing additional data relevant for the task at hand. Hence, the viewer sees the real world, plus a reflection of the video display screen which appears to float within the workspace.

Since the accommodation distance of the virtual overlay is about the same as the real object, the accommodation conflict is substantially reduced. On the other hand, the mirror between the viewer and the mixed environment may impede interaction. Moreover, the working volume is limited to the space behind the mirror.

14.4.8.3 FLOATS Display

The Fresnel-lens-based Optical Apparatus for Touchable-Distance Stereoscopy (FLOATS) system developed at the University of Tsukuba (Japan) is a free-viewing MR display allowing

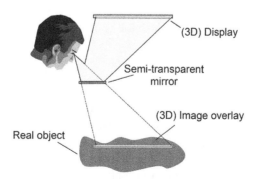

Figure 14.9 Virtual image overlay display concept. The image is perceived at approximately the same distance as the real object

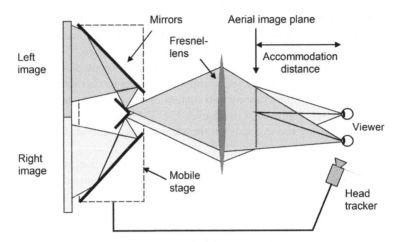

Figure 14.10 Top view of the head tracking FLOATS display using combined mirrors

unimpeded user interaction (Kakeya 2003). The display creates a stereoscopic image of virtual objects floating at a fixed location between the user and the display.

Figure 14.10 shows the basic optical arrangement. The Fresnel lens creates a real aerial image of the LCD panel showing the left- and right-eye images. Moreover, it creates images of the two small mirrors used to separate the views at the location of the user's eyes (field-lens principle, see Chapter 13).

Hence, the floating aerial image of the left/right view is visible only to the observer's left/right eye. The original approach proposed by the Communications Research Labs (Japan) used a large projection screen instead of the LCD and polarizing filters for stereo image separation (see Chapter 13).

14.4.8.4 Variable Accommodation MR Display

Another free-viewing MR display, where the workspace between the user and the virtual objects is free for direct interaction, has been developed at the Heinrich-Hertz-Institut (Germany). As shown in Figure 14.11, the basic optical concept of the Variable Accommodation MR (VAMR) display is similar to the FLOATS concept.

Two video projectors in a stereo projection arrangement are used to create a very bright stereo image pair. The stereo projector is focused on an aerial plane floating in front of (or behind) a large convex lens. The aerial image constantly changes its location according to the fixation point of the user. Hence, the viewer perceives an image of the currently observed virtual object in his/her accommodation distance (Boerger and Pastoor 2003).

This MR display creates holography-like images with a large depth range and supports the link of accommodation and convergence. On the other hand, it requires on-line measuring of the user's gaze direction and adaptation of the optical components of the display. Developed display prototypes for direct MR interactions within the user's arm reach are described in the following section.

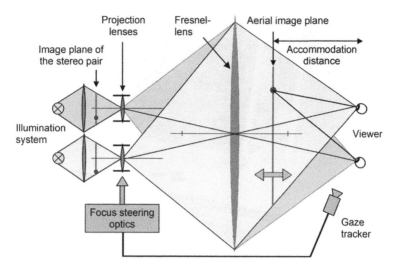

Figure 14.11 Top view of the Variable Accommodation MR (VAMR) display. The viewer accommodates on the aerial image plane and perceives a floating 3D object. The Fresnel lens images the exit pupils of the stereo projector at the locations of the left and right eye, respectively, and hence separates the constituent stereo images. The aerial image plane is dynamically aligned with the current position of the viewer's fixation point

14.5 EXAMPLES OF DESKTOP AND HAND-HELD MR SYSTEMS

14.5.1 Hybrid 2D/3D Desktop MR System with Multimodal Interaction

Figure 14.12 shows a recent prototype of the VAMR display, including a video-based device for finger tracking and simple hand-gesture recognition. The images of two high-resolution video projectors, with 1600×1200 RGB pixels each, are projected through a 30-inch filter panel containing the large convex Fresnel lens. As a result, computer-generated objects float above the desktop in very high quality (symmetric resolution, excellent stereo separation without cross-talk). Interaction is possible using the hand or a haptic device.

Normally, human stereoscopic vision is limited to a central binocular region where the visual fields of the left and right eye overlap. The binocular field is flanked by two monocular sectors within which objects are visible to one eye only. In these sectors, spatial vision relies on monocular depth cues such as shading, perspective and occlusion. The idea of the hybrid display (Figure 14.13) was derived from this basic concept of vision. The display shows 3D images in the centre part of a widescreen display. The surrounding area reproduces only monocular depth cues, but subtends a major portion of the user's visual field. The entire display area belongs to a joint data space. Hence, it is possible to make the 3D imagery merge seamlessly with the surrounding 2D region. Optionally, the 2D area may offer additional room where the user temporally places virtual objects and relevant tools.

The VARM display in the centre part of the workspace creates a stereo image hovering over the desk about 20 cm in front of the screen. The workplace offers various interaction

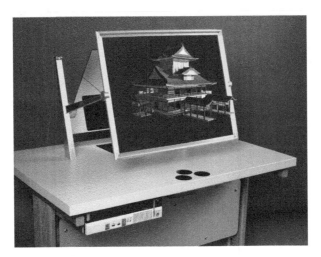

Figure 14.12 Prototype free-viewing VAMR display. The display creates high-resolution aerial 3D images which float 25 cm in front of the screen (filter panel) within the user's arm reach. A video-based hand tracker is embedded in the desktop at a location corresponding with the aerial image

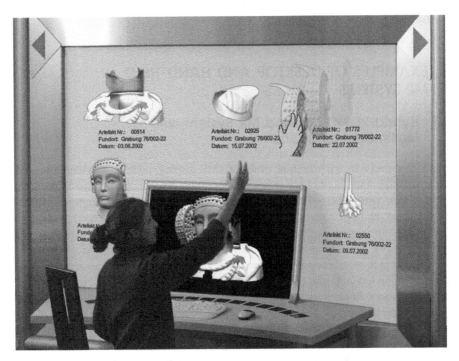

Figure 14.13 Combined 2D/3D MR display. This prototype workplace allows users to 'drag' computer-generated or scanned 3D objects from a high-resolution 2D rear-projection screen and 'drop' them into the centre (VAMR-type) 3D display for direct manipulation. Video-based hand-gesture recognition devices are embedded in the desktop. Cameras behind the front plate of the 3D display serve to recognize the viewer's gaze direction and head position. Both computer vision devices are used for multimodal interaction

modalities. Voice input combined with gaze tracking allows the user to visually and verbally address interactive objects. The display tracks the user's movements and adapts the optics as well as the 3D perspective to the current head position (motion parallax) . Hand gestures are recognized by devices embedded in the desk. When grabbing objects from the far 2D screen a realistic looking simulated hand is shown, virtually extending the user's arm; direct MR interaction with the real hand is possible if objects are within arm's reach. Optionally, tactile feedback when touching a virtual object is provided by a force-feedback device. Within the 3D area there is a perfect correspondence between the visual and tactile space. Corresponding spatialized audio is produced by an array of loudspeakers.

14.5.2 Mobile MR Display with Markerless Video-based Tracking

As opposed to the workspace constraints of desktop MR systems, mobile mixed reality (MMR) systems allow user interaction 'on the move'. Being mobile expands the applicability and acceptance of MR systems to their full potential. MMR systems can efficiently assist users in everyday tasks at home and in the office as well as in specialized applications in extended in-house and outdoor environments. Typical applications of MMR systems include virtual city and museum guides, architectural planning, industrial planning, maintenance and repair, annotation of product information, and entertainment.

To develop an adequate generic framework for MMR applications is a topic of ongoing research. Designers of MMR systems are confronted with various challenges. For example, all applications must be tailored to run with the limited resources of wearable or hand-held platforms. On the other hand, mobility requires real-time performance and robustness, while only few assumptions can be made about the actual environment.

As mentioned in Section 14.2, correct alignment of real and computer generated objects with respect to each other is a fundamental requirement of MR. Since mobile systems aim at applications in unprepared environments, sensors with a transmitter/receiver structure, such as electromagnetic and ultrasonic systems, are generally not adequate. Video-based registration has significant advantages over other techniques: it can be very accurate; it can use the very same image for registration and for display of the mixed environment; and it is suited for extended (virtually unlimited) workspace environments.

Major problems of the video-registration approach result from the fact that the camera is usually integrated in the mobile platform (PDA, Tablet PC, and HMD) worn by the user. Hence, even small movements of the user may produce large image motions which make the registration and tracking problem hard to solve. To simplify the registration problem and to handle user movements, a number of existing systems apply fiducial markers. Markers, however, limit the system's accuracy and require a 'controlled' environment.

Moreover, the so-called initialisation and recovery problems have to be solved. During initialization, the real target object(s) must be located. Once located, an object will often get lost during tracking — due to occlusion, large user movements (so that the target may be temporally not in view), or simply because the user puts down the mobile platform for a while to do something else. The system must recover (i.e., relocate the target object) as soon as the object is in sight again. Methods that usually perform well in stationary systems, such as the background subtraction and eigen-image methods, usually fail in mobile systems because neither the background nor the viewpoint remain constant. User-aided methods

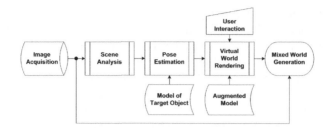

Figure 14.14   General flow diagram of a video- and model-based MMR system

requiring multiple interactions are also no adequate solution for the recovery problem, since most users find this process straining.

In the rest of this chapter, we describe two free-viewing prototypes based on a generic model-based framework for markerless video tracking developed at Fraunhofer Institute for Telecommunications, Heinrich-Hertz-Institut (Berlin, Germany). Model-based methods assume that a model of the target object consisting of features, such as points, lines, conic contours, texture, and colour, is known *a priori* known (Comport *et al.* 2003; Drummond *et al.* 2002).

A general flow diagram of the model-based approach is shown in Figure 14.14. A camera continually captures the environment, and the acquired images are analysed. Adequate features are extracted from the images and compared with the model in order to estimate the target object's pose with respect to the camera coordinate system. The estimated pose as well as the user's interaction is then used to render perceptively correct augmentations of the virtual world. The synthesized data are superimposed onto the live video image.

Figure 14.15 shows an *articulated video-MMR system*. A high resolution video display is mounted on the end effector of a passive kinematics chain with five revolving joints. The user can freely move the monitor within a large work envelope and observe the target object from different viewpoints. Stereo cameras fixed to the monitor view the object from almost the same perspective as the user and registrate the scene in real-time. Basically, the stereo information (epipolar constraints) makes registration more robust. Viewers see on the monitor the live video stream with additional information superimposed. Various display modes, such as 'show wireframes' and 'show hidden parts', are selectable. The cameras are the only sensors used for registration. The target object can be arbitrarily moved and repositioned, even during operation.

The system uses a generic model of the target object consisting of points, lines, conic contours and texture information. The object is first located via colour histograms and texture block-matching techniques. Once the target object is located, contours are extracted as the loci where the change of intensity is significant in one direction. Contours are known to be robust against lighting variations and can be localized accurately, since many measurements can be done along a contour. The extracted contour data are fitted to basic primitives such as lines and conics and brought in correspondence with the model. Pose estimation is then formulated as a minimization problem of the error vector between the extracted contours and the projected contours of the model, solved iteratively. If the object gets lost, the last valid patterns are used to recover registration.

Figure 14.16(a) shows a *handheld video-MMR system*. The system is implemented on a regular Tablet PC with a single webcam fixed on the back side of the device. The user

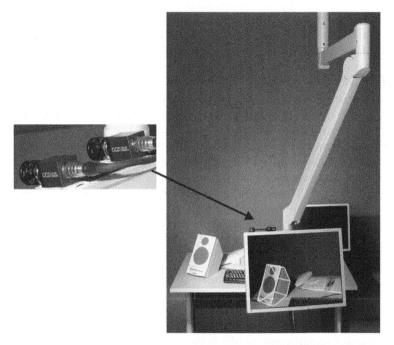

Figure 14.15 Articulated video-MMR system. The monitor mounted on the arm can be freely moved within a large working envelope, while it shows the scene captured with stereo cameras. Information about the target object (a loudspeaker) is superimposed in correct perspective onto the live scene (in this example the wireframe). The system enables a natural real-time user interaction

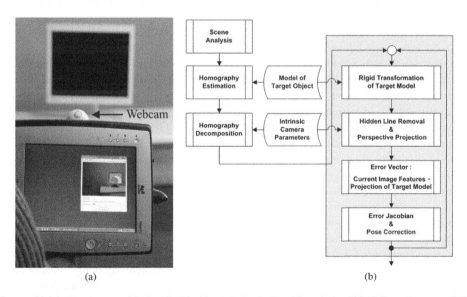

Figure 14.16 Freely movable handheld video-MMR device (a) and simplified flow diagram of the pose estimation (XE pose estimation) process (b) The webcam on the back side of the Tablet PC captures the real scene. The software recognizes the position and orientation of a known object (an LCD monitor) and correctly matches a virtual texture. All processing is done in real-time

can freely move the Tablet PC and observe the image of the target object from different viewpoints while various textures and information are superimposed onto the live image in real time. User interaction is possible via the Tablet PC's pen-sensitive screen. The registration framework shown in Figure 14.16 (b) is based on a coarse-to-fine model-based approach. Due to the limited resources of a Tablet PC target object registration is focused on the dominant planes of the target object (Simon *et al.* 2000; Sturm 2000). The perspective projections of different planes or of the same plane projected from different viewpoints are linked via plane projective transformations (homography). Homography provides on the one hand an adequate framework for tracking completely in the projective domain with less jitter. On the other hand, in case of known intrinsic camera parameters homography can be decomposed and recover metric information such as the three-dimensional position and orientation of the object. The provided pose estimate is then refined via an iterative pose estimation process similar to the one described.

An interesting extension of both MMR systems is to couple them with the free-viewing mixed reality displays described in Section 14.4.8 in order to enable comfortable 3D visualization of mixed environments.

14.6   CONCLUSIONS

One of the essential requirements of MR is the availability of adequate visualization technologies that allow users to perceive the mixed worlds and interact with them in a natural and familiar way. The challenge here is not only to correctly blend the virtual and the real scene components, but at the same time to support the natural perception mechanisms of human vision. The last aspect is of great importance in the context of comfortable user interaction and long-term visual perception.

Most of the MR applications reported in the literature are based on HMD technologies. HMDs have the great advantages that they support indoor and outdoor applications as well, since they are wearable, and that they allow hands-free working. However, HMDs require precise user-specific optical calibration and imply various limitations due to the limited display resolution, the limited field of view, and the reduced quality of the applied optics. A major drawback of current HMD-based solutions is that they are generally not suitable for long-term use due to the weight of the headset (inducing headache), imperfect support of human vision (causing visual fatigue), and imprecise dynamic registration (causing simulator sickness). Recently developed free-viewing displays overcome some of the limitations of HMDs. Existing solutions offer high-quality 3D scene rendering, support the natural mechanisms of visual perception, allow unhindered interaction, and are suitable for long-term use; however, the usable working volume of current free-viewing displays is significantly compromised compared to wearable-display solutions.

Although the existing results are very promising, developing adequate visualization technologies and concepts for full-scale MR applications is still a topic of ongoing research. On the other hand, the capabilities of the enabling technologies, particularly in the fields of high-brightness and high-resolution miniature displays, of tailored diffractive-optical elements, and of the graphics capabilities of PCs (with on-board devices for wireless connections to databases and networks) are continually improving and expanding. Hence we feel that selected MR applications will soon reach the point where it makes sense to leave the laboratories and to bring them to the mass market.

REFERENCES

Aiteanu D, Hillers B and Gräser A 2003 A Step forward in manual welding: demonstration of augmented reality helmet, *Proceedings of the 2nd IEEE and ACM International Symposium on Mixed and Augmented Reality,* Tokyo, 7–10 October 2003, 309–10.

ARCHEOGUIDE (2002) http://archeoguide.intranet.gr/project.htm

ARTHUR (2004) http://www.fit.fraunhofer.de/projekte/arthur/

Azuma R 1997 A survey of augmented reality. *Presence* **6** (4), 355–85.

Azuma R, Baillot Y, Behringer R, Feiner S, Julier S and MacIntyre B 2001 Recent advances in augmented reality. *IEEE Computer Graphics and Applications* **21** (6) 34–47.

Boerger G and Pastoor S 2003 Autostereoskopisches Bildwiedergabegerät (Autostereoscopic Display). Patent DE 195 37 499.

Blackwell M, Nikou C, DiGioia AM and Kanade T 1998 An image overlay system for medical data visualization. *Proceedings of the 1st International Conference on Medical Image Computing and Computer-assisted Intervention,* Cambridge MA, 232–40.

Blohm W, Beldie IP, Schenke K, Fazel K and Pastoor S 1997 Stereoscopic image representation with synthetic depth of field. *Journal of the SID,* **5**(33), 7–31.

Broll W, Augustin S, Krüger H, Krüger W, Maercker M, Grünvogel S and Rohlfs F 2004 The mixed reality stage — an interactive collaborative pre-production environment *Internationale Statustagung "Erweiterte und virtuelle Realität,* Leipzig, February 19–20.

Comport A. Marchand E and Chaumette F 2003 A real-time tracker for markerless augmented reality. *Proceedings of the IEEE/ACM International Symposium on Augmented Reality (ISAR 2003),* 36–45.

Damini D, Slomka PJ, Gobbi DG and Peters TM 2000 Mixed reality merging of endoscopic images and 3-D surfaces, *Proceedings of the 1st International Conference on Medical Image Computing and Computer-assisted Intervention,* Cambridge MA, 796–803.

Drascic D and Milgram P 1996 Perceptual issues in augmented reality. *SPIE* **2653,** 123–34.

Drummond T and Cipolla R 2002 Real-time visual tracking of complex structures. *IEEE Transactions on Pattern Analysis and Machine Intelligence,* **27**(7), 932–946.

Friedrich W and Wohlgemuth W 2002 ARVIKA — augmented reality for development, production and servicing. *Virtuelle und Erweiterte Realität,* Statustagung, Leipzig, 5–6 November 2002, http://informatiksysteme.pt-it.de/vr-ar-2/projekte/arvika/beitrag_ARVIKA.pdf

Furmanski C, Azuma R and Daily M 2002 Augmented-reality visualizations guided by cognition: perceptual heuristics for combining visible and obscured information. *Proceedings of IEEE/ACM International Symposium on Mixed and Augmented Reality (ISMAR 2002),* 215–224.

Holloway RL 1997 Registration error analysis for augmented reality. *Presence* **6**(4), 413–32.

Honda T, Shimomatsu M, Imai H, Kobayashi S, Nate H and Iwane H 2003 Second version of 3D display system by fan-like array of projection optics. *SPIE* **5006,** 118–27.

Hua H, Gao Ch, Brown LD, Ahuja N and Rolland JP 2001 Using a head-mounted projective display in interactive augmented environments. *Proceedings of IEEE/ACM International Symposium on Augmented Reality (ISAR 2001)* 217–23.

Inami M, Kawakami N, Sekiguchi D, Yanagida Y, Maeda T and Tachi S 2000 visuo-haptic display using head-mounted projector. *Proceedings IEEE Virtual Reality 2000.* 233–40.

Kakeya H 2003 Real-image-based autostereoscopic display using LCD, mirrors, and lenses. *SPIE* **5006,** 99–108.

Kasai I, Tanijiri Y, Endo T and Ueda H 2000 A forgettable near eye display. *4th International Symposium on Wearable Computers, IEEE Digest of Papers,* 115–118.

Kato H, Billinghurst M, Weghorst S and Furness T 1999 A mixed reality 3D conferencing application. *Technical Report R-99-1.* Seattle: Human Interface Technology Laboratory, University of Washington, http://www.hitl.washington.edu/publications/r-99-1/

Kato H, Tachibana K, Tanabe M, Nakajima T and Fukuda Y 2003 A city-planning system based on augmented reality with a tangible interface, *Proceedings of the 2nd IEEE and ACM International Symposium on Mixed and Augmented Reality,* Tokyo, 7–10 October 2003, 340–1.

Kiyokawa K, Kurata Y and Ohno H 2001 An optical see-through display for mutual occlusion with a real-time stereovision system. *Computers and Graphics* **25**(5), 765–79.

Lewis JR 2004 In the eye of the beholder, *IEEE Spectrum Online,* www.spectrum.ieee.org/WEBONLY /publicfeature/may04/0504reti.html.

Milgram P and Colquhoun H 1999 A taxonomy of real and virtual world display integration. In: Ohta Y and Tamura H (eds) *Mixed Reality — Merging Real and Virtual Worlds.* Ohmsha, Tokyo, and Springer, Berlin.

Milgram P and Kishino F 1994 A taxonomy of mixed reality visual displays. *IEICE Transactions on Information Systems,* **E77-D** (12) 1321–29.

Mon-Williams M and Wann JP 1998 Binocular virtual reality displays: when problems do and don't occur. *Human Factors* **40** (1), 42–9.

Ohshima T, Kuroki T, Yamamoto H and Tamura H 2003 A mixed reality system with visual and tangible interaction capability — application to evaluating automobile interior design, *Proceedings of the 2nd IEEE and ACM International Symposium on Mixed and Augmented Reality,* Tokyo, 7–10 October 2003, 284–5.

Ohta Y and Tamura H 1999 *Mixed Reality — Merging Real and Virtual Worlds.* Ohmsha, Tokyo, and Springer, Berlin.

Pettersen T, Pretlove J, Skourup C, Engedal T, and Løkstad T 2003 Augmented reality for programming industrial robots. *Proceedings of the 2nd IEEE and ACM International Symposium on Mixed and Augmented Reality,* Tokyo, 7–10 October 2003, 319–20.

POC (2004) http://www.poc.com/tech_summary/avionics/default.asp

Rolland JP, Krueger MW and Goon AA 1999 Dynamic focusing in head-mounted displays. *SPIE* **3639,** 463–9.

Schowengerdt BT, Seibel EJ, Kelly JP, Silvermann NL and Furness, TA 2003 Binocular retinal scanning laser display with integrated focus cues for accommodation. *SPIE* **5006,** 1–9.

Simon G, Fitzgibbon A and Zisserman A 2000 Markerless tracking using planar structures in the scene. *Proceedings of IEEE/ACM International Symposium on Augmented Reality (ISAR 2000),* 120–128.

Stapleton CB, Hughes CE and Moshell JM 2003 Mixed fantasy: exhibition of entertainment research for mixed reality, *Proceedings of the 2nd IEEE and ACM International Symposium on Mixed and Augmented Reality,* Tokyo, 7–10 October 2003, 354–5.

Sturm P 2000 Algorithms for plane-based pose estimation. *Proceedings of the International Conference on Computer Vision and Pattern Recognition (CVRP 2000),* 706–711.

Sugihara T and Miyasato T 1998 A lightweight 3-D HMD with accommodative compensation. *SID 1998 DIGEST,* 927–930.

Takagi A, Yamazaki S, Saito Y and Taniguchi N 2000 Development of a stereo video see-Through HMD for AR systems. *Proceedings of IEEE/ACM International Symposium on Augmented Reality (ISAR 2000),* 68–77.

Tamura H 2002 Steady steps and giant leap toward practical mixed reality systems and applications. *Proceedings of the International Status Conference on Virtual and Augmented Reality,* Leipzig, 3–12.

Wood RB 1992 Holographic head-up displays, in electro-optical displays (MA Karim ed.), Marcel Dekker, New York.

Yamazaki S, Inoguchi K, Saito Y, Morishima H and Taniguchi N 1999 A thin, wide field-of-view HMD with free-form-surface prism and applications. *SPIE* **3639,** 453–62.

15

Spatialized Audio and 3D Audio Rendering

Thomas Sporer and Sandra Brix

Fraunhofer Institute for Digital Media Technology, Ilmenau, Germany

15.1 INTRODUCTION

As long as objects are visible the human visual sense is dominant in normal environments to locate objects both in distance and direction. The visual sense is limited to objects in front, not occluded by other objects and with sufficient illumination. In contrast the auditory sense is able to detect objects in all directions. Objects emitting noise can be located, even if they are occluded by other objects. Actually the auditory sense of blind people is usually well trained, making it possible to locate objects because they change the sound field created by other objects. But even in situations where vision is dominant the auditory sense helps to analyse the environment and creates the feeling of immersion, the feeling of 'really being there'. As soon as immersion becomes an issue in audiovisual communication, that is if it is more than just recognizing the speech and who is talking, the correct, or at least plausible, reproduction of spatial audio becomes an important topic.

This chapter gives an overview about the basics of spatial audio perception, addresses the issue of multi-channel- and multi-object-based audio reproduction systems and discuss the problem of spatial coherence of audio and video reproduction.

15.2 BASICS OF SPATIAL AUDIO PERCEPTION

For the development of a special audio reproduction system it is essential to understand the behaviour of the human auditory sense. This section will explain the cues used to locate audio objects and discuss the major psychoacoustical effects important for audiovisual communication.

3D Videocommunication — Algorithms, concepts and real-time systems in human centred communication
Edited by O. Schreer, P. Kauff and T. Sikora   © 2005 John Wiley & Sons, Ltd

15.2.1 Perception of Direction

Similar to the visual sense the auditory sense is also based on two sensors. Due to the distance between the two ears and the shading by the head, sounds arriving from different angles in the horizontal plane arrive at different times (interaural time difference, ITD) and at different levels (interaural level difference, ILD, sometimes also called iteraural intensity difference, IID). Psychoacoustical experiments based on headphone reproduction have investigated the lateralization of sound depending on ILD and ITD. Figure 15.1 summarizes the results of such experiments. A lateralization of ±5 means that the sound is perceived near the entrance of ear channel, while 0 means it is perceived in the middle of the head. For headphone reproduction maximum lateralization is achieved at a level difference of about 12 dB and at a time difference of about 630 μs. Both values correspond approximately with the distance between the ears (Blauert 1974).

For frequencies below 1.5 kHz the interaural time difference is dominant. In this frequency range the ILD is very small due to diffraction of sound around the head. For higher frequencies the distance between both ears is too small to allow a definite estimation of the angle of incidence using the ITD, but diffraction is negligible and ILD becomes dominant.

Figure 15.2 shows all points in the horizontal plane where ILD and ITD are constant. It can be seen, that in the free field ILD and ITD are not sufficient to localize a sound object. In the presence of reflections the localization becomes definite. ILD and ITD do not give any hint about the vertical direction (elevation) of a sound source.

In addition to ILD and ITD the frequency response of the transmission channels from a sound object to each ear, the so-called head-related transfer functions (HRTF), is used for localizing objects. The HRTF of each individual is influenced by the shape of his head, pinna, shoulders and torso. For recording and measurement purposes dummy heads are sometimes used. These are designed to have averaged HRTF and usually include a torso part. The direction-dependent change of HRTF creates a change in the spectral envelope of sounds. If the spectrum of a sound is known to the listener and if the signal is sufficiently broadband the spectral difference between ears is sufficient to localize sound objects. Movements of the head change the spectral envelope of the received signal and enable the human brain to estimate the spectral content of the transmitted signal. To some extent this even enables one

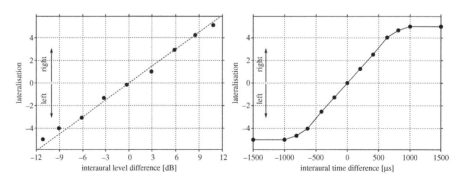

Figure 15.1 Lateralization dependent on the interaural level difference (left, test signal: broadband noise) and on the interaural time difference (right, test signal: impulses) respectively (data from Blauert 1974)

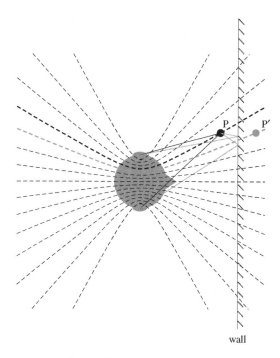

wall

Figure 15.2 Cone of confusion: dashed lines connect all potential positions of sound sources with same ILD and ITD. In the presence of a wall reflecting sound the mirror source P' of a sound source P helps to locate P

to localize sound sources if only one ear is working properly. The HRTF are also influenced by the vertical position of the sound objects: Depending on the elevation different parts of the spectrum are amplified. A discussion of these so-called directional bands can be found in Blauert (1997).

15.2.2 Perception of Distance

If sound objects are rather close to the ears the change of HRTF while moving the head is sufficient to estimate the distance of objects. For larger distances the angle of incidence for both ears is approximately the same and distance perception is not possible. If the distance of a sound object from the listener is changed, three parameters are used to perceive that movement: the change in level, the change of the spectral envelope and the Doppler shift. The change of the spectral envelope is due to the fact that there is more absorption in air for higher frequencies than for lower frequencies.

15.2.3 The Cocktail Party Effect

Localizing different sound objects enables one to separate different sound objects from each other. From a communication point of view this is equivalent of improving the signal-to-noise ratio. From a psychoacoustical point of view, this effect is called the binaural masking level difference (BMLD). The detection threshold of a sound signal (called the maskee) in the presence of a second sound signal (called the masker) in general is lower for binaural listening compared with monaural listening. A discussion of BMLDs in detail can be found in Blauert (1997).

15.2.4 Final Remarks

Perceptual experiments indicate that the localization performance the human auditory system is poor for frequencies below 100 Hz. However the perceptual impression of immersion is improved if the low-frequency content of sound signals presented to both ears are somehow decorrelated (Griesinger 1997).

15.3   SPATIAL SOUND REPRODUCTION

Over time, different methods for the reproduction of spatial sound have been developed. In general they can be grouped into discrete multi-channel loudspeaker setups, binaural reproduction and multi-object audio reproduction. The advantages and disadvantages of these methods will be explained in the following sections.

15.3.1   Discrete Multi-channel Loudspeaker Reproduction

In 1931 Blumlein filed a patent describing the basics of stereophonic recording and repro-duction (Blumlein 1931). In 1934 researchers at the AT&T Bell Laboratories introduced the concept of the acoustic curtain (many microphone, wired 1:1 with many loudspeakers). The group at Bell Labs also described two configurations of spatial sound reproduction with a reasonable number of transmission channels: binaural (two channels recorded using a dummy head) and multi-channel (Steinberg and Snow 1934). They proved that three loudspeakers provide superior quality in large rooms compared with two loudspeakers.

In 1953 Snow reported that 'the number of channels will depend upon the size of the stage and listening rooms, and the precision in localization required'. ... 'The effect of adding channels is not great enough to justify additional technical and economical efforts' (Snow 1953)

Two reproduction channels represent a minimal condition for spatial sound reproduction. In plain stereo, sound sources are mixed to a 'stage' in front of the listeners using mainly intensity cues. So-called phantom sources are created to place sound objects between the loudspeakers. If time cues are used in addition it is possible to create a limited perception of depth of the sound scene. Two-channel stereo is not able to recreate a spatial sound image of the real sound field, because sounds coming from the sides and back are missing. The optimum sound impression is limited to a very small portion of the reproduction room, the so-called sweet-spot. Adding a third loudspeaker in centre of the other two speakers the sweet-spot is enlarged. However the problem of proper mixing of sound objects is more difficult because sound objects mixed at the position of a loudspeaker sound differently than sound objects mixed to phantom sources.

Starting from plain stereo, quadrophony tried to enhance the naturalness of the sound field by adding two more loudspeakers at the back (Eargle 1971; 1972; Woodward 1977; Engebretson and Eargle 1981). Like plain stereo, the sweet-spot of quadrophony is very small.

Today in cinema and home cinema applications so-called 5.1 stereo is used (Figure 15.3). In addition to the left, centre and right speakers, two surround channels and a low-frequency enhancement channel (LFE) are used. In contrast to quadrophony the surround channels are placed at the sides of the listeners. The intention of this placement is to enable the reproduction of reflections from the walls of a (simulated) recording room. These reflections help to recreate the illusion of depth and help to localize sound sources in the front.

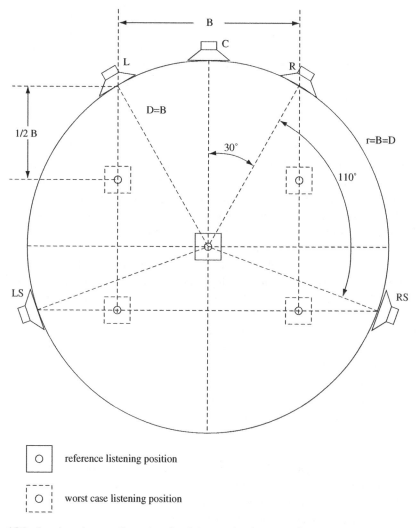

Figure 15.3 Loudspeaker configuration for 5.1 reproduction according to ITU-R BS.1116 (ITU-R 1994)

The LFE 0.1 channel was invented to give a surround system more bass headroom. In home applications the LFE often is misused for frequencies up to 200 Hz at the cost of a bad perception of immersion. Enhancements of 5.1 using an additional speaker at the back centre, or additional speakers in the front (7.1) are sometimes used in cinemas.

In a 5.1 system, side phantom imaging works well only for limited-bandwidth material, because different frequencies image differently on the sides, leading in turn to the break-up of sounds panned from front to rear. The 5.1 model does not take into consideration early first reflections, nor does it simulate a sense of height, for both these aspects of reproduction more channels are required. Compared with plain stereo the 5.1 stereo systems provide a larger sweet-spot (Harley *et al.* 2002). In Figure 15.3 the 'worst case listening positions' indicated how small it still is.

In the case of movie sound, dialogue, effects, atmosphere and music are usually mixed in 5.1. Dialogue is taken into the centre channel. Moving sound effects can be realized only when they are fast, so that the listener is not able to recognize the lack of phantom sources between the speaker positions. Lateral sources cannot be positioned, because of the huge audible gap between the front and surround speakers, so that objects can not move slowly from the front to back and vice versa. Surround speakers are set up in a diffuse array of speakers and consequently create a sound image that builds a kind of envelope from the listener's back. Precisely positioned sounds in the back are usually avoided, because of the sound interference they create.

Some of the limitations of 5.1 systems are addressed directly in Tom Holman's proposal of a 10.2 reproduction system (Holman 1999; 2001). Wide fronts are added to simulate early first reflections and by spreading out the sound in the front, clarity is also improved and more possibilities for envelopment are added. Separate side and front speakers are also added to eliminate the compromise between envelopment and front imaging, along with a front centre channel so that hard front pans are possible. Finally, two low-frequency channels are added for bass envelopment and improved spatial reproduction.

Another example in increasing channels to enhance spatial sound is the 22.2 reproduction system introduced by NHK (Nippon Hoso Kyo Kai — Broadcasting Corporation of Japan; Glasgal 2001). This system consists of 10 speakers at ear level, nine above and three below, with another two for low-frequency effects.

While 10.2 and 22.2 seem to solve only some of the problems already recognized in the early days by Blumlein, Steinberg and Snow, they add a lot of new problems on the side of signal acquisition and mastering: While recording for two- and five-channel stereo with simple microphone arrangements can provide reasonable results, for 10.2 and 22.2 special care has to be taken to allocate signals to loudspeakers. If not properly done new distortions such as sound colourations and comp-filtering appear.

At the same time some limitations remain:

- (virtual) sound sources can be placed at the loudspeaker position or between loudspeaker positions, but not between loudspeaker and listener (i.e., in the listening room);
- (virtual) sound sources placed between front and surround speakers are acoustically unstable, i.e., strongly dependent on the listener position;
- if there are additional loudspeakers for the third dimension the creation of content using these loudspeakers is very tricky;

- all formats store only final mixes of sound scenes, making interactive manipulation of content by the end user impossible;
- on the reproduction side the number and position of loudspeakers must be exactly as predefined in the recording/mastering;
- the sweet-spot is always a small fraction of the reproduction room.

15.3.2 Binaural Reproduction

Binaural techniques are a way to provide perfect spatial perception either using headphones or loudspeakers, as long as proper head-tracking is done, in the case of two-channel loudspeakers for just one person.

The binaural technology is based on the idea that it is sufficient to control the sound pressure at the listener's eardrums as the input to the human auditory system. Usually, recordings are done by means of artificial heads or by simulating HRTFs. For reproduction, headphones are typically applied. The spatial impression that can be achieved by this technology is limited by several facts:

Perfect localization and natural spatial perception cannot be achieved for all listeners because each person has an individual set of HRTFs. Hearing with the wrong set of HRTFs increase the probability of front–back confusion.

Naturally, the angle of sound incidence from a source relative to a listener is changed by head movements. With binaural reproduction by headphones without head-tracking, sound sources follow the movements of the head. The only plausible position of a sound source which follows head movement is to be within the head (in-head localization). Since binaural recording are obtained at one static position, reproduction will only be valid for that point. The listener cannot change his/her position within the sound field.

Recent research has shown that a more accurate localization can be achieved when the processed ear signals are dynamically adapted according to the actual orientation of the listener's head, which can be tracked by a sensor (Mackensen *et al.* 2000).

Even with head-tracking two limitations are still present: the direction of the sound sources appears elevated for some listeners compared with their real directions (Mackensen *et al.* 2000; Horbach *et al.* 1999) and the direction of a sound source and the room impression do not change when the listener is moving within the room. The method of head-tracked binaural reproduction is rather expensive and restricted to a single user, rather than groups of people. It can thus be used in multimedia workstations or professional applications only.

Due to complexity binaural reproduction today is used mainly in conjunction with channel-based audio formats. However it is also possible to render object-based spatial audio using binaural techniques.

15.3.3 Multi-object Audio Reproduction

Most reproduction systems that have been mentioned are based on the channel or track paradigm, i.e., the coding format defines the reproduction setup. That means it is not possible to make changes in parts of the master without doing the complete mixing again. In contrast to the channel oriented approach, multi-object audio reproduction offers an object-orientated approach. This means that audio sources are stored as audio objects, comprising certain properties such as position, size and audio (mono) track. A scene description shows how the individual objects are composed to build up an audio scene.

This section explains the concept of ambisonics and wave field synthesis, which are based on a parametric description of audio sound fields, enabling the manipulation of audio objects at the rendering site.

Ambisonics.   Ambisonics has been proposed as a successor of stereophony and was developed during the 1970s. Good overviews can be found in Furness (1990) and Gerzon (1992). This recording and reproduction method uses a different microphone and imaging technique from stereophony. The goal of this system is the reconstruction of the wavefront: a listener at the sweet-spot receives wavefronts that are copies of the original wavefronts. For that spherical harmonics are used to encode the direction of sound sources within a three-dimensional sound field.

For first-order ambisonics the analysis and reconstruction of the sound field are restricted to a small area, of the order of a wavelength. Therefore all incident waves in the original and reproduced field may be considered to be plane waves. For first-order ambisonics four microphones are arranged in the shape of a tetrahedron. Additional signal processing transforms these four components in the so called B-format: W (sound pressure), X (velocity component in the x-direction), Y (velocity component in the y-direction) and Z (velocity component in the z-direction).

For higher-order ambisonics either complicated microphones with many capsules are used or, audio objects are recorded using spot microphones and are recalculated to achieve an ambisonics representation.

An important advantage of the ambisonics format is that the number of loudspeakers is not limited to the number of transmission channels: it is possible to render any number of loudspeakers.

Lower-order ambisonics suffer from the sweet-spot problem, like discrete stereo reproduction. Increasing the order of the ambisonics increases the sweet-spot up to the full area of the reproduction room (Nicol and Emerit 1999). In contrast to discrete stereo reproduction ambisonics is able to create the impression of sound sources within the reproduction room.

Wave field synthesis.   Wave field synthesis (WFS) is a method to recreate an accurate replication of a sound field using wave theory and the generation of wavefronts. This method is able to generate a natural high-quality spatial sound image in a large listening area. It was invented by researchers at the Technical University of Delft who investigated the basic methods and did some prototype implementations of WFS (see e.g., Berkhout and de Vries 1989; Berkhout *et al.* 1993; Vogel 1993; Boone *et al.* 1995; de Vries 1996). The intuitive acoustic curtain concept is replaced here by a well funded wave theory.

The idea is based on the so-called Huygens principle (Huygens 1690): all points on a wavefront serve as individual point sources of spherical secondary wavelets. This principle is applied in acoustics by using a large number of small and closely spaced loudspeakers (so-called loudspeaker arrays). WFS-controlled loudspeaker arrays reproduce wave fields that originate from any combination of (virtual) sound sources, like an acoustic hologram. When driven properly, wavefronts are reproduced with perfect temporal, spectral and spatial properties in the whole room, overcoming the limitations of a narrow sweet-spot.

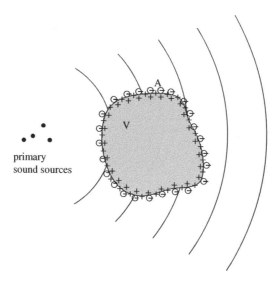

Figure 15.4 Monopoles and dipoles encircling a volume V create a sound field which is identical to the sound field of the primary sound sources within the volume, and zero outside the volume

From general linear acoustic theory an arbitrary sound field within a closed (fictive) volume can be generated with a distribution of monopole and dipole sources on the surface of this volume (Figure 15.4). The only restriction is that there are no additional acoustic sources within this volume. This can be expressed with the so-called Kirchhoff–Helmholtz integral, given by

$$P_A = \tfrac{1}{4\pi} \oint_S \left[\left(P\frac{1 + jk\Delta r}{\Delta r} \cos\phi \frac{\exp(-jk\Delta r)}{\Delta r} \right) + \right.$$
$$\left. + \left(j\omega\rho_0 v_n \frac{\exp(-jk\Delta r)}{\Delta r} \right) \right] dS \qquad (15.1)$$

where S is the surface of the volume, r is the coordinate vector of an observation point, Δr is the coordinate vector of the integrand functions on S. The sound pressure in the Fourier domain is given by P and k is the wave number ω/c, p_0 is the air density and v_n is the particle velocity in the direction normal of the normal vector n.

The first term in Equation (15.1) represents a distribution of dipoles that have a source strength, given by the sound pressure of the sound field at the surface. The second term represents a distribution of monopoles that have a source strength given by the normal velocity of the sound field. This predicts that by recording of $P(t)$ and $v_n(t)$ a sound field can be recreated exactly. The reproduction of this sound field is done by using a large number of monopole and dipole sources.

For practical purposes several simplifications are possible: by using either monopoles or dipoles it is possible to create the sound field within the encircled volume exactly, at the price of a mirrored sound field outside the target region. Furthermore the 3D integrals can be approximated using a finite number of identical loudspeakers placed in one plane. The approximations necessary for this step, going from 3D to 2D, are optimizing the sound

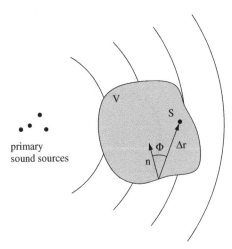

Figure 15.5 Illustration of the Kirchhoff–Helmholtz integral given in Equation (15.1)

field for a receiver line and using the assumption on an infinite long array of loudspeakers. Without any additional influence of the reproduction room the superposition of the sound fields generated by each loudspeaker composes the wave field in this plane up to the aliasing frequency. For an exact reproduction of all propagating waves, thereby neglecting near field effects, the spacing of the loudspeakers must be less than half of the shortest wavelength of the reproduced sound. In all these approximations air is regarded to be an ideal transport medium for sound propagation without any damping.

These approximations have some consequences:

- only monopoles are needed;
- the amplitudes of the sound reproduction are not absolutely correct in front or behind the receiver line;
- sound reproduction is correct only for wave field components in the horizontal plane;
- diffraction effects occur, due to finite or corner-shaped loudspeaker arrays.

Three representations of sound sources are possible in WFS (Figure 15.6): Virtual sound sources can be placed behind the loudspeaker arrays (so-called point sources) as well as in front of loudspeaker arrays (focused sources). In the case of sound sources in front of the array, the array radiates convergent waves toward a focus point, from which divergent waves propagate into the listening area. The third type are the so-called plane waves. Plane waves come from the same angular direction for all positions in the reproduction room and are often used for the reproduction of recorded or simulated room acoustics, or for ambient noise where exact placement plays a minor role.

WFS can easily be combined with room equalization. To lower the effect of the actual listening space on the perceived sound, partial cancellation of early reflections can be used (Kellermann 2001; Strobel *et al.* 2001). In addition, a virtual listening room can be added to the virtual sound sources by either synthetic acoustic spaces (e.g., as defined in the MPEG-4 standard Koenen 1999; Scheirer *et al.* 1999; Zoia and Alberti 2000) or by reproducing reverberation recorded in the room of the original performance (Horbach 2000).

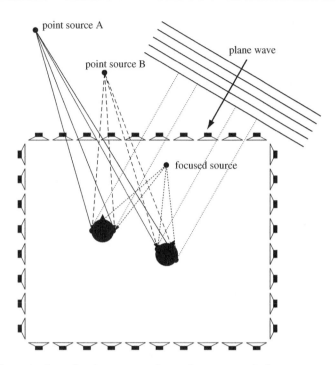

Figure 15.6 Reproduction of point sources, focused sources and plane waves using wave field synthesis

15.4 AUDIOVISUAL COHERENCE

Ideally, an audiovisual system should reproduce coherent visual and acoustical scenes. However there are audiovisual applications, where 2D (flat) video projection is combined with spatial sound reproduction, which provides depth information. This combination brings distortions in the audiovisual experiences of observers such as mismatch between perceived sound and visual information and loss of depth perception.

For the video part, distortions come from the fact that a natural representation of a (three-dimensional) scene is going to be represented as a two-dimensional image, so called 2D projection. Distortions are related to the loss of the binocular and motion parallax cues. So the main visual cues that remain for depth perception are size and positions of visual objects in the image relative to each other ('perspective'). Another aspect of using 2D projection is the observation of a 2D image from different observer points. In case of a 1:1 projection, there is only one point where the perceptual perspective of the picture coincides with the perspective of the original scene. For all other observer positions the perspective of the picture is distorted. A discussion of this problem can be found in Sedgwick (1991) among others.

In addition to the viewpoint problem the perception of a two-dimensional image cannot be reduced to just a geometrical problem because of other factors. One of them is the dual nature of pictures. Beside being a representation of a spatial layout existing in a three-dimensional virtual space that lies beyond the plane of the picture, a pictorial display is also a real object, consisting of markings of some sort, usually on a flat surface (Sedgwick 1991). This dual nature is one of the main reasons that the observer perceives the distortion to a smaller degree than the geometrical model suggests. Another point is that textures are used by observers to estimate the depth in a picture (Goldstein 1991; Sedgwick 1991). Additional problems are introduced by the use of different focal distances in modern movie production.

Spatial audio reproduction is usually limited to distance and direction, neglecting the perception of elevation. Due to the fact that the localization in the horizontal plane is much better than the perception of elevation, this does not influence the audiovisual experience. The use of spatial cues enables one to separate audio objects in the horizontal plane both in distance and direction.

Figure 15.7 illustrates the problem when flat video and spatial audio are used together. A visible object which is located in the middle of the screen is rendered at the correct spatial position **A'**. The point where the object is seen on the screen has also been drawn. Observer **A**, located at the ideal viewpoint, sees the object in the middle of the screen and hears the corresponding sound in the correct horizontal position of the true spatial auditory position. Observer **B**, located far on the side, sees the object in the middle of the screen as well, but does not hear the sound from the object in the expected position **B**. He also hears the sound from position **A'**. This causes a position error between the acoustical and the visual position of the object. The position error for a given position of the audio source depending on the observer position is calculated:

$$\varphi = \arccos\left(\frac{(\mathbf{I}_v - \mathbf{L})(\mathbf{I}_a - \mathbf{L})}{|\mathbf{I}_v - \mathbf{L}||\mathbf{I}_a - \mathbf{L}|}\right) \tag{15.2}$$

where $\mathbf{I}_v$ = vector of the image on the screen, $\mathbf{I}_a$ = vector of the corresponding sound source, with $\mathbf{L}$ = vector of the observer, φ = position error between image and sound of an object.

For practical applications the maximal position error from a perceptual point of view has to be found in order to calculate the maximal spatial depth, which can be used for the sound

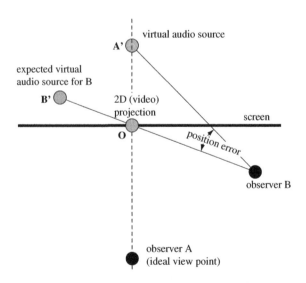

Figure 15.7 Perception of an audiovisual scene in case of different observer positions

sources positioning. Perceptual experiments performed using 2D video projection and wave field synthesis report a maximum acceptable position discrepancy between audio and visual images from 5° to 7° for untrained listeners (Melchior *et al.* 2003).

15.5 APPLICATIONS

Today video communication is focused on video quality and speech intelligibility. Spatial audio does not play an important role; the need to transmit non-speech audio signals is often regarded as unnecessary. Boone and de Bruijn (2003) show that WFS can improve speech intelligibility for teleconferencing.

Other application areas where spatial audio reproduction is important are themed environments, caves, cinemas and live performances (Start 1997). Themed environments today use multi-channel systems with eight and more channels. Over time the number of loudspeakers has increased more and more and the cost of audio production increased, too. The shift from channel-based audio to object-based audio has already started. For themed environments this brings additional benefits: with object-based audio (and video) it is easy to manipulate scenes in real-time, overcoming the limitation of pre-recorded fixed content. Similar trends are foreseen for caves. In cinemas spatial audio today is mainly represented as 5.1 or 7.1 stereo systems. In 2003 the first cinema based on WFS started daily operation, in 2004 the first WFS system was installed in a sound stage in Studio City, CA. While WFS can reproduce channel-based content in better quality, using the concept of a virtual loudspeaker (Boone *et al.* 1999) and thus increasing the sweet-spot, it can only play out its real capabilities when native, object-based content is available. For live performances, such as concerts, the content is created on-site. The first concerts with object-based reproduction systems have already taken place.

15.6 SUMMARY AND OUTLOOK

Spatialized audio rendering is currently in transition from channel-based formats to object-based formats. Current state-of-the-art systems can provide excellent spatial quality in the horizontal plane. Current research is focusing on audiovisual coherence, especially when spatial audio is combined with flat video, and on the extension of spatial audio to the third dimension.

REFERENCES

Berkhout AJ and de Vries D 1989 Acoustic Holography for Sound Control, *Proceedings of the 86th AES Convention,* Preprint 2801

Berkhout AJ, de Vries D and Vogel P 1993 Acoustic control by wave-field Synthesis, *Journal of the Acoustic Society of America* **93**.

Blauert J 1974 *Räumliches Hören.* Hirzel, Stuttgart, Germany.

Blauert J: *Spatial Hearing* (revised edition) (MIT, Cambridge, MA, 1997). Original edition published as *Räumliches Hören* (Hirzel, Stuttgart, Germany, 1974).

Blumlein A: Improvements in and relating to sound-transmission, sound-recording and sound-reproduction systems, Patent No. 394325, December 14th, 1931

Boone M *et al.* Spatial sound-field reproduction by wave-field synthesis - *Journal of Audio Engineering Society,* **43**(12), 1995.

Boone MM, Horbach U; de Brujin W: Virtual surround speakers with wave-field synthesis, Preprint 4928, *Proceedings of the 106th AES Convention,* Munich, 1999.

Boone MM, de Bruijn WPJ: Improving speech intelligibility in teleconferencing by using wave field synthesis, Preprint 5800, *Proceedings of the 114th AES Convention,* Amsterdam (March 2003)

de Vries D: Sound reinforcement by wave-field synthesis: adaptation of the synthesis operator to the loudspeaker directivity characteristics - *Journal of the Audio Engineering Society,* Vol.44, No. 12, 1996.

Engebretson M and Eargle JM: State-of-the-art cinema sound reproduction systems: technology advances and system design considerations, preprint 1799, 69th AES Convention, Los Angeles, May 1981.

Eargle J: Multi-channel stereo matrix systems — an overview, *Journal of the Audio Engineering Society* 19(7), pp. 552–559 (1971).

Eargle JM: 4-2-4 matrix systems: standards, practice and interchangeability preprint 871, 43th AES Convention, New York, September 1972.

Furness RK: Ambisonics: an overview paper presented at AES 8th International Conference, Washington, May 1990

Gerzon MA: general metatheory of auditory localization preprint 3306, 92nd AES Convention (1992)

Glasgal R: Ambiophonics. Achieving physiological realism in music recording and reproduction preprint 5426, 111th AES Convention, New York, 2001

Goldstein EB: Perceived orientation, spatial layout and the geometry of pictures. In: *Pictorial Communication in Virtual and Real Environments,* Edited by Ellis SR (ed.) Taylor & Francis 1991

Griesinger D: Spatial impression and envelopment in small rooms, preprint 4638, 103rd AES Convention, New York, September 1997

Harley R, Holman T: Home theatre for everyone: a practical guide to today's home entertainment systems Acapella Publishing; 2nd edn (June 2002)

Holman T: *5.1 Surround Sound — Up and Running,* Focal Press, Woburn, MA, (December 1999)

Holman T: The number of loudspeaker channels 19th International Conference of the AES, Elmau, Germany (June 2001)

Horbach U, Karamustafaoglu A, Boone MM: Practical implementation of a data-based wave-fields reproduction system preprint 5098, 108th AES Convention, Paris, 2000.

Horbach U, Karamustafaoglu A, Pellegrini R, Mackeusen P, Thiele G: Design and applications of a data-based auralisation system for surround sound preprint 4976, 106th AES Convention, Munich, 1999.

Huygens C: Trait de la lumire. Ou sont expliqués les causes de ce qui luy arrive dans la reflexion, et dans la refraction. Et particulierement dans l'étrange refraction du cristal d'Islande. Par C.H.D.Z. Avec un discours de la cause de la pesanteur", published by Pierre van der Aa, Leiden, 1690

ITU-R BS.1116: Methods for the subjective assessment of small impairments in audio systems including multi-channel sound systems - International Telecommunication Union, Geneva, Switzerland, 1994.

Kellermann W: Acoustic echo cancellation for microphone arrays, In: D. Ward and M. Brandstein (Eds.), *Microphone Arrays: Techniques and Applications,* Springer, Berlin, 2001, pp. 231–256.

Koenen R: MPEG-4. Multimedia for our times, IEEE Spectrum, vol. 36, no. 2, pp. 26–33, 1999.

Mackensen P et al.: Head-tracker based auralization systems: additional consideration of vertical head movements preprint 5135, 108th AES Convention, Paris, 2000.

Melchior F, Brix S, Sporer T, Röder T, Klehs B: Wave field synthesis in combination with 3D video projection 24th International Conference of the AES, Banff, Canada (June 2003)

Nicol R and Emerit M: 3D-sound reproduction over an extensive listening area: a hybrid method derived from holophony and ambisonic, paper presented at AES 16th International Conference, Rovaniemi, Finland (April 1999)

Scheirer ED, Väänänen R, Huopaniemi J: Audio BIFS: describing audio scenes with the MPEG-4 multimedia standard IEEE Transactions on Multimedia, vol. 1, no. 3, pp. 237–250, 1999.

Sedgwick HA: The effects of viewpoint on the virtual space of pictures. In: *Pictorial Communication in Virtual and Real Environments,* Ellis SR (ed.), Taylor & Francisi, London 1991

Start E: Direct sound enhancement by wave field synthesis, *PhD. Thesis,* Technical University Delft, 1997.

Snow WB: Basic principle of stereophonic sound Journal of SMPTE, volume 61, pages 567–589, 1953

Steinberg JC, Snow WB: Auditory perspectives — physical factors, "Electrical Engineering", volume 53, pages 12–17, 1934

Strobel N, Spors S, Rabenstein R: Joint audio — video signal processing for object localization and tracking In: M. Brandstein, D. Ward (Eds.) 'Microphone Arrays: Techniques and Applications', Springer, Berlin, 2001, pp. 197–219.

Vogel P: "Application of Wave-Field Synthesis in room acoustics", *PhD. Thesis,* Technical University Delft, 1993.

Woodward JG: Quadraphony — a review Journal of the Audio Engineering Society Vol.25, No.10, pp. 843–854, October 1977.

Zoia G, Alberti C: A virtual DSP architecture for MPEG-4 structured audio, Proceedings of the COST-G6 Conference on Digital Audio Effects DAFX'00, Verona, Italy, December 2000.

16

Sensor-based Depth Capturing

João G.M. Gonçalves and Vítor Sequeira

Institute for the Protection and Security of the Citizen (IPSC), European Commission – Joint Research Centre, Ispra, Italy

16.1 INTRODUCTION

Videocommunication, as the real-time transmission of sequences of images representing a real scene, requires sensing devices to capture the moving reality. Traditionally, real-world scenes were sensed projecting by means of optical lenses the visual aspect of the scene onto a plane. There, a sensor (e.g., an electronic tube or CCD array) would convert the light intensity and colour representing the scene into an electrical signal. This conversion was two-dimensional as there was no means of knowing the distance of the object to the sensory device. Though these techniques worked well, there are limitations in the quality and quantity of the information to transmit. These limitations restrict the visualization at the reception and in particular users' interactivity.

It is considered that the key to users' interactivity is the availability of three-dimensional (3D), or depth, information. Given two objects and a sensory device, this information can be either qualitative, i.e., knowing that one object is behind the other as seen from the sensor, or quantitative, i.e., knowing the precise location (and hence distances) of the objects in a world coordinate frame (e.g., having the X, Y, Z origin at the sensory device).

This quest for realistic 3D representations of real-world scenes raises two major challenges:

1. Completeness — how complete should a 3D representation be? Real-world scenes can be quite complex in the sense that objects can, and normally do, hide others. Further more, it is important to know the level of detail, i.e., depth accuracy and texture mapping, required for modelling each object;

3D Videocommunication — Algorithms, concepts and real-time systems in human centred communication
Edited by O. Schreer, P. Kauff and T. Sikora © 2005 John Wiley & Sons, Ltd

2. Time resolution — how often should a 3D representation be updated? This is particularly important when moving objects exist in the scene.

The answer to these challenges is not straightforward as there are many factors and technologies involved, including the sensory devices and display mechanisms used, not to mention the transmission aspects (e.g., compression level). All these factors will ultimately determine the level of interactivity for a given application area. To give a practical example, 3D videoconferencing may require a medium level of interactivity on a specific and limited volume of space, whereas, a user requesting 3D navigation maps on her/his mobile is less demanding. At the other extreme, full 3D television is heavily demanding on interactivity.

This chapter focuses on active techniques to acquire range or depth images. These are the most common and reliable method of acquiring range images, i.e., images representing explicitly the geometric structure of a scene (as opposed to inferring the 3D structure from 2D images). Active techniques can be used in almost any environment. There are two limitations, however: (a) objects should have a minimum reflectance in order that enough energy is reflected to the sensor — for the largest majority of objects this does not constitute a problem as in practical terms the reflectance is sufficient; (b) objects surfaces should be Lambertian in nature, i.e., they should diffuse incident light in the wavelength region of the illuminating light, normally in the red or near infrared spectra — i.e., surfaces should not act as mirroring devices.

It should be noted that, whatever the sensor or technique used in acquiring 3D data, the two challenges above require specific solutions for a given application area.

Completeness. 3D implies the possibility to interactively and realistically change the viewpoint of the scene being visualized. When changing a viewpoint, the model either provides the extra spatial information or does not. In the latter case, occlusions will be displayed and this can be rather disturbing for visualization. It should be made clear that to build 3D complete models, whatever the sensor or technique used, it is necessary to capture the reality from different locations in order to resolve occlusions. This means that different datasets must be registered — finding a common coordinate frame, and fused — selecting the best data from each set and discarding the remaining (Sequeira *et al.* 1999; 2002).

Time resolution. This attribute is most important for comfortable viewing of moving objects. Given that in most scenes not all objects normally move (e.g., in indoor scenes only people move, the walls and furniture do not) it is possible to separate static from moving objects. This suggests the differentiation of sensors and model accuracies for the objects in each category. Further to this, it is well known from 2D video transmission that fast objects do not necessarily require as much visual detail as static ones. This human factor can be used to set the 3D acquisition instrumentation and parameters to achievable values.

The techniques described in this chapter fall in two categories: triangulation and time-of-flight. Two new techniques are described at the end: focal plane arrays and fast image shutters. The text discusses the advantages/limitations of each technique, sources of error as well as the problems associated with data calibration and missing data. A section at the end attempts to provide guidelines on the most suitable techniques for specific applications. There is one common point: no technique provides real-time data – Focal plane arrays can provide real-time data, but the overall technique still needs to be improved in terms of spatial and depth resolutions. This current technological limitation suggests the combined use of active techniques for 3D modelling of static objects and of passive techniques for moving objects. These latter techniques are described in Chapter 7.

16.2 TRIANGULATION-BASED SENSORS

The basis for triangulation-based methods is the triangle formed by a distinct point on the surface of the object, one laser pointer and a camera. A laser stripe is scanned in one dimension across the scene. Commonly, a cylindrical lens transforms a low-power visible laser beam into a plane, which is projected onto the object in the field of view of the instrument (Figure 16.1a). This plane marks a profile-line at the object.

All the points of the line are thus marked optically. Looking to the profile-line from the viewpoint of the camera, the line appears to be curved in the image. The lateral displacement shows the elevations and indentations of the object's surface. If the position of the light source, the orientation of the light plane, and the position and orientation of the camera are all known, a vision system can calculate the position of each of the points on the bright line in three dimensions.

Figure 16.1(b) details the geometry for triangulation-based sensors. Considering that h is the height of the camera above the laser light plane, r is the distance between the camera and the object, y is the y-coordinate of the projection of the laser light on the object as seen on the camera focal plane bitmap and f is the focal length of the camera, Equation (16.1) is derived based on the similarity of triangles.

$$\frac{y}{f} = \frac{h}{r+f} \tag{16.1}$$

Since f and h are known, the range r can be determined as a function of y Equation (16.2)

$$r = \frac{fh}{y} - f \tag{16.2}$$

For each profile-line one picture must be captured. In order to improve the measurement accuracy and acquisition speed it is possible to use multiple cameras and light patterns (Battle *et al.* 1998; Blais 2004).

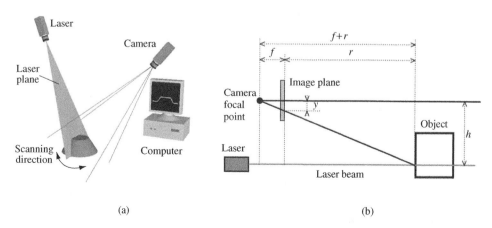

(a)   (b)

Figure 16.1   Optical triangulation: (a) principle of operation; (b) geometry of triangulation

Table 16.1   Measurement precision for different object distances

Object Distance (cm)	2.5	8	17	36	120
Precision (μm)	10	25	45	55	210

Table 16.1 gives an idea of the precision of the distance measurements when a triangulated laser striping technique is used. It should be noted that, overall, the precision decreases with the object distance and increases with the image resolution of the CCD sensor used in the camera.

A typical sensor provides a single profile of distances. In order to have a complete range image the sensor must be mounted on a scanning mechanism. Typical arrangements for the scanning mechanism are a rotational or translation stage. Figure 16.2 shows a commercially available triangulation-based 3D laser measurement device mounted on a computer controlled linear translation stage (2000 steps/mm).

Figure 16.3 (a) displays a grey level representation of a typical range image acquired with the sensor in Figure 16.2, where darker areas correspond to shorter distances. In addition to the distance measurement, the sensor gives, at each point, the energy reflected by the laser

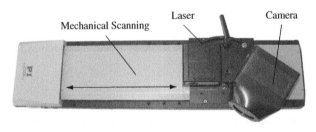

Figure 16.2   Triangulation-based laser measurement device mounted on a linear translation stage

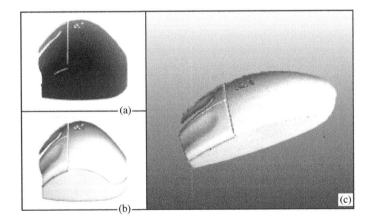

Figure 16.3   Range image from a computer mouse: (a) grey level; (b) reflectance; (c) 3D geometrically corrected perspective view

beam, generating simultaneously two registered images: range (Figure 16.3a) and reflectance (Figure 16.3b) image. When displayed with a regular grid, range images suffer from deformation associated with the geometry of the scanning system. Therefore, a geometrical correction must be performed to obtain data in a format independent of the acquisition system. A 3D Cartesian view of the same range image with geometrically corrected data is shown in Figure 16.3(c).

16.3 TIME-OF-FLIGHT-BASED SENSORS

The second main approach to measure distances is a direct consequence of the propagation delay of a wave. Distance is extracted by measuring the time-of-flight between an emitted signal and its corresponding echo. The most common sensors are ultrasonic and laser range finders (Figure 16.4). Ultrasonic sensors provide good depth accuracy and resolution, but suffer from poor spatial resolution due to the associated beam divergence. As a result, ultrasonic sensors are better suited to object detection than to 3D modelling and will not be further discussed.

A laser range finder measures distances along the orientation of the laser beam. To measure a horizontal distance profile, or to map the distances to the complete environment (i.e., range image), it requires the deflection of the laser beam. This is achieved either by moving the instrument itself (e.g., by means of pan-and-tilt device) or by deflecting the laser beam with a computer-controlled mirror.

Time-of-flight range sensors have limitations due to the technique used for emitting and detecting the received pulse and thus determining the distance. The emitted laser beam is not an ideal line, but rather a cone with a given, normally small, divergence (e.g., 2 mrad). It is common in commercially available systems to average earlier arrivals that have a strong signal-to-noise radio and report that as a single range. Distance measurement errors may occur when the laser footprint lies in a region where depth changes rapidly (e.g., a range roof edge or range discontinuity, or step edge). In this case, the range measurements are a combination of distances (Figure 16.5) and may have no physical meaning. This problem is commonly referred as to the *mixed point* problem.

Time-of-flight laser range sensors can be broadly divided into two main types: continuous wave modulated in amplitude or frequency; and pulsed wave or time delay. Sensors using this principle have recently been manufactured with an external scanning mechanism or without scanning by using an array of receivers (see Section 16.4).

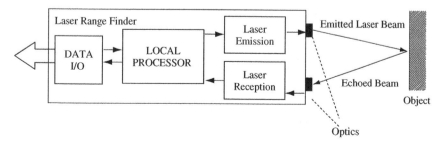

Figure 16.4 Time-of-flight principle

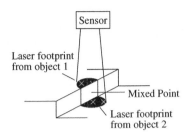

Figure 16.5 Mixed point problem

16.3.1 Pulsed Wave

Pulsed wave techniques are based on the emission and detection of a pulsed laser beam. Range is determined by directly measuring the time elapsed between the emission and the received echo (Figure 16.6). Distance is given by Equation (16.3), where c is the speed of light and Δt is the pulse travel time (i.e., from pulse emission to the detection of the corresponding 'echo')

$$d = \frac{c\Delta t}{2} \tag{16.3}$$

Figure 16.7 shows a commercially available pulsed wave scanner. Figure 16.8 shows an example of a typical scan taken with a Riegl Z210 scanner.

The best accuracies for this type of sensor are of the order of 10–20 mm, for distances of a few hundred metres.

16.3.2 Continuous-wave-based Sensors

AM phase-shift. A continuous laser beam modulated in amplitude with a reference waveform, normally a sine wave, is emitted and the range is determined by measuring the phase

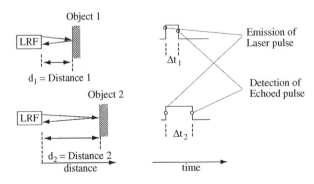

Figure 16.6 Pulsed wave principle

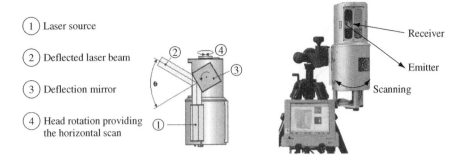

1. Laser source
2. Deflected laser beam
3. Deflection mirror
4. Head rotation providing the horizontal scan

Receiver

Emitter

Scanning

Figure 16.7 Pulsed wave scanner

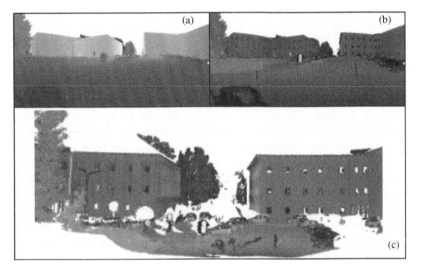

(a)

(b)

(c)

Figure 16.8 3D scan of an urban area using a Riegl Z210 scanner: (a) grey level; (b) reflectance; (c) 3D perspective view (888 × 2400 points, for a field of view of 60 × 260° and acquisition time of about 6 minutes)

shift between the emitted and received laser beams. By amplitude modulating a laser beam at frequency $f_a = c/\lambda_a$, as illustrated in Figure 16.9, and measuring the phase difference $\Delta\phi$

$$\Delta\phi = 2\pi f_a \Delta t = 4\pi f_a \frac{d}{c} \tag{16.4}$$

between the returning and the outgoing signal, the range is obtained as

$$d(\Delta\phi) = \frac{c}{4\pi f_a}\Delta\phi = \frac{\lambda_a}{4\pi}\Delta\phi \tag{16.5}$$

Considering that the phase measurement $\Delta\phi$ is periodic in time, the distance measurement is valid for range values shorter than one wavelength of the modulated signal. This value determines the *ambiguity* interval λ_a. If there are no ambiguity resolving mechanisms (such as

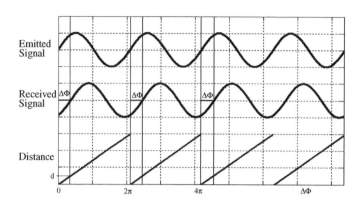

Figure 16.9 Amplitude modulation laser ranging

instructions from an operator, or heuristic rules), the *depth of field* must be constrained to fit into a single wavelength λ_a, or ambiguity interval.

The frequency of the modulating signal is used for trade-off between depth accuracy and the ambiguity interval (i.e., field of view). If higher modulating frequencies are used, better depth resolution is obtained, and if lower modulating frequencies are used a larger depth of field is obtained. The range of distances that can be measured is limited by the wavelength of the modulating signal. In order to overcome this trade-off and increase the unambiguous distance range, some manufacturers modulate the laser beam with multiple frequencies. In this case the lower modulating frequency determines the field of view, i.e., the ambiguity interval whereas longer frequencies improve the depth resolution. The best accuracies for this type of sensor are of the order of 3–5 mm, for distances below 100 metres.

Figure 16.10 shows a commercially available AM-Phase shift scanner using a double frequency (Langer *et al.* 2000). Figure 16.11 shows an example of a typical scan from an industrial area.

FM frequency shift. A continuous laser beam modulated in frequency with a reference waveform is emitted and the range is determined as a function of the frequency shift between the emitted and received laser beams. If the optical frequency of the transmitted signal varies linearly between $(v - \Delta v/2)$ and $(v + \Delta v/2)$ during the period $(1/f_m)$, a reference signal can

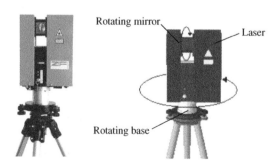

Figure 16.10 AM phase shift scanner

Figure 16.11 Scan from an industrial area using a Z+F Imager 5003: (a) grey level; (b) reflectance; (c) 3D perspective view (1700 × 1700 points, for a field of view of 60 × 60° and acquisition time of about 1 minute)

be coherently mixed with the returning signal to create a beat frequency f_b that is a function of the distance. The distance is determined from Equation (16.6).

$$d(f_b) = \frac{cf_b}{4f_m \Delta v} \tag{16.6}$$

Since this method is not based on phase measurements there are no problems associated to an ambiguity interval. Figure 16.12 illustrates the frequency modulation principle.

Unlike the AM phase shift the accuracy for a FM frequency shift scanner is controlled not only by the modulating frequency, but also by the signal-to-noise ratio at the detector and the number of samples acquired per measurement (Stone *et al.* 2004). This technique is currently the most accurate of the time-of-flight type. Typical accuracies for this method range from 50 to 300 μm at ranges of 25 m.

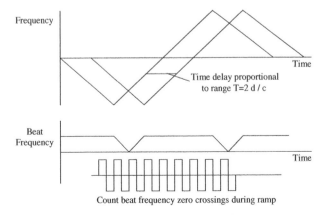

Figure 16.12 Frequency modulation laser ranging

16.3.3 Summary

The advantage of AM phase shift is usually better range resolution and linearity. The phase can be measured during the whole wave modulation rather than during pulse edges only, reducing the requirements for the bandwidth of the electronics. Modulation, however, generates ambiguity interval problems. Multiple-frequency modulation of the light beam is used to minimize this ambiguity, but absolute maximum range is still lower when compared with pulse systems. The use of FM frequency shift allows even better range accuracy.

Range accuracies for pulsed wave systems vary from 10 to 20 mm, for distances of a few hundred metres. AM modulation generally has a better accuracy, of order of 3–5 mm, but covers a maximum distance of about 50 m. FM frequency shift can have very high accuracies, varying from 50 to 300 μm at ranges of 25 m, but the cost is extremely high. Due to their cost FM frequency sensors are normally used for quality control of large mechanical objects (e.g., planes). Table 16.2 summarizes the performance of time-of-flight-based-sensors.

16.4 FOCAL PLANE ARRAYS

New miniaturized systems capable of measuring an array of distances in real-time and without any mechanical scanning parts ('scannerless') are now entering the market. One way of achieving this is by using a custom solid-state area image sensor, allowing the parallel measurement of the phase, offset and amplitude-modulated light field that is emitted by the system and reflected back by the camera surroundings. Depth measurement is based on the time-of-flight principle.

Focal plane array sensors (FPA) require the scene to be actively illuminated. In the case of the commercially available SwissRanger (Oggier *et al.* 2004), the CMOS array is part of an assembly comprising an array of 48 infrared LEDs (light emitting diodes) emitting light modulated with a signal of 20 MHz. This AM modulated light is diffused by the objects in the scene and projected back onto the CMOS sensor by means of the optical lens provided. Each element in the CMOS array sensor is able to measure the phase shift between the emitted light and the light focused onto itself. Each element of the array thus computes an individual phase difference measurement, which is proportional to distance according to Equation (16.5). A range map is thus built representing the distances of the objects in the scene to the sensor.

Figure 16.13 shows the SwissRanger featuring an FPA sensor with an area array of 124×160 elements. It can run at up to 30 frames/s with accuracies of about 10 mm. The ambiguity interval for this sensor is 7.5 m. Figure 16.14 shows an example of a typical scan.

Table 16.2 Performance of time-of-fight-based sensors

Type	Maximum range (m)	Accuracy (mm)	Speed (3D points/s)
Pulsed wave	100–800	4–20	1000–10 000
AM phase shift	55–80	2–5	250 000–500 000
FM frequency shift	2–25	0.05–0.3	2–1000

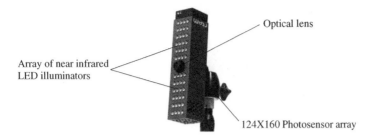

Optical lens

Array of near infrared
LED illuminators

124X160 Photosensor array

Figure 16.13 SwissRanger (Oggier *et al.* 2004)

Figure 16.14 Snapshot of a real time (20 frames/s) scan of a person: (a) grey level; (b) reflectance;
(c) 3D perspective view (124 × 160 points, for a field of view of 42 × 46°)

16.5 OTHER METHODS

Another way of determining the range without any moving parts is by using a fast image
shutter in front of the CCD chip and blocking the incoming light (Figure 16.15). The concept
of operation is based on generating a square laser pulse of short duration having a field of
illumination equal to the field of view of the lens. As the light hits the objects in the field
of view it is reflected back towards the camera. The collected light at each of the pixels is
inversely proportional to depth of the specific pixel.

This technology enables real-time (i.e., frame rate) depth keying and finds applications in
video production.

16.6 APPLICATION EXAMPLES

All sensors and systems described in this chapter provide a direct depth measurement output.
The use of each system depends obviously on the requirements of each application area.
Several parameters are to be considered, among which: (a) spatial resolution (the number of
samples per solid angle); (b) depth resolution (related to the sensor's accuracy and signal to
noise ratio); and (c) total acquisition time.

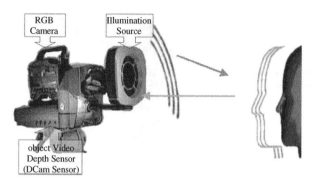

Figure 16.15 ZCAM (Iddan and Yahav 2001; courtesy of 3DV Systems)

Spatial and depth resolutions have a large impact on the quality of the final range image. Laser scanners, either pulsed or continuous wave, are the best candidates for achieving high-quality 3D images over a wide range of distances. On the other hand, and in spite of the tremendous progress in the last five years, all laser scanners are far from providing 3D images at frame rate (i.e., better than 10 frames per second). One can conclude that laser scanners are better suited for off-line applications, i.e., those requiring the creation of static, high-quality, dimensionally accurate 3D models.

An application in videocommunication which can profit from these high-quality 3D models is 3D virtual studios. The fact that all the information is in 3D enables a wide range of media capturing and production capabilities: (a) multiple TV cameras (different viewpoints); (b) moving cameras; (c) insertion of other off-line 3D virtual or reconstructed material, including high-quality animated characters; (d) independent 3D viewer-dependent visualization.

The technical pressure is transferred from media generation and production (including capture) to the graphical visualization engine. These techniques are widely used off-line in the most recent film productions and there have been many experiments in TV broadcasting.

Triangulation-based scanners are mainly used for high-accuracy object (in particular, surface) modelling, with accuracies between 10 and 100μm. The fact that depth error increases sharply with the distance of the object to the scanner prevents the utilization of this technique for distances beyond 1–2 m.

Focal plane Array sensors constitute a recent technological advance, in the sense that they enable real-time 3D capturing. Their image quality, defined by the spatial and depth resolutions, is still too limited to be widely used in the video communication industry. Nevertheless, it is certainly a technology to track in the following years, as it may bring important developments creating significant application niches (e.g., 3D teleconferencing or 3D remote surveillance). Manufacturers are working towards improved spatial and depth resolutions as well as increasing the sensor's field of view.

3D modelling of urban areas, achieved with laser scanning techniques, is an application area that has recently attracted much interest from mobile phone operators and services. Indeed, bringing georeferenced, photo-realistic and dimensionally accurate 3D urban models to mobile terminals (some of which already equipped with global positioning systems) will improve city navigation, as well as providing added-value information and services to tourists and professionals.

16.7 THE WAY AHEAD

Due to their chaotic nature, it is difficult to predict which will be the most influential technologies in 3D videocommunications including 3D sensing. With respect to real-time 3D sensing, advances in focal plane array technologies are to be expected. In the R&D pipeline are improved spatial and depth resolutions as well as increased field of view. One way to increase spatial resolution is to increase the number of elements in the solid-state detection array. This is not straightforward, as smaller array elements sizes decrease the sensor sensitivity. This can be overcome within certain limits with more powerful illumination. Increasing the field of view and depth resolution is probably easier by modulating the illumination with more than one frequency.

For laser scanning devices, manufacturers are at work on better detection technologies, including multiple frequency modulations, capable of improving depth resolution. Spatial resolution is still, and will probably continue to be, dependent on the mechanical precision of scanners. These technologies will still probably furnish the most accurate 3D models in the future.

In terms of applications, 3D technologies are here to stay. Indeed, more and more application areas will find the benefits of photo-realistic, dimensionally accurate 3D models. 3D videocommunications will increase along two main lines. The first is generic and includes 3D videoconferencing, interactive networked video games or full 3D television. Increased production costs will be divided by wide mass audiences.

The second line of development will be based on professional, highly specific applications. The scope of the application is to increase the ability of operators (and hence their productivity) in dealing with complex tasks, including remote teleoperation and/or telepresence. 3D models will be used to 'anchor' (i.e., to act as a reference) multiple and varied datasets aiming at easy interpretation of real-time, multi-variable complex phenomena or processes. Mixed (or augmented) reality technologies (see Chapter 17) will play an important role in these new human interfacing applications. Sensors of particular importance include real-time, highly accuracy positioning sensors, as well as all new types of human-computer interaction devices. Research efforts will concentrate on specific real-time software to process the multiple and various input data streams (some of which generated on-line, some others retrieved from off-line repositories) and have them seamlessly registered, fused and displayed. Many applications will result from the successful integration of existing technologies, which will constitute the building blocks for further developments.

It can be said that, however uncertain the new technologies may be, there are certainly interesting times ahead.

16.8 SUMMARY

The range of distances to measure determines the technologies to be used. For short distances, up to 100 cm, triangulation-based equipment with short baselines is more appropriate, with accuracies ranging from 10 to 100μm. For intermediate distances, between 1 and 10 m, there is an overlap between triangulation-based equipment with wide baselines and time-of-flight sensors with accuracies between 30μm and 2–3 mm. For larger distances, up to several kilometres time-of-flight sensors are preferably used. Accuracies can be as fine as 3–4 mm

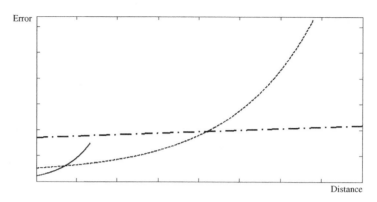

Figure 16.16   Scanner accuracy (dashed dot line — time-of-flight systems; solid line — triangulation with a short baseline; dashed line — triangulation with wide baseline)

for the best time-of-flight sensors Figure 16.16 gives an overview of the errors for different techniques.

Present commercial time-of-flight range sensors cost 25 000–400 000€, with 100 000€ being an average cost. Sizes vary but almost none have volumes less than 15 000 cm³ and weigh around 15 kg.

A list of the commercially available range sensors, both triangulation and time-of-flight, can be found in Boehler and Marhs (2002) and an exhaustive survey for time of flight sensors in Stone *et al.* (2004).

Presently there is a lot of research by several different companies in the development of scanerless sensors; their the biggest drawback of these sensors is their resolution. The best commercially available sensor has a 124 × 160 pixel array, yet the largest research system is a 256 × 256 pixel array. These are promising results in the quest to achieve a truly 3D Video camera.

REFERENCES

Battle J, Mouaddib E and Salvi J, Recent progress in coded structing Light as a technique to solve the correspondence Problem: survey *Pattern Recognition* **31**(7) 963–982, 1998.

Blais F, 'Review of 20 years of range sensor development', Journal of Electronic Imaging, 13(1), 231–240, Jan. 2004.

Boehler W, Marbs A,3D Scanning instruments.CIPA, heritage documentation - International Workshop on Scanning for Cultural Heritage Recording, pp. 9–12, Corfu, Greece, September 2002.

Iddan GJ, Yahav G; '3D Imaging in the studio', SPIE Vol. 4298, pp. 48, January 2001.

Langer D, Mettenleiter, M, Härtl F and Fröhlich C, Imaging Ladar for 3-D surveying and CAD modeling of real world environments, International Journal of Robotics Research, Vol.19, No.11, Sage Publications, Inc., pp. 1075–1088, November 2000.

Oggier T, Lehmann M, Kaufmann R, Schweizer M, Richter M, Metzler P, Lang G, Lustenberger F, and Nicolas Blanc, An all-solid-state optical range camera for 3D real-time imaging with sub-centimeter depth resolution (SwissRangerTM), Proceedings of SPIE – Volume 5249 Optical Design and Engineering, Laurent Mazuray, Philip J. Rogers, Rolf Wartmann, Editors, February 2004, pp. 534–545

Sequeira V, Ng K, Wolfart E, Gonçalves JGM and Hogg D, 1999 Automated reconstruction of 3D models from real environments. *ISPRS Journal of Photogrammetry and Remote Sensing.* **54**, 1–22.

Sequeira V, Gonçalves JGM, 3D reality modelling: photo-realistic 3D models of real world scenes", *Proceedings of the 1st International Symposium on 3D Data Processing Visualization and Transmission (3DPVT 2002)*, pp. 776–783, IEEE CS Press, 2002.

Stone WC, (BFRL), Juberts, M, Dagalakis N, Stone, J, Gorman J. (MEL) 'Performance analysis of next-generation Ladar for manufacturing, construction, and mobility', NISTIR 7117, National Institute of Standards and Technology, Gaithersburg, MD, May 2004.

17
Tracking and User Interface for Mixed Reality

Yousri Abdeljaoued, David Marimon i Sanjuan, and Touradj Ebrahimi

Ecole Polytechnique Fédérale de Lausanne (EPFL), CH-1015 Lausanne, Switzerland

This chapter introduces the field of mixed reality that merges the real and virtual worlds seamlessly. To develop a mixed reality system, two fundamental problems must be addressed: tracking and user interface. Tracking in mixed reality is defined as the task of estimating the user's viewing pose. The goal of tracking is to achieve an accurate registration between real and virtual objects. In order to allow the user to interact efficiently with the mixed reality scene, interaction techniques must be designed. A wide range of mixed reality applications, such as medical and entertainment applications, have been identified. We will focus on the new trends and applications in the area of mixed reality.

17.1 INTRODUCTION

Motion tracking in mixed reality (MR) is defined as the task of measuring in real-time the position and orientation of the human head, limbs or hand-held devices in order to interact with the MR environment. Three types of interaction have been identified (Foxlin 2002): view control, navigation and manipulation. For example, the user's head pose allows an accurate registration of virtual and real objects which gives the user the impression that the virtual objects are part of the real world. Figure 17.1 shows a typical block diagram of a MR system. It consists of a set of sensors which serve for sensing the environment. They deliver the input data for motion tracking. Examples of these sensors are magnetic, optical, and inertial sensors. The camera, which is an optical sensor, has an additional role in motion tracking. It serves for the acquisition of the real scene where the virtual objects will be inserted. Based on the measurements obtained by the sensors, the position and the

3D Videocommunication — Algorithms, concepts and real-time systems in human centred communication
Edited by O. Schreer, P. Kauff and T. Sikora   © 2005 John Wiley & Sons, Ltd

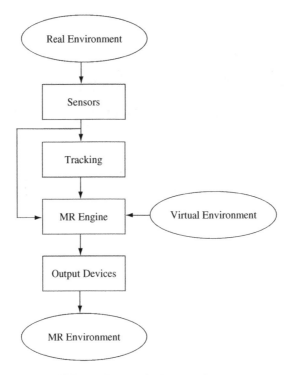

Figure 17.1   Typical block diagram of a MR system

orientation of the human head are estimated. These are sent to the MR engine where the seamless integration between real and virtual objects, as well as rendering, is performed. The MR engine also interprets human gestures or motion of hand-held devices and adapts the MR environment to allow for the interaction between the user and the MR scene. Output devices, such as head-mounted displays (HMD), are used for viewing the merged virtual and real environments. The output devices of MR are the focus of Chapter 14.

MR applications are diverse and range from the traditional medical area, such as computer aided surgery, to recent industrial applications. Several applications of 3D videocommunication make use of MR technology such as content creation in film production (Chapter 3) and immersive videoconferencing (Chapter 5).

This chapter is organized as follows. Section 17.2 introduces the requirements that a tracker should fulfil and presents the different tracker categories based on several physical laws. Section 17.3 is dedicated to the design of MR interfaces. Section 17.4 presents the most recent MR applications.

17.2 TRACKING

The following categories of trackers can be identified based on their physical sensing technologies: magnetic, optical, inertial, mechanical, and vision-based. Each category is based on a physical principle to measure the position and the orientation of the object. Many surveys on motion tracking consider vision-based tracking as a subclass of optical tracking

(Foxlin 2002). We treat it as a separate class because there is an increasing interest in this area and the main modality addressed in this book is video. We will briefly introduce magnetic, optical, and mechanical tracking, but we will focus mainly on vision-based tracking.

To evaluate a tracker the following performance parameters have been proposed (Burdea and Coiffet 2003):

- *accuracy* represents the difference between the object's actual and measured 3D position;
- *jitter* is equivalent to the change in tracker output when the tracked object is stationary;
- *drift* corresponds to the steady increase in tracking error with time;
- *latency* is the time between the change in object position or orientation and detection of this change by the sensor detects.

17.2.1 Mechanical Tracking

Mechanical trackers measure displacement or bending of rigid or semi-rigid pieces. These pieces can convert their bending or mechanical pressure into electric signals. These signals can be interpreted as relative position or rotation. Specifically, the active pieces are poten-tiometers, bend sensors, and optical fibres. Potentiometers produce a voltage proportional to their torsion or displacement. A bend sensor is composed of tiny patches of carbon or plastic that change resistance, depending on the bending level. Optical fibres conduct more or less light, also depending on bending level.

The study of body movements has increased the performance level in sports. Mechanical devices are often used for this purpose as well as for hand motion analysis. Different types of equipment are available, from body-based to hand haptic devices. Although in principle they have an unlimited range, the former may be an impractical solution if the user needs large free movement. Force-feedback gives the latter additional touch perception in a mixed or virtual reality experience.

Commercial examples of such technology are the Haptic Workstation (two-handed Cyber-Force systems for virtual prototyping) and the CyberGlove, both from Immersion Corporation (Immersion 2004).

17.2.2 Acoustic Tracking

Acoustic trackers sense sound waves to measure distance. A transmitter–receiver pair is needed to calculate the speed of sound. Two techniques are commonly used: time-of-flight and phase coherence. Commercial acoustic ranging systems use the time-of-flight of a pulse, typically of ultrasonic waves, from the source to the destination. Phase coherence systems are based on the principle that certain types of waves, e.g., ultrasonic, conserve phase during their lifetime. Knowing the frequency and phase at the origin, and the shift at the end, position can be estimated from the difference within a wavelength cycle. Isolating phase may be inconsistent as the phase difference could exceed a period when motion is too fast. As a result, acoustic sensors perform best for slow motion. They also have to deal with several properties of sound waves. The main disadvantage of acoustic trackers is the slow speed of sound, which creates a delay. Also, the environment can create occlusion or wave repetitions by reflecting the signal (Vallidis 2002). An example of such technology is the Logitech's 3D Mouse created by FakeSpace Labs, Inc., consisting of three ultrasonic speakers set in triangular position (FakeSpaceLabs 2004).

17.2.3 Inertial Tracking

Inertial trackers calculate position and orientation using accelerometers and gyroscopes. They are based on the well-known physical relation between force and acceleration

$$F = m \; a.$$

The displacement of the sensor mass inside the sensor housing is measured by an accelerometer. This displacement is due to the force applied to the housing. Acceleration is calculated by translating this displacement into a measurable force. Finally, position is obtained by double-integration of the acceleration.

A mechanical gyroscope consists of a rapid rotation wheel set inside a frame that permits it to turn freely in any direction. The conservation of angular momentum causes it to maintain the same attitude (direction). The secondary spinning movement induced by external rotation is measured. From its speed, it is possible to calculate orientation in one direction.

Originally, inertial navigation systems (INS) used gimbals as rotating elements. Their size and weight were inappropriate for human transportation and so, for mobile MR. The use of such devices in mixed and augmented reality did not start until microelectronic mechanical systems (MEMS) technology appeared in the 1990s. Recent advances led to the strapdown INS type. It contains three orthogonal accelerometers for 3D position (x, y, and z) and three orthogonal gyroscopes for 3D orientation (roll, pitch, and yaw).

Inertial trackers are self-dependent (i.e., there is neither a transmitter nor a receiver system). This is an advantage over acoustic or optic rangers, where there is a need for a direct line-of-sight. Being chip-implementable, passive, with low latency, and without environment requirements, makes them one of the most efficient tracking systems. Drawbacks arise from the double-integrator needed to calculate position and the single-integrator for orientation. A bias in the accelerometers induces an incrementing error (i.e., drift) in position estimation. At low frequency (i.e., slow motion), noise in measurement is confused with acceleration (Allen *et al.* 2001). If there is really slow motion or the user is still, the tracker could produce the same estimation. Gyroscopes are affected by similar problems. Another drawback is the effect of gravity. Inertial trackers compensate it by estimating its direction. A divergence of the estimation can produce an increasing error in position. Their good performance in a fast moving environment though, makes them a good candidate for a hybrid solution. A system that is drift-free can solve the main drawback of inertial trackers by applying a periodical reset of their coordinates.

Examples of commercial products are: InterSense's InterTrax2, InterCube2 (Figure 17.2), and IS-300 Pro. Recent tracking systems use inertial sensors to deal with occlusion (Livingston *et al.* 2003) and body motion (Kourogi and Kutata 2003; Roetenberg *et al.* 2003).

17.2.4 Magnetic Tracking

Magnetic trackers employ a magnetic field generated by a stationary transmitter to determine the real-time position of a moving receiver (Raab *et al.* 1979). A calibration algorithm is used to determine the position and orientation of the receiver relative to the transmitter. Since the magnetic field penetrates dielectrics, this tracking category overcomes the problem of line-of-sight. Other advantages of magnetic tracking are convenience of use and high accuracy.

Figure 17.2 The InertiaCube 2 inertial tracker. © InterSense Inc. Reprinted by permission

However, this accuracy is seriously degraded by magnetic fields due to ferromagnetic or conductive material in the environment.

The transmitter consists of three coils wound orthogonally around a common core. These coils are activated sequentially either by alternating currents (AC tracker) or direct currents (DC tracker) to produce three orthogonal magnetic fields. For the AC tracker, the receiver consists of three small orthogonal coils which measure the nine elements (three measurements for each of the sequentially activated transmitter coils). Figure 17.3 shows a representative AC tracker, Fastrack from Polhemus.

For the DC tracker, fluxgate magnetometers or Hall effect sensors are used as the receiver. By inducing a short time delay between the excitation of the transmitter and the measurement

Figure 17.3 The Fastrack AC tracker. © Polhemus. Reprinted by permission

of the resulting field in the receiver, eddy currents induced in nearby metal objects die out. In this way, the distortions in the transmitter magnetic fields due to eddy currents are minimized.

17.2.5   Optical Tracking

An optical tracker uses optical sensing to determine the real-time position and orientation of an object. This tracker consists of two components: targets and optical sensors. Because of the high speed of light, these trackers have a small latency. Other advantages are the high update rate and the large range. However, optical trackers require direct line-of-sight.

Optical trackers can be classified into outside-looking-in and inside-looking-out systems. In an outside-looking-in tracker, the sensors are fixed on the walls or ceiling looking in the workspace. These sensors measure the bearing angles of the targets attached to the object or user. The position of the target is computed using triangulation. The orientation is estimated from the position data. One of the drawbacks of this class is that the orientation resolution is degraded as the distance between the targets attached to the object or user decreases. This category of optical tracker is suited for motion capture for animation and biomechanics.

In an inside-looking-out arrangement, the sensors are attached to the tracked object or user and the targets are paced at fixed locations on the ceiling or walls. This arrangement has better orientation resolution and a larger working volume compared with outside-looking-in systems. These advantages make the inside-looking-out arrangement attractive for MR applications.

Another way to classify the optical trackers is based on the use of active or passive targets. An example of an active target is the infrared light emitting diode (ILED). Typical sensors used to detect these active targets are the lateral-effect photodiode (Wang *et al.* 1990) and quad cells (Kim *et al.* 1997). They determine the centroid of the light on the image plane. Passive targets (e.g., markers) and sensors (e.g., CCD cameras) are treated separately in the vision-based category. An example of an optical tracker is HiBall (Figure 17.4).

17.2.6   Video-based Tracking

Video-based tracking is a major topic within the motion tracking research area. Different advances have been achieved, most of them following one of the two main paths: marker- or fiducial-based tracking, and markerless tracking. The first, introduced a leap for MR applications. The second one, dealing especially with tracking in unprepared environments, attempts to remove the limitations of marker-based tracking. In this section we describe both tendencies.

Marker-based tracking.   Fiducial markers, such as squares and circles, are easily tracked, thanks to their contrasted contours. The pose of the marker can be recovered from their geometric properties. Another advantage of fiducial markers is their coding capability. Once perspective distortion is recovered from imaged shapes, pattern recognition algorithms can be applied. This enlarges the possibilities of mixed realities as each pattern can be associated with a different information or virtual object.

Figure 17.4 The HiBall optical tracker. © 3rdTech. Reprinted by permission

We discuss in the following representative marker-based tracking techniques that propose square markers. A conferencing system was presented in Kato and Billinghurst (1999) that came to be the well-known ARToolKit technology. First, the image is binarized, and regions similar to squares (surrounded by four lines) are selected. These regions are then normalized and matched to off-line registered patterns. To resolve the normalization problem, the algorithm estimates the camera projection matrix P using the imaged position of the four corners and the four lines.

In Zhang *et al.* (2001) the same principle with a more advanced square coded shape is used. Their work is applied to a large environment inside an industrial plant. The markers are used as a map of the plant to guide the user and also to give extra information of the surroundings of the user (e.g., the description of the machine in front). This framework is composed of a mobile computer with a camera connected via wireless network to a server. The user can browse the location-relevant information.

Another marker-based system is the one developed by Liu *et al.* (2003). This work targets architecture and urban planning using a round table. Since the collaborative work is over a plane (the table), homography-based pose estimation is used for each marker. Similarly to what is described in next section, a transformation from the plane of the camera to the plane of the table is sufficient to estimate the position of the camera. Additionally, this technique presents re-configurable markers based on colour segmentation.

The work of Naimark and Foxlin (2002) describes the creation of circular markers and compares them with square ones. It presents the former to be more robust to template matching, especially for a large set of markers. The approach consists of two steps: acquisition and tracking. Acquisition deals with new detected markers, resolving distortion and matching their code. Then tracking can be achieved faster, using only their centroid position and assuming the code from previously tracked frames.

Marker-based tracking has proven to be an easy-to-use tool in MR applications, especially in indoor environments. However, in order to recognize fiducial markers, they must always be in view; the method fails rapidly with occlusions or in cases where the marker partially disappears. The number of markers in the whole environment plays a key role in this technique. Different studies (Kawano *et al.* 2003; Naimark and Foxlin 2002; Zhang *et al.* 2002) reveal that creating identification codes for fiducial markers has limitations in scalability. This becomes a problem if the environment asks for hundreds or thousands of uniquely coded markers, such as industrial applications.

Markerless Tracking.   Markerless tracking overcomes the limitations of active environments and markers by using only natural features. Though all techniques in this area point to unprepared environments, geometrical constraints or a few references such as models or key views of a scene are necessary for registration. Several solutions can be classified based on the references taken from the environment: planar structures, non-planar structures and model-based techniques.

Move-matching techniques simultaneously estimate camera pose and the 3D structure of the imaged scene using *planar structures* in the sequence. In some cases they are highly accurate and almost jitter-free. As a counterpart, they require great computing power that makes them unsuitable for real-time applications. The algorithm presented in Neumann and You (1999) uses natural features such as corners or edges, and regions which group co-planar feature points. In Simon *et al.* (2000), the algorithm computes motion tracking using homographies. This work assumes that the region is always visible, which is generally the case for an indoor environment (ceiling or floor are commonly visible). First, the projection is calculated for a selected plane. After this step, pose is tracked using homographies. In Prince *et al.* (2002), a picture of the surface to augment is used as a reference. The position for augmentation inside the surface is noted prior to real-time execution. The picture could be seen as a marker in the scene. However, the algorithm is based on frame-to-frame feature tracking. Consequently, we consider this approach as a markerless technique rather than a marker-based one.

A different approach uses *non-planar structures* and takes advantage of the epipolar constraint. The work of Chia *et al.* (2002) uses one or two *a priori* calibrated key images of a scene and the fundamental matrix to track the relation between the reference(s) and the actual frame. MR is assisted in this case by the two- and three-view constraints, respectively.

The drawback of these two classes of techniques is that in most cases Kalman-type filters are applied. Human head motion, which is widely used in mobile MR, is hardly modelled by this type of filters. Additionally, tracking usually recovers pose from previous estimates. The length of the sequence critically increases error as it is accumulated over time (Lepetit *et al.* 2003).

Another solution to register the scene is to use a rough CAD model of parts of the real environment, generally an object. This is known as *model-based tracking*. In Lepetit *et al.* (2003), in addition to the CAD model, pose estimation is assisted with key fames created off-line.

The work of Najafi and Klinker (2003) presents a combination of these techniques. It applies a 3D model together with the epipolar constraint in a stereo camera system. These two techniques prove to be robust to occlusions. The obvious drawback is the need for a CAD model, which can be cumbersome if the environment changes.

Markerless tracking has not yet proven to be an all-purpose solution. For the most robust systems, it still needs a little preparation of the environment. And what is more important, it is not suitable for fast movement, since the applied techniques generally assume small inter-frame displacement. Nevertheless, they deal with many limitations of other trackers.

Markerless tracking is unlimited in range and almost free of scene preparation. As will be seen in next section, the accuracy in slow motion makes these methods a good candidate for combination with other tracking solutions.

17.2.7 Hybrid Tracking

None of the tracking techniques discussed above represent the proverbial *silver bullet* (Welch and Foxlin 2002). However, their limitations or weaknesses are complementary in some ways. Synergies have proven to be necessary for some applications. These synergies are referred to as *hybrid systems*. The fusions commonly developed are inertial–acoustic and inertial–video (Allen *et al.* 2001). As seen in Section 17.2.3, high performance of inertial sensors is achieved for fast motion. On the other hand, in order to compensate for drift, an accurate tracker is needed for periodical correction. A fusion with trackers that compensate for this poor performance at slow motion are the best candidates.

A well known inertial–acoustic solution is the InterSense IS-900 (InterSense 2000), which was first presented in Foxlin *et al.* (1998). The inertial tracker (accelerometers and gyroscopes) extracts 6DOF, and an ultrasonic tracker provides ranging correction to compensate for the drift of the first. The correction is achieved within a Kalman filter framework. This work is ongoing since 1999. Nowadays, the system presents jitter-free robust tracking that is immune to metallic, acoustic (even to blocking of line-of-sight) or optical interference. Different solutions have been developed targeting diverse cases ranging from desktop to large broadcast studio applications. The IS-900 is also presented as wireless equipment.

Video and inertial tracking have been presented as complementary technologies. Video-based tracking performs better at slow motion and fails with rapid movements. Other solutions such as inertial–acoustic tracking still have the environment preparation constraint while video-based tracking is suitable for both indoor and outdoor mobile applications. During the last few years, robust trackers have been implemented combining these two techniques. As presented in You *et al.* (1999), the fusion of gyros and markerless tracking shows that the computational cost of the latter can be decreased by 3DOF orientation self-tracking. It reduces the search area and avoids tracking interruptions due to occlusion.

A marker-based tracker was presented in Kanbara *et al.* (2000b). Three blue markers served as reference for pose estimation with a stereoscopic camera using the epipolar constraint. In addition, this system correctly rendered mutual occlusion between real and virtual objects. Virtual objects were segmented by real objects in front. However, this work has the limitations described in Section 17.2.6 (i.e., limited number of markers, and markers must be imaged). This led to a hybrid solution, combining inertial and marker-based tracking, which enables the algorithm to estimate the position of markers (especially when those were out of scope). This fusion was presented in Kanbara *et al.* (2000a).

InterSense have also presented an inertial–video hybrid (Foxlin and Naimark 2003). A core framework is developed to deal with different tracking sensors in a defined timing scheme. This framework performs autocalibration, automapping and tracking. The first sets up the system's accuracy. The second initializes the system coordinates in space. The user walks around the scene and the system maps the marker's positions. The tracking uses Kalman filtering algorithms to merge both sensors.

Another video–inertial sensing technique is presented in Klein and Drummond (2003). In this case model-based tracking is used in conjunction with inertial tracking. The fast sampling rate of the inertial sensors compensates blurring of images with rapid motion.

Hybrid systems have demonstrated far better performance than single solutions in all cases. Although separate research in each tracking topic is necessary, tracking in MR in the near future will still be linked to fusion algorithms.

17.3 USER INTERFACE

Many MR systems focused on adapting traditional interfaces, such as text-based (e.g., keyboard) and graphical user interfaces (e.g., mouse), for interacting with virtual objects. However these interfaces are inconvenient because they cannot easily provide 6DOF inputs. To allow direct and natural interactions with MR environment, tangible and gesture-based interfaces have been proposed.

17.3.1 Tangible User Interfaces

Tangible user interfaces (TUI) support physical interactions through the registration of virtual objects with physical objects. Manipulating and grasping a virtual object is equivalent to interacting with the corresponding physical object. Typically, a tracker is integrated in the physical object, that serves as the tangible interface, to determine its position and orientation.

Many TUIs have been proposed. Their design depends on the targeted applications. In Ishii and Ullmer (1997), a prototype system, called metaDESK (Figure 17.5) is presented. A possible application scenario of this system is the manipulation of 2D and 3D graphical maps. By exploiting the widespread desktop metaphor in GUI interfaces, metaDESK transforms the elements of a GUI interface into graspable objects. For example, the window and the icon are physically instantiated by an arm-mounted flat-panel display called active-LENS, and a physical icon referred to as a phicon, respectively. Optical, mechanical, and magnetic trackers are embedded within the metaDESK system to determine the position and orientation of the graspable physical objects.

The personal interaction panel (PIP) is a two-handed interface used to control a collaborative augmented reality system (Szalavari and Gervautz 1997). It consists of two lightweight

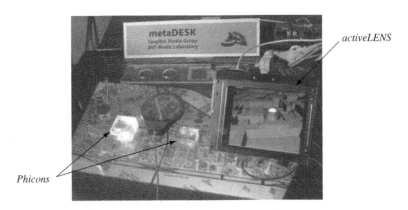

Figure 17.5 The metaDESK system. From (Ishii and Ullmer 1997). © 1997 ACM Inc. Reprinted by permission

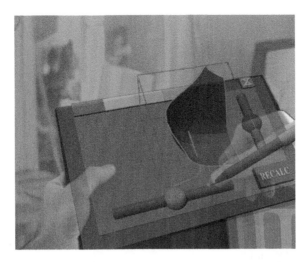

Figure 17.6 The personal interaction panel. From (Schmalstieg *et al.* 2002). © 2002 Massachusetts Institute of Technology. Reprinted by permission

hand-held props, a pen and a panel, both with embedded trackers. The props are augmented with virtual objects, transforming them into application-dependent interaction tools (Figure 17.6).

Hauber *et al.* (2004) developed a tangible teleconferencing system which simulates a traditional face-to-face meeting. In the shared environment remote users, represented by virtual avatars, are arranged around a virtual table on which virtual objects are placed. Based on ARToolKit (2003), several tangible interfaces are designed to load, translate, rotate, and scale the virtual objects. For example, the translation tool consists of an ARToolKit marker attached to a handle. Through the motion of the physical tool, the virtual object is translated.

17.3.2 Gesture-based Interfaces

Similarly to the use of hands by humans to interact with the real world, gesture-based interfaces enable intuitive and natural interactions with the MR environment. Compared with the traditional interfaces and TUI's, gesture-based interfaces also allow additional degrees of freedom through finger motion. The gesture interfaces are based on the tracking of the position of fingers. According to Quek (1994;1995), gestures are classified into two main classes: communicative and manipulative. The communicative gestures allow the user to express thoughts and to convey information. This class of gestures is influenced by cultural and educational factors. On the other hand, manipulative gestures allow the user to interact with objects. Since manipulative gestures prevail within the area of MR, we will focus on this class. The reader who is interested in interpretation of hand gestures in the context of the more general area of human–computer interaction (HCI) is referred to Pavlovic *et al.* (1997).

In recent years many gesture-based interfaces for MR applications have been developed. In order to make the interactions as natural as possible, there are a number of criteria a

Figure 17.7 Manipulation of virtual objects by finger tracking. From (Dorfmuller-Ulhaas and Schmalstieg 2001). © 2001 IEEE.

gesture-based interface should fulfil, such as haptic and visual feedback, multi-fingered and two-handed interactions, and robustness to occlusion.

In Dorfmuller-Ulhaas and Schmalstieg (2001) a finger tracking system for three-dimensional input in MR environments is presented (Figure 17.7). To allow robust tracking of the user's index finger in real-time, a glove fitted with retroreflective markers is used. Infrared light is reflected by the retroreflective markers towards a stereo camera pair. The following procedures are applied to estimate the pose of the joints of the user's index finger. First, the 2D locations of the markers in both images are extracted using infrared filters. Then, the extracted 2D locations of the markers are correlated by employing epipolar constraints and a kinematic model of the finger. Finally, a Kalman filter is used to smooth and predict the motion of the user's finger. Since only the index finger interaction is supported, natural grabbing is not possible.

Walairacht *et al.* (2002) developed a two-handed, multi-finger haptic interface (Figure 17.8). A user can manipulate virtual objects with eight fingers (four on the left hand and four on the right hand). Each finger is connected to three strings which are attached to the edge of a rectangular frame. The tension of these strings is controlled by motors to allow the user to perceive force feedback when manipulating virtual objects. Visual feedback is also provided by registering real and virtual images with correct geometrical occlusions. The positions of the fingertips are estimated from the length of the strings. One of the drawbacks of this system is that the interactions are limited within the working space of the rectangular frame.

Buchmann *et al.* (2004) presented an urban design system which supports several types of gesture interactions, such as grabbing, dragging, rotating, dropping, pushing and pointing (Figure 17.9). This system also provides haptic feedback, correct occlusion cues and allows two-fingered input. The ARToolKit video marker-based tracking library is used for tracking both the user's pose and fingers. In order to reliably track the user's viewpoint, 24 markers are placed on the board. That is, tracking will be maintained even when some markers are occluded by the user's hand. For finger tracking, a glove with small markers attached to the tips of the thumb and the index finger, and to the hand, is used. To support haptic feedback, a buzzer is attached to each of these fingertips. The recognition of the gestures is based on

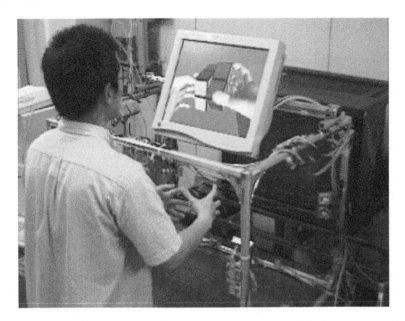

Figure 17.8   The two-handed, multi-finger haptic interface. From (Walairacht *et al.* 2002). © 2002 Massachusetts Institute of Technology. Reprinted by permission

(a)   (b)

Figure 17.9   The urban design system. From (Buchmann *et al.* 2004). © 2004 ACM Inc. Reprinted by permission

the tracked position of the fingertips. Despite the versatility of the system, the use of only two fingers to manipulate virtual objects makes the interaction less natural.

A promising new research direction in the area of user interfaces is the fusion between different mono-modal interfaces to improve the recognition performance and the accuracy of the interactions. For example, in Olwal *et al.* (2003) speech- and gesture-based interfaces are integrated for the selection of objects.

17.4 APPLICATIONS

A large number of applications are reported in the literature (Azuma 1997). One may cite medical, design, and entertainment applications. We will focus on the most recent applications, namely mobile applications, collaborative, and industrial applications.

17.4.1 Mobile Applications

The MR interface has the attractive features for providing a navigation aid that guides the user to a target location and an information browser that displays location-based information. Both features support the development of mobile applications such as a tourist guide.

The early versions of MR mobile systems are composed mainly of a notebook and a head-mounted display (HMD) (Hoellerer *et al.* 1999; Julier *et al.* 2000; Kalkusch *et al.* 2002). However, these systems have a large form factor and do not provide the required ergonomic comfort. In order to overcome these limitations, personal digital assistant (PDA) based systems have been proposed (Gausemeier *et al.* 2003; Newman *et al.* 2001; Regenbrecht and Specht 2000). Due to the limited capabilities of PDAs in terms of computation and display, the main tasks (i.e., tracking, composition and rendering) are offloaded to a computing server and the PDAs are used mainly as displays. The drawbacks of such configuration are a limited work area and waste of the available bandwidth.

Wagner and Schmalstieg (2003) proposed a self-contained PDA-based system (Figure 17.10), where all the tasks are performed by the client. In some selected locations, called *hotspots*, the tracking task is outsourced to the computing server. ARToolKit(2003) and a commercial outside-looking-in system are used for tracking. The system dynamically and transparently switch between the three modes, namely ARToolKit, outside-looking-in, and tracking offloaded to server mode. To test the effectiveness of the developed system, a 3D navigation application that guides a user through an unknown building to a chosen location is presented.

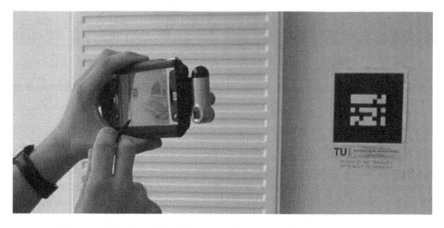

Figure 17.10 A mobile MR system based on a PDA. © 2003 IEEE.

17.4.2 Collaborative Applications

In face-to-face collaboration, multiple communication channels are at work to convey the information. In addition to speech, humans use accompanying gestures to communicate (Teston 1998). Examples of accompanying gestures are gaze, hand movements, head movements, and body posture. Unlike conventional collaborative interfaces, such as screen-based interface, MR collaborative interfaces are suited for collaboration because they support the perception of the accompanying gestures and allow seamless interactions. That is, when people collaborate, they can see each other and the objects of interest simultaneously. Therefore the task space, containing the objects, is a subset of the interpersonal communication space (Billinghurst and Kato 2002).

In Kobayashi *et al.* (2003) a MR interface for IP network simulation was presented. It supports face-to-face collaborative design and simulation of an IP network by a group of network designers and their customers. This system consists mainly of a sensing table, allowing users to directly manipulate network topologies.

MR interfaces are also suited to support remote collaboration. In Billinghurst and Kato (2000), MR conferencing interfaces are developed, where a remote user appears attached to a real card as a life-sized, live virtual window (Figure 17.11). In this way the sense of presence for the remote user is higher and the accompanying gestures are better perceived compared with traditional audio- and videoconferencing systems.

17.4.3 Industrial Applications

There is an increasing interest from industry to improve its current workflow by adopting MR technology. In order to ensure user acceptance, MR researchers need to deeply understand each industrial process and its requirements, to make a feasibility study, and to consider marketing and commercialization aspects of potential MR solutions (Navab 2004).

Figure 17.11 Remote users appear attached to real cards in the AR conferencing system. From (Billinghurst and Kato 2002). © 2002 ACM Inc. Reprinted by permission

An example of a manufacturing procedure is the *intelligent welding gun* (Echtler *et al.* 2004). In order to evaluate new car concepts and validate product functionality, car prototypes are built. These processes rely heavily on manual work rather than on automation because of the limited number of car prototypes. MR technology can be used to improve the efficiency of the stud welding process. The intelligent welding gun guides the welder directly to the specified positions where it is possible to shoot studs with high precision. The setup consists of a regular welding gun with an attached display. Tracking is achieved using an outside-looking-in system with a set of stationary cameras that track highly reflective markers on the welding gun and on the car frame (Figure 17.12). Different visualization schemes, such as the *notch and bead metaphor*, are proposed for finding and aiming at stud positions.

Another example of industrial applications is presented in Navab (2004). In many cases, virtual models, manufacturing information, and part specifications of industrial sites are available. However, this source of valuable information is not efficiently exploited for monitoring and control procedures. The MR interface allows the worker to have easy access to engineering, monitoring, and maintenance data. Figure 17.13 shows the augmentation of a waste water plant by the associated floor map and the virtual model of pipework. The ultimate MR maintenance system (Goose *et al.* 2003) aims at integrating virtual models, industrial drawings, factory images, vision-based localization and tracking, wireless communication and data access, and speech-based interaction to provide end-to-end solutions that empower mobile workers.

Figure 17.12   The intelligent welding gun. From (Echtler *et al.* 2004). © 2004 Springer.

Figure 17.13 Augmentation of a waste water plant by the associated floor map and the virtual model of pipework. From (Navab 2004). © 2004 IEEE. Reprinted by permission

17.5 CONCLUSIONS

MR is still in its infancy. It includes a number of hard-related issues such as tracking and interface design. Markerless tracking increases the range and supports outdoor applications. Hybrid tracking seems to be a promising solution to achieve high accuracy. Multimodal interfaces with haptic feedback allow natural and efficient interaction between the user and the MR environment.

Because MR interfaces support mobility and collaborative work, a tremendous number of MR applications are reported in the literature, covering diverse areas. However, few applications have come out of the laboratories to be used in real industrial settings (Klinker *et al.* 2004). This is due to the lack of collaboration between industry and academic research to identify real scenarios where MR could be adopted. ARVIKA (2003) is one of the few consortia that includes many industrial partners and has succeeded in identifying a number of real scenarios for MR technology.

REFERENCES

Allen B, Bishop G and Welch G 2001 Tracking: Beyond 15 minutes of thought *Course Notes, Annual Conference on Computer Graphics and Interactive Techniques (Siggraph)*.
ARToolKit 2003 www.hitl.washington.edu/artoolkit/.
ARVIKA 2003 www.arvika.de.
Azuma R 1997 A survey of augmented reality. *Presence: Teleoperators and Virtual Environments* **6**(4), 335–385.
Billinghurst M and Kato H 2002 Out and about: Real world teleconferencing. *British Telecom Technical Journal* **18**(1), 80–82.

Billinghurst M and Kato H 2002 Collaborative augmented reality. *Communications of ACM* **45**(7), 64–70.

Buchmann V, Violich S, Billinghurst M and Cockburn A 2004 Fingertips: gesture based direct manipulation in augmented reality *Proceedings of the 2nd International Conference on Computer graphics and interactive techniques in Austalasia and SouthEast Asia (Graphite)*, pp. 212–221.

Burdea G and Coiffet P 2003 *Virtual Reality Technology* 2 edn. Wiley.

Chia KW, Cheok A and Prince S 2002 Online 6 dof augmented reality registration from natural features *Proceedings of the International Symposium on Mixed and Augmented Reality*, pp. 305–313.

Dorfmuller-Ulhaas K and Schmalstieg D 2001 Finger tracking for interaction in augmented environments *Proceedings of the IEEE and ACM International Symposium on Augmented Reality*, pp. 55–64.

Echtler F, Sturm F, Kindermann K, Klinker G, Stilla J, Trilk J and Najafi H 2004 The intelligent welding gun: Augmented reality for experimental vehicle construction *Virtual and Augmented Reality Applications in Manufacturing*, In Chapter 17 (Ong SK and Nee AYC eds.) Springer Verlag.

FakeSpaceLabs 2004 www.fakespacelabs.com/.

Foxlin E 2002 Motion tracking requirements and technologies In (ed.) Stanney K *Handbook of Virtual Environment Technologies* Lawrence Erlbaum Publishers, Hillsdale, N.J. pp. 163–210.

Foxlin E and Naimark L 2003 Vis-tracker: a wearable vision-inertial self-tracker *Proceedings of IEEE Virtual Reality*, pp. 199–206.

Foxlin E, Harrington M and Pfeifer G 1998 Constellation: A wide-range wireless motion-tracking system for augmented reality and virtual set applications *Proceedings of Computer Graphics (SIGGRAPH)*, pp. 371–378. ACM Press, Addison-Wesley, Orlando(FL), USA.

Gausemeier J, Fruend J, Matyszok C, Bruederlin B and Beier D 2003 Development of a real time image based object recognition method for mobile ar-devices *Proceedings of the 2nd International Conference on Computer graphics, Virtual Reality, Visualisation and Interaction in Africa*, pp. 133–139.

Goose S, Sudarsky S, Zhang X and Navab N 2003 Speech-enabled augmented reality supporting mobile industrial maintenance. *IEEE Transactions on Pervasive Computing* **2**(1), 65–70.

Hauber J, Billinghurst M and Regenbrecht H 2004 Tangible teleconferencing *Proceedings of the Sixth Asia Pacific Conference on Human Computer Interaction (APCHI 2004)*, pp. 143–152.

Hoellerer T, Feiner S, Terauchi T, Rashid G and Hallaway D 1999 Exploring mars: Developing indoor and outdoor user interfaces to a mobile augmented reality system. *Computers and Graphics* **23**(6), 779–785.

Immersion 2004 www.immersion.com/3d/.

InterSense 2000 *Technical Overview IS-900 Motion Tracking System InterSense* Inc. www.isense.com.

Ishii H and Ullmer B 1997 Tangible bits: Towards seamless interfaces between people, bits and atoms *Proceedings of the Conference on Human Factors in Computing Systems (CHI '97), ACM Press*, pp. 234–241.

Julier SJ, Baillot Y, Lanzagorta M, Brown D and Rosenblum L 2000 Bars: Battlefield augmented reality system *NATO Information Systems Technology Panel Symposium on New Information Processing Techniques for Military Systems*.

Kalkusch M, Lidy T, Knapp M, Reitmayr G, Kaufmann H and Schmalstieg D 2002 Structured visual markers for indoor pathfinding *Proceedings of the 1st IEEE International Workshop of ARToolKit*. p. 8.

Kanbara M, Fujii H, Takemura H and Yokoya N 2000a A stereo vision-based augmented reality system with an inertial sensor *Proceedings of the IEEE and ACM International Symposium on Augmented Reality*, pp. 97–100.

Kanbara M, Okuma T, Takemura H and Yokoya N 2000b A stereoscopic video see-through augmented reality system based on real-time vision-based registration *Proceedings of IEEE Virtual Reality*, pp. 255–262.

Kato H and Billinghurst M 1999 Marker tracking and hmd calibration for a video-based augmented reality conferencing system *Proceedings of the 2nd IEEE and ACM International Workshop on Augmented Reality*, pp. 85–94.

Kawano T, Ban Y and Uehara K 2003 A coded visual market for video tracking system based on structured image analysis *Proceedings of the 2nd IEEE and ACM International Symposium on Mixed and Augmented Reality*, pp. 262–263.

Kim D, Richards SW and Caudell TP 1997 An optical tracker for augmented reality and wearable computers *Proceedings of the IEEE Virtual Reality Annual International Symposium*, pp. 146–150.

Klein G and Drummond T 2003 Robust visual tracking for non-instrumented augmented reality *Proceedings of the 2nd IEEE and ACM International Symposium on Mixed and Augmented Reality*, pp. 113–122.

Klinker G, Najafi H, Sielhorst T, Sturm F, Echtler F, Isik M, Wein W and Truebswetter C 2004 Fixit: An approach towards assisting workers in diagnosing machine malfunctions *Proceedings of the International Workshop exploring the Design and Engineering of Mixed Reality Systems - MIXER 2004, Funchal, Madeira, CEUR Workshop Proceedings.* (http://sunsite.informatik.rwth-aachen.de/Publications/CEUR-WS/Vol-91/paper Ell.pdf)

Kobayashi K, Hirano M, Narita A and Ishii H 2003 Ip network designer: interface for ip network simulation *Proceedings of the 2nd IEEE and ACM International Symposium on Mixed and Augmented Reality*, pp. 327–328.

Kourogi M and Kurata T 2003 Personal positioning based on walking locomotion analysis with self-contained sensors and a wearable camera *Proceedings of the 2nd IEEE and ACM International Symposium on Mixed and Augmented Reality*, pp. 103–112.

Lepetit V, Vacchetti L, Thalmann D and Fua P 2003 Fully automated and stable registration for augmented reality applications *Proceedings of the 2nd IEEE and ACM International Symposium on Mixed and Augmented Reality*, pp. 93–102.

Liu Y, Storring M, Moeslund T, Madsen C and Granum E 2003 Computer vision based head tracking from re-configurable 2d markers for an *Proceedings of the 2nd IEEE and ACM International Symposium on Mixed and Augmented Reality*, pp. 264–267.

Livingston M, Swan JE I, Gabbard J, Hollerer T, Hix D, Julier S, Baillot Y and Brown D 2003 Resolving multiple occluded layers in augmented reality *Proceedings of the 2nd IEEE and ACM International Symposium on Mixed and Augmented Reality*, pp. 56–65.

Naimark L and Foxlin E 2002 Circular data matrix fiducial system and robust image processing for a wearable vision-inertial self-tracker *Proceedings of the International Symposium on Mixed and Augmented Reality*, pp. 27–36.

Najafi H and Klinker G 2003 Model-based tracking with stereovision for an *Proceedings of the Second IEEE and ACM International Symposium on Mixed and Augmented Reality*, pp. 313–314.

Navab N 2004 Developing killer apps for industrial augmented reality. *Proceedings of IEEE Computer Graphics and Applications*, pp. 16–20.

Neumann U and You S 1999 Natural feature tracking for augmented reality. *IEEE Transactions on Multimedia* **1**(1), 53–64.

Newman J, Ingram D and Hopper A 2001 Augmented reality in a wide area sentient environment *Proceedings of the IEEE and ACM International Symposium on Augmented Reality*, pp. 77–86.

Olwal A, Benko H and Feiner S 2003 Senseshapes: using statistical geometry for object selection in a multimodal augmented reality *Proceedings of the 2nd IEEE and ACM International Symposium on Mixed and Augmented Reality*, pp. 300–301.

Pavlovic V, Sharma R and Huang T 1997 Visual interpretation of hand gestures for humancomputer interaction: a review. *IEEE Transactions on Pattern Analysis and Machine Intelligence* **19**(7), 677–695.

Prince S, Xu K and Cheok A 2002 Augmented reality camera tracking with homographies. *Computer Graphics and Applications* **22**(6), 39–45.

Quek F 1994 Toward a vision-based hand gesture interface *Proceedings of the Virtual Reality Software and Technology Conference*, pp. 17–31.

Quek F 1995 Eyes in the interface. *Image and vision Computing*, **13**(6), 511–525.

Raab FH, Blood EB, Steiner TO and Jones HR 1979 Magnetic position and orientation tracking system *IEEE Transactions on Aerospace and Electronic Systems*, **AES-15**, 709–718.

Regenbrecht HT and Specht RA 2000 A mobile passive augmented reality device - mpard *Proceedings of the IEEE and ACM International Symposium on Augmented Reality*, pp. 81–84.

Roetenberg D, Luinge H and Veltink P 2003 Inertial and magnetic sensing of human movement near ferromagnetic materials *Proceedings of the 2nd IEEE and ACM International Symposium on Mixed and Augmented Reality*, pp. 268–269.

Schmalstieg D, Fuhrmann A, Hesina G, Szalavari Z, Encarnao L, Gervautz M and Purgathofer W 2002 The studierstube augmented reality project. *Presence: Teleoperators and Virtual Environments* **11**(1), 33–54.

Simon G, Fitzgibbon A and Zisserman A 2000 Markerless tracking using planar structures in the scene *Proceedings of the IEEE and ACM International Symposium on Augmented Reality*, pp. 120–128.

Szalavari Z and Gervautz M 1997 The personal interaction panel: A two-handed interface for augmented reality. *Computer Graphics Forum* **16**(3), 335–346

Teston B 1998 L'observation et l'enregistrement des mouvements dans la parole: Problemes et methodes In Santi S *et al.* (eds): *Oralit et gestualit. Communication multimodale, interaction*: L'Harmattan, pp. 39–58.

Vallidis NM 2002 *WHISPER: A Spread Spectrum Approach to Occlusion in Acoustic Tracking* PhD thesis University of North Carolina at Chapel Hill, Department of Computer Science.

Wagner D and Schmalstieg D 2003 First steps towards handheld augmented reality *Proceedings of the 7th IEEE International Symposium on Wearable Computers*, pp. 127–135.

Walairacht S, Yamada K, Hasegawa S and Sato M 2002 4 + 4 fingers manipulating virtual objects in mixed-reality environment. *Presence: Teleoperators and Virtual Environments* **11**(2), 134–143.

Wang JF, Azuma R, Bishop G, Chi V, Eyles J and Fuchs H 1990 Tracking a head-mounted display in a room-sized environment with head-mounted cameras *Proceedings of SPIE Helmet-Mounted Displays II*, vol. 1290.

Welch G and Foxlin E 2002 Motion tracking: no silver bullet, but a respectable arsenal. *Computer Graphics and Applications* **22**(6), 24–38.

You S, Neumann U and Azuma R 1999 Hybrid inertial and vision tracking for augmented reality registration *Proceedings of IEEE Virtual Reality*, pp. 260–267.

Zhang X, Fronz S and Navab N 2002 Visual marker detection and decoding in an systems: A comparative study *Proceedings of the International Symposium on Mixed and Augmented Reality*, pp. 97–106.

Zhang X, Genc Y and Navab N 2001 Taking an into large scale industrial environments: Navigation and information access with mobile computers *Proceedings of the IEEE and ACM International Symposium on Augmented Reality*, pp. 179–180.

Index

Printed and bound in the UK by
CPI Antony Rowe, Eastbourne

Printed and bound by CPI Group (UK) Ltd, Croydon, CR0 4YY

27/10/2024

14580292-0002